Maxwell's Equations

Maxwell's Equations

Paul G. Huray

A John Wiley & Sons, Inc., Publication

Published by John Wiley & Sons, Inc., Hoboken, New Jersey
Published simultaneously in Canada

Library of Congress Cataloging-in-Publication Data:

Huray, Paul G., 1941–
Maxwell's equations / Paul G. Huray.
p. cm.
Includes bibliographical references.
ISBN 978-0-470-54276-7 (cloth)
1. Maxwell equations. I. Title.
QC670.H87 2010
530.14'1–dc22

2009019340

Printed in the United States of America

10 9 8 7 6 5 4 3 2 1

This book is dedicated to the author's lifelong partner
Susan Lyons Huray

Contents

Acknowledgments

The author owes a debt of gratitude to Dr. Yinchao Chen of the Electrical Engineering department at the University of South Carolina (USC). Dr. Chen has published articles with the author and has had many discussions on the techniques and meaning of the solutions to Maxwell's equations and their applications. Other USC professors who contributed to the physical and chemical understanding of printed circuit board (PCB) materials were Michael Myrick of the Chemistry Department and Richard Webb of the Physics and Astronomy Department.

The author, Dr. Chen, and three Signal Integrity engineers (Brian Knotts, Hao Li, and Richard Mellitz) from the Intel Corporation in Columbia, SC, created the first graduate Signal Integrity program in 2003, which has since produced more than 80 practicing Signal Integrity engineers, many of whom read and corrected early drafts of this text.

The author conducts industrial research part time with the Intel Corporation in the area of high-speed electromagnetic signals. In this work, he has had the privilege to work closely with Richard Mellitz and Stephen Hall, on applications of electromagnetism for practical use. It was their penetrating questions that prompted many of the explanations in this text. Another Intel employee, Dan Hua, provided a sequence of exchanged papers on the evaluation of scattering and absorption in the language of vector spherical harmonics; it was through these discussions that the sections on absorption by small good conducting spheres arose. Gary Brist taught the author (and many of his graduate students) about the process of manufacturing PCB stack-ups and stimulated many of the questions that are sprinkled throughout the book. Anusha Moonshiram and Chaitanya Sreerema conducted many of the high-frequency vector network analyzer (VNA) measurements in this text. Femi Oluwafemi conducted many of the numerical simulations on phase analysis to identify time-dependent fields inside good conductors and provided many of the final comparisons to the VNA data. Guy Barnes and Paul Hamilton provided the Fabry–Perot measurements of permittivity. Brandon Gore helped work on magnetic losses, and David Aerne assisted the analysis of spherical composition profiles and near neighbor interference effects. Peng Ye was a sounding board for arguments about the analytical analysis associated with electromagnetic field dynamics. Kevin Slattery introduced the author to near-field scanning electromagnetic probes and helped direct the work of two USC graduate students, Jason Ramage and Christy Madden Jones, whose work on proof of Snell's law at microwave frequencies and absorption by impurities appears in the text. Intel engineers such as Howard Heck, Richard Kunze, Ted Ballou, Steve Krooswyk, Matt Hendrick, David Blakenbeckler, and Johnny Gibson passed through USC during the writing of this text to present lectures to the author's Signal Integrity classes and to build richness into the intellectual atmosphere. Mark Fitzmaurice was always ready to help make the Signal Integrity program at USC a success through his support for measurement equipment, student internships, and common sense.

Many USC undergraduate and graduate students contributed to the testing and writing of this book. Steven Pytel worked with the author on scanning electron microscope (SEM) and analysis measurements at the Oak Ridge National Laboratory in Oak Ridge, TN, and, while working for Intel, was the sounding board for many of the arguments presented here. After receiving his PhD, he became an employee of the Ansoft Corporation of Pittsburg, PA, where he became an applications engineer for Signal Integrity tools. He is primarily responsible for the material in Chapter 8 of the author's book *The Foundations of Signal Integrity* on numerical simulations. Ken Young helped with editing, Fisayo Adepetun provided assistance with figures and the reduced magnetic moment analysis, and David London supported web pages for testing and transmittal of the chapters. Tom McDonough gave lectures to the Signal Integrity classes on the use of Synopsys Corporation of Mountain View, CA, HSpice software and helped in the analysis of ceramic capacitor fields.

John Fatcheric of the Oak Mitsui Corporation of Camden, SC, assisted the presentation on copper surface production. Bob Helsby, Charles Banyon, Arien Sligar, and Zol Cendes of the Ansoft Corporation of Pittsburg, PA, supported the use of forefront numerical solutions to Maxwell's equations. James Rautio of Sonnet Software in Syracuse, NY, assisted on the history Maxwell and the use of his portrait. Mike Resso of Agilent Corporation of Santa Clara, CA, supported a joint Intel–Agilent VNA donation. Lee Riedinger, Harry M. Meyer III, Larry Walker, and Marc Garland of the Oak Ridge National Laboratory assisted in making qualitative and quantitative measurements of PCB components by SEM and Auger analysis. José E. Rayas Sánchez of the Instituto Technológico y de Estudios Superiores de Occidente (ITESO) at the Jesuit University of Guadalajara, Mexico, James Gover of Kettering University in Flint, MI, and John David Jackson of University of California-Berkeley and Lawrence Berkeley National Laboratory, in Berkeley, CA, provided discussions on Maxwell's interpretations and Signal Integrity of high-speed circuits.

Introduction

James Clerk Maxwell

Maxwell's Equations[i]

Differential form	Integral form
$\vec{\nabla}\times\vec{E} = -\frac{\partial\vec{B}}{\partial t}$	$\oint_C \vec{E}\cdot d\vec{l} = -\frac{d\Phi_B}{dt}$
$\vec{\nabla}\times\vec{H} = \vec{J} + \frac{\partial\vec{D}}{\partial t}$	$\oint_C \vec{H}\cdot d\vec{l} = I + \oint_S \frac{\partial\vec{D}}{\partial t}\cdot d\vec{s}$
$\vec{\nabla}\cdot\vec{D} = \rho_V$	$\oint_C \vec{D}\cdot d\vec{s} = Q$
$\vec{\nabla}\cdot\vec{B} = 0$	$\oint_S \vec{B}\cdot d\vec{s} = 0$

BIOGRAPHY

James Clerk Maxwell was born on June 13, 1831, in Edinburgh, Scotland, and was educated at his country home until he was 8 years old, when his mother died. At age 10, he attended Edinburgh Academy, where he was given the nickname[ii] "Dafty." At age 16, he attended the University of Edinburgh, and, at age 19, he went to Peterhouse Cambridge but moved to Trinity to obtain a fellowship. In 1856, he moved to Marischal College in Aberdeen to be near his father, who shortly thereafter died. Maxwell married Katherine Mary Dewar, the daughter of the Principal of Marischal College, in 1859. In 1860, Maxwell was appointed to the chair of Natural Philosophy at King's College in London, where he did his most productive work, and, from 1871 until his death on November 5, 1879, he was the first Cavendish Professor at Cambridge, where he produced his two-volume treatise,[iii] created a working laboratory, and edited Henry Cavendish's researches for publication. Albert

Einstein once described Maxwell's work[iv] as the "most profound and the most fruitful that physics has experienced since the time of Newton."

ELECTROMAGNETIC CONCEPTS IN THE MID-1800s

John Tyndall wrote a series of essays, addresses, and reviews entitled *Fragments of Science*,[v] which began with the copy of a 1865 essay on "The Constitution of Nature,"[vi] in which the perspective of "modern" science described electromagnetics as follows:

> *From the phenomena of sound, as displayed in the air, "men's minds" ascend to the phenomena of light, as displayed in the ether; which is the name given to the interstellar medium. The notion of this medium must not be considered as vague or fanciful conception on the part of scientific men. Of its reality most of them are as convinced as they are of the existence of the sun and the moon. The luminiferous ether has definite mechanical properties. It is almost infinitely more attenuated than any know gas, but its properties are those of a solid rather than of a gas. It resembles jelly rather than air.*

It was the revolutionary thinking of Einstein that changed that description to a relativistic philosophy that there is no luminiferous ether, but, rather, the speed of light is the same for all observers, independent of their own velocity relative to the source. Einstein was one of the first scientists to adopt the rubric of the observer for logically deducing outcomes, but it was an extension of one of the fundamental principles of science[vii] that this text has used to develop a process of explaining physical processes: "Upon every particle there exists an observer and that observer is you." Students in this endeavor are asked to "think like an electromagnetic wave" for the purpose of finding a logical solution.

In an article on "Scientific Materialism"[viii] in 1868, Tyndall also noted

> *Mathematics and physics have been long accustomed to coalesce, and here they form a single section. No matter how subtle a natural phenomenon may be, whether we observe it in the region of sense, or follow it into that of imagination, it is in the long run reducible to mechanical laws. But the mechanical data once guessed or given, mathematics are all-powerful as an instrument of deduction. The command of Geometry over the relations of space, and the far-reaching power which Analysis confers, are potent both as a means of physical discovery, and of reaping the entire fruits of discovery.*

In this book, we follow one of the most beautiful and fruitful applications of mathematics that has ever been written. It is the culmination of many individual mathematicians and scientists that has led us to the solutions of Maxwell's equations. Most of these early mathematicians were not interested in applications, and most did their work prior to the development of Maxwell's equations. French, German, and English academicians such as Bessel, Cauchy, Clebsch, Dirichlet, Fourier, Gauss, Green, Hankel, Helmholtz, Hermite, Hilbert, Laguerre, Laplace, Laurent, Legendre, Lorentz, Lorenz, Neumann, Poisson, Rayleigh, Schmidt, Stokes and Wronski built a foundation of solutions for partial differential equations that stood alone. When Sturm and Liouville categorized their work into a common framework, they were able to demonstrate that a set of complete, orthogonal, functions could uniquely describe a set of eigenfunctions that guaranteed a solution to Maxwell's equations. This framework is one of the most elegant sets of solutions ever to be devised, and we are the beneficiaries of their work. We shall adopt their efforts with reference to their work. Relevant texts for this systematic approach are given by Arfken,[ix] and Jackson,[x] and a self-study set of outlines is available from Spiegel.[xi]

Rather than focus on mathematical proofs of solutions, this book is intended to be a foundation for the discipline of electricity and magnetism upon which measurements, simulations, and "rules-of-thumb" are built. In that sense, it is intended to be a book that takes theory to practice. It is written in the language of an electrical engineer rather than a mathematician or physicist and is intended to support engineering practice.[xii]

Practical work has provided a dilemma for scientists since the time of Maxwell. For example, following an essay "On Prayer as a Form of Physical Energy," Tyndall includes a comment on the popular interest in understanding electromagnetics:

> *It is the custom of the Professors in the Royal School of Mines in London to give courses of evening lectures every year to working men. The lecture-room holds 600 people; and tickets to this amount are disposed of as quickly as they can be handed to those who apply for them. So desirous are the working men of London to attend these lectures that the persons who fail to obtain tickets always bear a large proportion to those who succeed. Indeed, if the lecture-room could hold 2,000 instead of 600, I do not doubt that every one of its benches would be occupied on these occasions. The information acquired is hardly ever of a nature which admits of being turned into money. It is therefore, a pure desire for knowledge, as a good thing in itself, and without regard to its practical application, which animates the hearers of these lectures.*

Tyndall concludes his Fragments of Science with the following insight:

> *Two orders of minds have been implicated in the development of this subject; first, the investigator and discoverer, whose object is purely scientific, and who cares little for practical ends; secondly, the practical mechanician, whose object is mainly industrial. It would be easy, and probably in many cases true, to say that the one wants to gain knowledge, while the other wishes to make money; but I am persuaded that the mechanician not infrequently merges the hope of profit in the love of his work.*

EXTENSION OF ELECTROMAGNETIC THEORY INTO THE 2000s

One of the great ironies of history is that Maxwell's equations were not written by Maxwell, at least not in their vector form stated on the first page. Maxwell was convinced that the laws of electromagnetism would be best formulated in the form of a quaternion, which had been invented by the Irish mathematician Sir William Rowan Hamilton in 1843 because they worked in four dimensions and could therefore include three-dimensional space and time. The original form of Maxwell's equations was thus in the form of 20 quaternion expressions that will be discussed in the vector chapter, which included eight equations dealing with electromagnetic fields that include the magnetic vector potential and 12 that deal with the magnetic scalar potential, magnetic mass, and magnetic conductivity. Some scientists today[xiii] say that the vector-reformulated equations by Heavyside, Gibbs, Fitzgerald, Lodge, and Hertz are insufficient to describe some physical outcomes such as the Aharonov–Bohm effect, the Josephson effect, the quantum Hall effect, the De Hass Van Alphen effect, and the Sagnac effect, all of which make the magnetic scalar and vector potentials (sometimes called the A_μ fields with $\mu = 0, 1, 2, 3$) physically meaningful constructs. The Heavyside formulation (and this book) treats the magnetic vector field as a mathematical convenience. As Lord Kelvin said in 1892, "Quaternions came from Hamilton after his really good work had been done; and, though beautifully ingenious, have been an unmixed evil to those who have touched them in any way, including Clerk Mawell." Maxwell's first formulation of a magnetic charge density and the possible existence of magnetic monopoles were forgotten for half a century until P. A. M. Dirac again speculated on their existence in 1931. In deference to my former colleague, Yakir Aharonov, these potentials are mentioned extensively in this book but usually in the vein of an engineer's interpretation that they give the correct mathematical result but show few practical macroscopic properties and do not interact directly with particles. A few scientists like H. Harmuth and K. Meyl go a step further to state that there exist no kind of monopoles, electric or magnetic; that the alleged electric monopoles (charges) are only secondary effects of electric and magnetic fields. We will see that this description of fields as wavelike entities is contained in the dual nature of matter as either particle or wavelike, depending on the interpretation for a given application. Both descriptions have a disconcerting problem: the magnetic vector potentials require "action at a distance" and that electromagnetic history influences current properties, whereas the vector field interpretation requires that waves propagate in luminiferous ether that has no physically measurable properties.

INTENT OF THE BOOK

It is the intent of this book to first clarify the concept of Tyndall's jelly, which transmits "action at a distance" (as it was called in 1865). This portion of the book is a mathematical and physical treatment that enables "working men" (and women) to understand the greatest set of equations ever devised,[1] now called Maxwell's equations. It focuses on our "pure desire for knowledge" and is intended to permit the readers to convince themselves that Maxwell's equations provide a framework for a multitude of application. In the current book, techniques that show how to obtain analytic solutions to Maxwell's equations for ideal materials and boundary

[1] In 2004, Physics Web readers voted Maxwell's equations and Euler's equation to be the greatest equations of all time.

conditions are presented. These solutions are then used as a benchmark for the student to solve "real world" problems via computational techniques, first confirming that a computational technique gives the same answer as the analytic solution for an ideal problem. A subsequent book, *The Foundations of Signal Integrity*,[xiv] concentrates on the solutions to Maxwell's equations in a variety of media (various flavors of jelly) and with a variety of boundary conditions.

This information is presented to the twenty-first-century students[2] in the hope that they will consider mathematical and physical concepts as *integral*. The students are challenged to not accept uncertainty but to be honest with themselves in appreciating and understanding the derivations of the electromagnetic giants. After the mathematical solution has been obtained, we hope that the students will ask, "What are these equations telling me?" and "How could I use these in some other application?" Perhaps the students will delve even deeper to ask, "What are the physical phenomena that cause fields to exist, to move, to reflect, or to transmit through materials?" With such an armada of knowledge, the students can take these electromagnetic concepts to further applications and to further "stand on the shoulders of giants"[3] (perhaps for monetary gain).

And while the students may criticize the concept of luminiferous ether and make fun of the ancient practice of including an essay on *Prayer* in the context of natural law, they should review the beauty of such a set of symmetric equations named for Maxwell. It is worthy of note that, while we call these equations Maxwell's equations, a student might ask, "*Why* do these equations describe nature in such a simple form?" "Is it possible that the mind of man is incapable of understanding or postulating a more complex set of equations?" or "Is it not possible that nature is so mathematically beautiful because God made it that way for our pleasure?"

NOTES

i. James Clerk Maxwell, "A Dynamical Theory of the Electromagnetic Field," *Philosophical Transactions of the Royal Society of London* 155 (1865): 459–512. Oliver Heaviside reformulated Maxwell's equations (originally in quaternion format) to this asymmetric vector form. In Chapter 7, concepts of electric vector potential **and** magnetic vector potential are shown to make the equations fully symmetric.

ii. http://www-groups.dcs.st-and.ac.uk/~history/Mathematicians/Maxwell.html

iii. James Clerk Maxwell, *A Treatise on Electricity & Magnetism*, Vol. 1, unabridged 3rd ed. (New York: Dover, 1954), Vol. 2 (La Verne, CA: Merchant Books, 2007).

[2] One reader from the Physics Web polled that rated Maxwell's equations as the most beautiful equations ever derived recalled how he learned Maxwell's equations during his second year as an undergraduate student, "I still vividly remember the day I was introduced to Maxwell's equations in vector notation," he wrote. "That these four equations should describe so much was extraordinary... For the first time I understood what people meant when they talked about elegance and beauty in mathematics or physics. It was spine-tingling and a turning point in my undergraduate career."

[3] The quote "If I have seen farther than others, it is because I have stood on the shoulders of giants" was attributed to Sir Issac Newton because it appeared in a letter he wrote to Robert Hooke in 1675 but was also used by an eleventh-century monk named John of Salisbury, and there is evidence he may have gotten it from an older text while studying with Abelard in France.

iv. Paul Arthur Schilpp, ed., *Albert Einstein: Philosopher-Scientist* (La Salle, IL: Open Court, 1951), 63. Einstein once wrote, "The special theory of relativity owes its origin to Maxwell's equations of the electromagnetic field."
v. John Tyndall, *Fragments of Science*, Vol. 1 (New York: D. Appleton, 1897), 4.
vi. *Fortnightly Review* (1865), Vol. 3, p. 129.
vii. A lighthearted quip by James Baird, a PhD student of Norman Ramsey, from a lecture at the Oak Ridge National Laboratory in 1967.
viii. President's Address to the Mathematical and Physical Section of the British Association at Norwich.
ix. Hans J. Wever and George B. Arfken, *Mathematical Methods for Physicists*, 6th ed. (Burlington, MA: Elsevier Academic Press, 2005).
x. John David Jackson, *Classical Electrodynamics*, 3rd ed. (Danvers, MA: John Wiley & Sons, 1999).
xi. Murray R. Spiegel, *Schaum's Outline Series: Advanced Mathematics for Engineers and Scientists* (McGraw-Hill, 1999).
xii. Textbooks that support practical design practices are Stephen H. Hall, Garrett W. Hall, and James A. McCall, *High-Speed Digital System Design* (New York: John Wiley & Sons, 2000), Stephen H. Hall and Howard L. Heck, *Advanced Signal Integrity for High-Speed Digital Designs* (Hoboken, NJ: John Wiley & Sons, 2009), and Howard Johnson and Martin Graham, *High-Speed Signal Propagation: Advanced Black Magic* (Upper Saddle River, NJ: Prentice Hall, 1993).
xiii. T. W. Barrett, *Topological Foundations of Electromagnetism*, World Scientific Series in Contemporary Chemical Physics, Vol. 26, 2008, begins with "Electromagnetic Phenomena Not Explained by Maxwell's Equations."
xiv. Paul G. Huray, *The Foundations of Signal Integrity* (Hoboken, NJ: John Wiley & Sons, 2009).

Chapter 1

Foundations of Maxwell's Equations

LEARNING OBJECTIVES

- Review selected chronological developments of electromagnetic concepts
- Appreciate the role of electromagnetic theory in electrical engineering
- Use fundamental electromagnetic field quantities, units, and universal constants
- Use statistical concepts for determining the precision of a measured number
- Understand and apply principles of complex variables and phasor notation

1.1 HISTORICAL OVERVIEW

Some credit the existence of electric charge to a discovery more than two and a half thousand years ago by a Greek astronomer and philosopher, Thales of Miletus. He found that an amber (ἤλεκτρον) rod, after being rubbed with silk or wool, would attract straw and small pieces of parchment. The Greek word for amber is *éléktron*, from which the words *electron*, *electronics*, *electricity*, *electromagnetic*, and *electrical engineer* are derived.

The discovery of the magnetic polarities of lodestone (μάγνηζ), a natural material found in the Thessalian Magnesia, from which we derive[i] the name *magnetic*, by Pierre de Maricourt occurred around 1269. From that time through the early seventeenth century, progress in the study of magnetism was slow, but, during the seventeenth century, there were notable contributions by a number of scientists toward understanding magnetism. A. Kirchner demonstrated that the two poles of a magnet have equal strength, and Newton attempted to formulate the laws governing the forces between bar magnets.

The inverse square law of electric and magnetic forces was not postulated until John Michell proposed it in 1750 and Coulomb confirmed it in 1785. Coulomb's

Maxwell's Equations, by Paul G. Huray

law may be said to be the starting point of modern electromagnetic theory. Subsequent landmark developments in electromagnetic theory include the derivation by Laplace in 1782 and Poisson in 1813 of the famous equations that bear their names. Gauss published the divergence theorem, often called Gauss's law, in the same year.

Experiments with electric current could be performed only after invention of the battery by Volta in 1800. Having a source for generating a continuous current, Oersted, in 1820, was able to demonstrate the production of magnetic fields by electric currents. His discovery prompted others to investigate the relationship between electric current and magnetic fields. In 1820, Ampere announced a discovery relating to the forces between electric current-carrying conductors and magnets and the mutual attraction or repulsion of two electric currents. These experiments led to the formulation of what is now called Ampere's law. In 1820, Biot and Savart repeated Oersted's experiment to determine a law of force between current carrying conductors, giving us the so-called Biot–Savart law.

During the period of Oersted and Ampere, Faraday was also experimenting on the interaction between current-carrying conductors and magnetic fields and developed an electric motor in 1821. Faraday's experiments on developing induced currents by changing magnetic fields led to the law of electromagnetic induction in 1831. Faraday also proposed the concept of magnetic lines of force and laid the foundation of electromagnetic field theory.

In 1864, Maxwell proposed[ii] *A Dynamical Theory of the Electromagnetic Field* and thus unified the experimental researches of over a century through a set of equations known as Maxwell's equations. These equations were verified by Hertz in 1887 in a brilliant sequence of demonstrations. It is now generally accepted that all electromagnetic phenomena are governed by Maxwell's equations.

1.2 ROLE OF ELECTROMAGNETIC FIELD THEORY

Electromagnetic field theory is the study of the electric and magnetic phenomena caused by electric charges, q, at rest or in motion. There are two kinds of electric charges, positive and negative, following a definition given by Benjamin Franklin. Both positive and negative charges are sources of an electric field intensity,[1] $\boldsymbol{E}$ (or $\vec{E}$). Moving charges produce a current that can further give rise to a magnetic field intensity, $\boldsymbol{H}$ (or $\vec{H}$). A vector field is defined as a spatial distribution of a vector quantity, which may or may not be function of time. A time-varying electric field intensity is always accompanied by a magnetic field intensity and vice versa. In other words, time-varying electric and magnetic field intensities are intrinsically coupled and result in an electromagnetic field intensity. Time-dependent electromagnetic field intensities produce waves that radiate from their source toward an observation point. Many authors call this the causality principle because, they argue, the phenomenon does not work in the opposite direction. But, in our study of electromagnetic

[1] Other authors often use a bold type or a capital letter with an overhead vector to represent a vector quantity and interchange the two designations freely. This book will also color-code the electric field intensity and magnetic field intensity to make their representation clear in equations and drawings.

fields propagating in a waveguide, we will see that boundary conditions at a conducting surface boundary require a charge density distribution to support the electric and magnetic field intensities defined by Maxwell's equations. Because conduction electrons do not travel at velocities comparable with field propagation velocities, we can argue that the surface charges on the conductors must be induced by the field intensities. Such a picture of the physical universe gives symmetry to nature as defined by a "*principle of equivalence*": It is equivalent to view charges and currents as the source of electromagnetic fields or to view electromagnetic fields as the source of induced charges and currents.

The concept of propagating fields and waves is essential in the explanation of action at a distance. Satellite and mobile communications demonstrate that electric fields and magnetic fields propagate; that electromagnetic waves move in free space or in a medium such as air, water, resin fiberboard, or any other material. As we will see, they propagate *without* the presence of a luminiferous ether or "jelly."

Electromagnetic field theory is important in that it can explain many phenomena and solve complicated problems that conventional circuit theory cannot address. For instance, a mobile antenna can receive signals transmitted from base stations, where there are no physical connections between the transmitter and receiving antennas, and no free-space currents or voltages defined as in circuit theory. Another good example is the strong coupling that may exist between components printed some distance apart on circuit boards even though there are no identifiable resistance, capacitance, or inductance elements between them. By using computer techniques and electromagnetic theory, however, the intentional coupling between widely separated antennas and the unintentional coupling between nearby circuit components can be accurately predicted. In the discipline of Signal Integrity, the phenomenon is called "cross talk."

1.3 ELECTROMAGNETIC FIELD QUANTITIES

Historically, quantities in electromagnetic field theory are divided into two categories: source quantities and field quantities. The *source* of an electromagnetic field usually refers to electric charges at rest or in motion, while field quantities are usually observed or computed at an *observation* or *field* point. In this chapter, we will distinguish between classical view of cause and effect, at least for the purpose of discussion by routinely displaying source coordinates with a prime, for example, (x', y', z'); and field or observation coordinates as unprimed, for example, (x, y, z). However, we are mindful that it is equivalent to take the view that fields induce charges or charges induce fields and we shall see in the case of field propagation in a transmission line or waveguide that this duality can lead to a more complete understanding of power loss.

Electric Charges and Charge Densities

The symbol q or Q is used to denote electric charge, which is a fundamental property of matter and exists only in positive or negative integral multiples of the charge on an electron, $-e$, where

$$e = 1.60217653(14) \times 10^{-19}\ \mathrm{C}. \tag{1.1}$$

C is the abbreviation for the meter–kilogram–second (or International System of Units [SI]) unit of charge, coulomb.[2] A coulomb is a very large unit for charge because it takes $1/1.60 \times 10^{-19}$ or 6.25×10^{18} electrons to make up 1 C. The quantity in parenthesis (14) is the standard deviation in all measurements that have been compiled by the National Institute of Standards and Technology to obtain an average of the measured values of e. This and other quantities described below can be found at http://physics.nist.gov/cgi-bin/cuu/Value?e.

The principle of conservation of electric charge is a fundamental postulate. The statement that electric charge is *conserved* simply means that it can neither be created nor destroyed. The principle of conservation of electric charge must be satisfied at all times and in all situations in electrical engineering.

Next, we define a volume charge density, ρ_v, as a source quantity as follows:

$$\rho_v = \lim_{\Delta v \to 0} \frac{\Delta q}{\Delta v} = \frac{dq}{dv}\left(\mathrm{C/m^3}\right), \tag{1.2}$$

where Δq is the amount of charge in a very small volume Δv. In many cases, an amount of charge Δq may be identified with an element of surface, Δs, or an element of line, Δl. In such cases, it will be more appropriate to define a surface charge density, Σ_s, or a line charge density, λ_l:

$$\Sigma_s = \lim_{\Delta s \to 0} \frac{\Delta q}{\Delta s} = \frac{dq}{ds}\left(\mathrm{C/m^2}\right) \tag{1.3}$$

$$\lambda_l = \lim_{\Delta l \to 0} \frac{\Delta q}{\Delta l} = \frac{dq}{dl}\left(\mathrm{C/m}\right). \tag{1.4}$$

In general, all charge densities are point functions of space coordinates and may also be time dependent. In some texts, the surface charge density, Σ_s, may be labeled σ_s or ρ_s, and the line charge density, λ_l, may be labeled ρ_l. Alternate labeling is necessary in preventing confusion when, in the same section or publication, we discuss electrical conductivity, traditionally labeled σ; and/or scattering cross section, traditionally labeled σ_s. Likewise, we often refer to the distance to the z-axis in cylindrical coordinates by the symbol ρ.

Current and Current Density

Electric current is the rate of transfer of charge across a reference surface with respect to time;[3] that is,

[2] One of the oddities of science and technology is that we traditionally do not capitalize the written unit that represents a person's name (like Coulomb) unless the symbol (e.g., C) is used for that unit.

[3] As mentioned earlier, a time-varying electric field intensity is always accompanied by a magnetic field intensity and vice versa. In this book, charges and electric field intensities will be colored red and currents and magnetic field intensities blue. Thus, a time derivative of a red quantity produces a blue quantity, as shown in Equation 1.5 and vice versa.

$$I = \lim_{\Delta t \to 0} \frac{\Delta q}{\Delta t} = \frac{dq}{dt} \text{ (C/s or A)}, \tag{1.5}$$

where the unit of current is a coulomb per second (C/s) or ampere (A). A physical current must flow through a finite area; hence, it is not a point function but may be time dependent. However, in electromagnetic field theory, we define a vector point function, current density, $\vec{J}$, which measures the amount of current flowing through a unit area normal to the direction of current flow. The current density $\vec{J}$ is a vector whose magnitude and direction are the current per unit area (A/m^2), and the direction of current flow at a point in space, respectively. $\vec{J}$ may also be a time-dependent quantity.

Electromagnetic Field Quantities

An electromagnetic field can be described by four field quantities:

Electric field intensity	$\vec{E}$ (V/m);
Electric flux density or displacement	$\vec{D}$ (C/m^2);
Magnetic field intensity	$\vec{H}$ (A/m); and
Magnetic flux density	$\vec{B}$ (Wb/m^2 or T).

Here, the unit T stands for the tesla or volt-second per square meter and is named in honor of Nikola Tesla (1857–1943), who helped the understanding of rotating field poles in electric motors and transformers. The electric field intensity $\vec{E}$ is the vector field used in electrostatics when charge is at rest in free space and is defined as the electric force on a unit test charge. The electric displacement vector $\vec{D}$ (also called the electric flux density or displacement flux) is a vector field used in studying the electric fields inside material objects. Similarly, magnetic field intensity $\vec{H}$ is a vector needed in discussing magnetic phenomonen, that is the field generated at a point in free space by steady or time-varying electric currents in a source; it is related to the magnetic force acting on a moving charge. The magnetic flux density $\vec{B}$ is useful in the investigation of the magnetic fields within material objects where the material modifies the field intensity.

When there is no time variation in field quantities, the electric field quantities ($\vec{E}$, $\vec{D}$) are independent from the magnetic field quantities ($\vec{H}$, $\vec{B}$). In time-dependent cases, however, the electric and magnetic fields are coupled; that is, time-varying ($\vec{E}$, $\vec{D}$) will give rise to ($\vec{H}$, $\vec{B}$) and vice versa. The electromagnetic properties of materials are governed by the so-called *constitutive* relations between $\vec{E}$ and $\vec{D}$, and $\vec{H}$ and $\vec{B}$. The equations that represent these constitutive relations are called Maxwell's equations.

1.4 UNITS AND UNIVERSAL CONSTANTS

In this book, as in most contemporary engineering texts, we will adhere to the SI, often called the meter–kilogram–second system built from seven basic units, as shown in Table 1.1. All derived units can be expressed in terms of these quantities.

In the SI system, the speed of light is an exact quantity as a consequence of the definition of the meter adopted in 1983, the definition of the kilogram adopted in 1889, the definition of the second adopted in 1967:

1. **Meter** is the length of the path traveled by light in a vacuum during a time interval of 1/299,792,458 of a second.
2. **Kilogram** is the unit of mass; it is equal to the mass of the international prototype of the kilogram.
3. **Second** is the duration of 9,192,631,770 periods of the radiation corresponding to the transition between the two hyperfine levels of the ground state of the cesium-133 atom.
4. **Ampere** is that constant current that, if maintained in two straight parallel conductors of infinite length, of negligible circular cross section, and placed 1 m apart in vacuum, would produce between these conductors a force equal to 2×10^{-7} newton per meter of length.
5. **Kelvin**, the unit of thermodynamic temperature, is the fraction 1/273.16 of the thermodynamic temperature of the triple point of water.
6. **Mole** is the amount of substance of a system that contains as many elementary entities as there are atoms in 0.012 kg of carbon 12; its symbol is "mol." When the mole is used, the elementary entities must be specified and may be atoms, molecules, ions, electrons, other particles, or specified groups of such particles.
7. **Candela** is the luminous intensity, in a given direction, of a source that emits monochromatic radiation of frequency 540×10^{12} Hz and that has a radiant intensity in that direction of 1/683 W per steradian.

Table 1-1. Seven Basic Units

Quantity	Unit	Abbreviation
Length	Meter	m
Mass	Kilogram	kg
Time	Second	s
Current	Ampere	A
Temperature	Kelvin	K
Amount of substance	Mole	mol
Luminous intensity	Candela	cd

In electromagnetic field expressions, we frequently encounter three constants: the speed of light (and all other electromagnetic waves) in free space, c, the dielectric permittivity of free space, ε_0, and the magnetic permeability of free space, μ_0. Note that there is no uncertainty (no standard deviation) in any of these terms because they are defined exactly.

We define

$$c \equiv 299{,}792{,}458\,(\mathrm{m/s}) \approx 3\times10^{8}\,(\mathrm{m/s}). \tag{1.6}$$

In addition, the magnetic permeability of free space, μ_0, is defined as

$$\mu_0 \equiv 4\pi\times10^{-7}\,(\mathrm{H/m})\left(\mathrm{or\ N/A^2}\right)(\mathrm{or\ \Omega s/m}). \tag{1.7}$$

Thus, using an equality that we will later derive for free space that includes the electric permittivity of free space, ε_0, we can deduce the exact value

$$\varepsilon_0 \equiv 1/\mu_0 c^2 = 8.854187817\ldots\times10^{-12}\,(\mathrm{F/m})\left(\mathrm{or\ C^2/N\,m^2}\right)(\mathrm{or\ s/\Omega m}) \tag{1.8}$$

where the units H/m and F/m stand for henry per meter and farad per meter, respectively. We again note that, because they are defined, there is no uncertainty in any of the constants c, ε_0, or μ_0.

For convenience, we will often use the value 3×10^8 m/s for the speed of light because it is easier to recall than the defined figure, and, consistent with this approximation and Equation 1.8, we will often use the approximation $1/36\pi\times10^9\,\mathrm{C^2/N\,m^2}$ for ε_0. This practice is common in the study of electromagnetic fields and seldom leads to significant error. Nonetheless, in critical computations, the more accurate values of c and ε_0 may be required.

In free space, the constants ε_0 and μ_0 are the proportionality constants between the electric field intensity, $\vec{E}$, and the electric flux density, $\vec{D}$, and the magnetic field intensity, $\vec{H}$, and the magnetic flux density, $\vec{B}$, respectively, such that

$$\vec{D} = \varepsilon_0\vec{E}\ \text{(in free space)} \tag{1.9}$$
$$\vec{B} = \mu_0\vec{H}\ \text{(in free space)}. \tag{1.10}$$

Finally, we note that the force, $\vec{F}_{12}$, between two charges, q_1 and q_2, is given by the experimentally confirmed Coulomb's law, which is expressed as

$$\vec{F}_{12} = k_e\frac{q_1q_2}{r_{12}^2}\hat{a}_{12} = \frac{1}{4\pi\varepsilon_0}\frac{q_1q_2}{r_{12}^2}\hat{a}_{12}, \tag{1.11}$$

where k_e is Coulomb's constant and is approximately equal to $9\times10^9\,\mathrm{N\,m^2/C^2}$.

From Equation 1.11, we can see that a measurement of $\vec{F}_{12}$ (in kg m/s^2) and of r_{12}^2 (in m^2), with the derived quantity for ε_0 of Equation 1.8 for two identical charges, q, leads to a measured value of charge, q, in units of mass, length, and time.

The measured value of the ampere as defined by the SI is found from the force created by two parallel wires of length dl_1-carrying current I_1 and dl_2-carrying current I_2 respectively, by the Biot–Savart force law:

$$\vec{F}_{12} = \frac{\mu_0}{4\pi}\frac{(I_1dl_1)(I_2dl_2)}{r_{12}^2}\hat{a}_{12}. \tag{1.12}$$

As we will see, the inverse square "law" of Coulomb (Equation 1.11) and the inverse square "force law" of Biot–Savart (Equation 1.12) also lead us to Maxwell's equations. Many researchers have tried, unsuccesfully to date, to measure any deviation from the inverse square law for these quantities. It is worthy of note that the gravitational force between two masses, m_1 and m_2, separated by r_{12} follows a mathematical expression similar to that of Coulomb's law or the Biot–Savart law:

$$\vec{F}_{12} = G\frac{m_1 m_2}{r_{12}^2}\hat{a}_{12} = \frac{1}{4\pi\varepsilon_g}\frac{m_1 m_2}{r_{12}^2}\hat{a}_{12}, \tag{1.13}$$

where $G = 6.673(10) \times 10^{-11}\ \mathrm{m^3/kg\,s^2}$ is the gravitational constant. In this equation, the author has chosen to define a new constant, ε_g, so that the gravitational force law looks the same as Coulomb's law.

The symmetry of the equations leads the casual observer to postulate[4] another force due to the *mass current* K_1 and K_2 in two parallel lengths dl_1 and dl_2, respectively:

$$\vec{F}_{12} = \frac{\mu_g}{4\pi}\frac{(K_1 dl_1)(K_2 dl_2)}{r_{12}^2}\hat{a}_{12}. \tag{1.14}$$

Here, another constant, μ_g, has been defined in order to make the force due to *mass currents* symmetric to the Biot–Savart force law. This force has been postulated by others and a measurement of it is being attempted by a group of researchers from Stanford and NASA.[iii]

Many researchers have also tried to measure deviations from an inverse square law for Equation 1.13 as well. It is partly the similarity of these forces that gives us confidence that the "laws" are correct. However, we should note that more powerful forces within the nucleus, the weak and strong forces, do not obey an inverse square law, so we should leave open the possibility that a future correction may need to be made to any one or all of these "laws."

EXERCISES

1.1 Compare the gravitational and electric forces[5] between a proton and an electron if they are separated by the same distance, as shown in Figure 1.1.

SOLUTION Suppose an electron at $\vec{r}_1$ and a proton at $\vec{r}_2$, so the distance between them is $r_{12} = |\vec{r}_{12}|$ where $\vec{r}_{12} = \vec{r}_2 - \vec{r}_1$.

We know from Coulomb's law that $\vec{F}_{12} = k_e(q_1 q_2/r_{12}^2)\hat{r}_{12}$ is the electrostatic force between the electron and the proton and from Newton's law

[4] Areas of speculation are often used in this text in shaded boxes; they are intended to stimulate thinking of the student on a topic she might not otherwise have considered.

[5] Gravitational force $\ll$ weak force $\ll$ electromagnetic force $\ll$ strong forces.

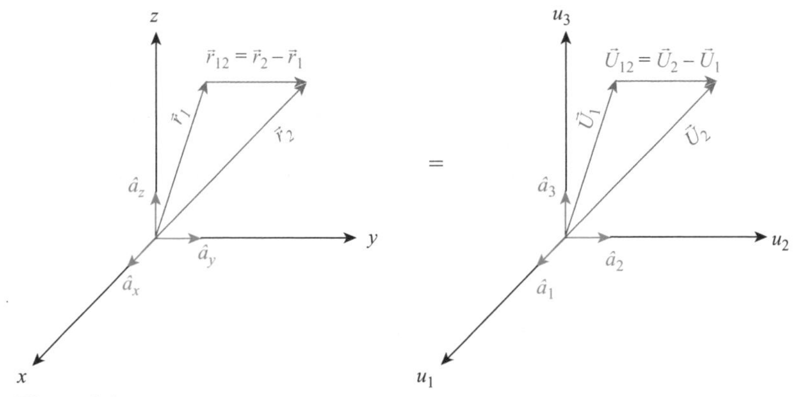

Figure 1.1 Vector representations of physical locations in space.

that $\vec{F}_{12} = G(m_1 m_2/r_{12}^2)\hat{r}_{12}$ is the gravitational force between the electron and the proton.

NOTE Both terms are attractive, both are proportional to the product of two measured quantities (charge and mass, respectively), and both are proportional to the inverse square of their separation. The gravitational constant G is $6.67 \times 10^{-11}\,\mathrm{m^3/kg\,s^2}$ and the electric constant k_e is $8.99 \times 10^9\,\mathrm{N\,m^2/C^2}$. The measured masses and charges are $m_e = 9.11 \times 10^{-31}\,\mathrm{kg}$, $q_e = -1.60 \times 10^{-19}\,\mathrm{C}$, $m_p = 1.67 \times 10^{-27}\,\mathrm{kg}$, $q_p = +1.60 \times 10^{-19}\,\mathrm{C}$. Thus,

$$\vec{F}_{\text{due to electrostatic charges}} = \left(8.99\times10^9\,\mathrm{N\,m^2/C^2}\right)\left(1.60\times10^{-19}\,\mathrm{C}\right)\left(1.60\times10^{-19}\,\mathrm{C}\right)\left(\hat{r}_{12}/r_{12}^2\right)$$

$$\vec{F}_{\text{due to gravity}} = \left(6.67\times10^{-11}\,\mathrm{N\,m^2/kg^2}\right)\left(1.67\times10^{-27}\,\mathrm{kg}\right)\left(9.11\times10^{-31}\,\mathrm{kg}\right)\left(\hat{r}_{12}/r_{12}^2\right)$$

and the ratio of these two forces is 2.27×10^{39}, independent of their separation.

SOLUTION The electrostatic force between an electron and a proton is so much larger than the gravitational force between an electron and a proton that we may ignore the gravitational forces.

1.2 Using classical arguments for an electron bound to a proton in a hydrogen atom with a circular radius of 1 Å, determine its tangential velocity.

SOLUTION From Exercise 1.1, $\vec{F}_{\text{due to electrostatic charges}} = 2.30 \times 10^{-28}\,\mathrm{Nm^2}(\hat{r}_{12}/r_{12}^2)$.

For $r_{12}^2 = (1\,\text{Å})^2 = 10^{-20}\,\mathrm{m^2}$, $\vec{F}_{\text{due to electrostatic charges}} = 2.30 \times 10^{-8}\,\mathrm{N}\hat{r}_{12}$.
This force seems small until you use it to compute the acceleration of an electron:

$$|a_e| = \left|\vec{F}_{\text{due to electrostatic charges}}\right|/m_e = 2.3\times10^{-8}\,\mathrm{N}/9.11\times10^{-31}\,\mathrm{kg}$$
$$= 2.52\times10^{22}\,\mathrm{m/s^2} = 2.58\times10^{21}\,\mathrm{g},$$

where g is the acceleration of gravity.

From our knowledge of centripetal forces, for an electron circling a proton at a radius of 1 Å, $a_e = v_t^2/r$, where v_t is the electron's tangential velocity. Thus

$$v_t^2 = ra_e = \left(10^{-10}\,\mathrm{m}\right)\left(2.52\times10^{22}\,\mathrm{m/s^2}\right) \text{ or } v_t = 1.58\times10^6\,\mathrm{m/s}.$$

CONCLUSION The tangential velocity of an electron circling a hydrogen nucleus (a proton) is approximately 0.5% the speed of light.[6] For larger Z atoms, the accelerations and tangential velocities will be even closer to the speed of light, so we must take into account relativistic effects when computing electron velocities for heavy atoms.

Electrical engineers prefer to express the properties of an electromagnetic wave[iv] via its frequency, f (in Hz), or its wavelength, λ (in m), which are related by

$$c = \lambda f. \tag{1.15}$$

Physicists and astronomers often express the properties of an electromagnetic wave via its energy, E, or its temperature, T, which are related by

$$E = hf, \tag{1.16}$$

where

$$h = 6.62606876(52) \times 10^{-34}\ \mathrm{J\,s}\ \left(\text{or } 4.13566727(16) \times 10^{-15}\ \mathrm{eV\,s}\right)$$

is Plank's constant.

$$E = k_B T, \tag{1.17}$$

where $k_B = 1.3806503(24) \times 10^{-23}$ J/K is Boltzmann's constant.

1.3 Find the wavelength, energy, and temperature of a 2.4-GHz wave.[7]

SOLUTION

$$\lambda = \frac{c}{f} = \frac{3.00 \times 10^8\ \mathrm{m/s}}{2.40 \times 10^9\ 1/\mathrm{s}} = 0.125\,\mathrm{m} = 12.5\,\mathrm{cm}$$

$$E = hf = \left(4.14 \times 10^{-15}\ \mathrm{eV\,s}\right)\left(2.40 \times 10^9\ 1/\mathrm{s}\right) = 9.94 \times 10^{-6}\ \mathrm{eV} = 1.59 \times 10^{-24}\ \mathrm{J}$$

$$T = \frac{E}{k_B} = \frac{1.59 \times 10^{-24}\ \mathrm{J}}{1.38 \times 10^{-23}\ \mathrm{J/K}} = 0.115\,\mathrm{K}$$

1.4 Find the wavelength, frequency, and characteristic temperature of a 1 keV x-ray.[8]

SOLUTION

$$T = \frac{E}{k_B} = \frac{(10^3)(1.60 \times 10^{-19}\ \mathrm{C})V}{1.38 \times 10^{-23}\ \mathrm{J/K}} = 11.6 \times 10^6\ \mathrm{K}$$

$$f = \frac{E}{h} = \frac{(10^3)\mathrm{eV}}{4.14 \times 10^{-15}\ \mathrm{eV\,s}} = 2.42 \times 10^{17}\ \mathrm{Hz}$$

[6] Had our answer come out closer to the speed of light, we would need a recalculation using the special theory of relativity for mass rather than the classical theory for rest mass.

[7] This frequency is common in computer central processing units (CPUs), cell phones, and microwave ovens.

[8] This energy is common at the face of a cathode ray tube (CRT) if electrons are accelerated by a 1 kV potential.

$$\lambda = \frac{c}{f} = \frac{3.00\times10^{8}\ \text{m/s}}{2.42\times10^{17}\ 1/\text{s}} = 1.24\times10^{-9}\ \text{m} = 12.4\ \text{Å}$$

1.5 Because $1/\sqrt{\mu_0\varepsilon_0} = c$ for electromagnetic waves, it is not unreasonable to postulate that $1/\sqrt{\mu_g\varepsilon_g} = c$ for gravomagnetic waves. With this postulate, find the value of the constants ε_g and μ_g and compare the magnitude of the force caused by a 1 C/s electrical current with that of a 1 kg/s mass current if they are in the same lengths and have the same distance of separation.

SOLUTION
If $G = 6.673 \times 10^{-11}\ \text{N m}^2/\text{kg}^2 = 1/4\pi\varepsilon_g$, then $\varepsilon_g = 1.193 \times 10^{9}\ \text{kg s}^2/\text{m}^3$.
If $1/\sqrt{\mu_g\varepsilon_g} = c$, then $\mu_g = 9.317 \times 10^{-27}$ m/kg.

$$\frac{F_{\text{charge current}}}{F_{\text{mass current}}} = \frac{\mu_0}{\mu_g} = 1.35\times10^{20}$$

1.6 If there is one "free electron" (conduction electron) per Cu atom,[9] compute the number of free electrons in a 1-m^3 block of Cu and find the average velocity (drift velocity) of electrons needed to produce a current of 1 C/s in one direction and the mass current of those same electrons.

SOLUTION The number of "free electrons" in a block of copper is

$$N = \frac{\text{density}}{\text{molar mass}}\text{Avogadro's number} = \frac{8.93\times10^{3}\ \text{kg/m}^3}{64\times10^{-3}\ \text{kg/mol}}\left(6.023\times10^{23}\ \text{e/mol}\right)$$

and

$$q_N = N\left(1.60\times10^{-19}\ \text{C/e}\right) = 1.34\times10^{10}\ \text{C/m}^3.$$

is the "free electron" charge in a block of copper. If this charge in a 1-m^3 block is moving across one of the 1-m^2 faces at a velocity of 1 m/s, then it will produce a current I = 1.34 × 10^{10} C/s. Thus, to produce an electric current of 1 C/s, $\langle v \rangle$ need be only 7.44 × 10^{-11} m/s.

The mass of the "free electrons" in the block of copper is $m_N = N$ (9.11 × 10^{-31} kg/e) = 0.0766 kg/m^3. If this mass is moving at an average velocity of 7.44 × 10^{-11} m/s, then the mass current across a 1-m^2 face of the block will be 5.70 × 10^{-11} kg/s.

[9] The designation "free electrons" is given by those electrons outside the bound core of an ion; these electrons interact with their neighbors to such an extent that they lose track of which one was their parent and thus are "free" to move in the conductor.

1.5 PRECISION OF MEASURED QUANTITIES

Standard Uncertainty and Relative Standard Uncertainty

Definition

The **standard uncertainty** σ_y of a measurement result, y, is the estimated standard deviation of y.

Meaning of Uncertainty

If the probability distribution characterized by the measurement result y and its standard uncertainty σ_y is approximately normal (Gaussian), and σ_y is a reliable estimate of the standard deviation of y, then the interval from $y - \sigma_y$ to $y + \sigma_y$ is expected to encompass approximately 68.26% of the distribution of values that could reasonably be attributed to the value of the quantity Y of which y is an estimate. This implies that it is believed with an approximate level of confidence of 68.26% that Y is greater than or equal to $y - \sigma_y$ and is less than or equal to $y + \sigma_y$, which is commonly written as $Y = y \pm \sigma_y$.

Use of Concise Notation

If, for example, $y = 1234.56789\,\text{U}$ and $\sigma_y = 0.00011\,\text{U}$, where U is the unit of y, then $Y = (1234.56789 \pm 0.00011)$ U. A more concise form of this expression, and one that is in common use, is $Y = 1234.56789(11)$ U, where it understood that the number in parentheses is the numerical value of the standard uncertainty referred to the corresponding last digits of the quoted result.

Appendix A contains a review of statistical definitions, examples, and interpretations. See http://physics.nist.gov/cuu/Uncertainty/index.html for additional information.

1.6 INTRODUCTION TO COMPLEX VARIABLES

Complex numbers are frequently used in the applications of electromagnetic applications. In this section, the definition and fundamental operations of complex numbers and complex variables will be reviewed.

A complex number, z, can be written as

$$z = \text{Re}(z) + j\,\text{Im}(z) = x + jy, \tag{1.18}$$

where x and y are both real numbers; x is said to be the real (Re) part of z, y is said to be the imaginary (Im) part of z, and $j = \sqrt{-1}$. z can be also be expressed in polar form by

$$z = |z|e^{j\theta}, \tag{1.19}$$

where $|z|$ and θ are both real and are called the *amplitude* and *phase* of z. With the use of Euler's identity

$$e^{j\theta} = \cos\theta + j\sin\theta, \tag{1.20}$$

we obtain

$$z = |z|\cos\theta + j|z|\sin\theta. \tag{1.21}$$

Comparing Equations 1.20 and 1.21, we conclude

$$x = |z|\cos\theta, \tag{1.22a}$$
$$y = |z|\sin\theta, \tag{1.22b}$$

and, inversely,

$$|z| = \sqrt{x^2 + y^2} \tag{1.23a}$$

$$\theta = \tan^{-1}\left(\frac{y}{x}\right) \quad 0 \le \theta \le 2\pi. \tag{1.23b}$$

The above relations can be graphically represented as shown in Figure 1.2.

The complex conjugate of z, designated with an asterisk (*), is a complex number that replaces j with $-j$ in all places; that is,

$$z^* = (x + jy)^* = (x - jy) = |z|e^{-j\theta}. \tag{1.24}$$

The magnitude of z is the square root of the product of z and its complex conjugate:

$$|z| = \sqrt{z \cdot z^*} = \sqrt{(x + jy)(x + jy)^*} = \sqrt{x^2 + y^2} \tag{1.25a}$$

or

$$|z| = \sqrt{|z|e^{j\theta}|z|e^{-j\theta}} = \sqrt{x^2 + y^2}. \tag{1.25b}$$

Arithmetic Operations with Complex Numbers

Arithmetic with complex numbers is tedious when carried out by hand but otherwise is very much like arithmetic with real numbers.

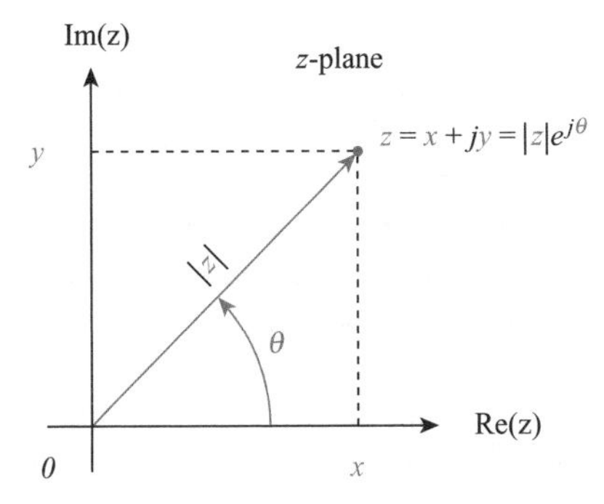

Figure 1.2 The relation between rectangular and polar coordinates.

1. Addition and subtraction:

$$z_1 + z_2 = (x_1 + jy_1) + (x_2 + jy_2) = (x_1 + x_2) + j(y_1 + y_2) \tag{1.26a}$$

$$z_1 = z_2 = (x_1 + jy_1) - (x_2 + jy_2) = (x_1 - x_2) + j(y_1 - y_2) \tag{1.26b}$$

2. Multiplication:

$$z_1 \cdot z_2 = (x_1 + jy_1)\cdot(x_2 + jy_2) = (x_1 \cdot x_2 - y_1 \cdot y_2) + j(x_1 \cdot y_2 + x_2 \cdot y_1) \tag{1.27a}$$

In the polar form, the multiplication of two complex numbers can be written as

$$\begin{aligned} z_1 \cdot z_2 &= |z_1| e^{j\theta_1} |z_2| e^{j\theta_2} = |z_1 z_2| e^{j(\theta_1 + \theta_2)} \\ &= |z_1 z_2| [\cos(\theta_1 + \theta_2) + j\sin(\theta_1 + \theta_2)] \end{aligned}. \tag{1.27b}$$

3. Division: For any $z_2 \neq 0$,

$$\begin{aligned} \frac{z_1}{z_2} &= \frac{(x_1 + jy_1)}{(x_2 + jy_2)} = \frac{(x_1 + jy_1)(x_2 - jy_2)}{(x_2 + jy_2)(x_2 - jy_2)} \\ &= \frac{(x_1 x_2 + y_1 y_2) + j(x_2 y_1 + x_1 y_2)}{(x_2^2 + y_2^2)} \end{aligned} \tag{1.28a}$$

or

$$\begin{aligned} \frac{z_1}{z_2} &= \frac{|z_1| e^{j\theta_1}}{|z_2| e^{j\theta_2}} = \frac{|z_1|}{|z_2|} e^{j(\theta_1 - \theta_2)} \\ &= \frac{|z_1|}{|z_2|} [\cos(\theta_1 - \theta_2) + j\sin(\theta_1 - \theta_2)] \end{aligned}. \tag{1.28b}$$

4. Power: For any positive or negative integer n, we have

$$\begin{aligned} z^n &= \left[|z| e^{j\theta}\right]^n = |z|^n e^{jn\theta} \\ &= |z|^n [\cos(n\theta) + j\sin(n\theta)] \end{aligned}. \tag{1.29}$$

Arithmetic Functions of Complex Numbers (Complex Variables)

A function of a complex number could be a combination of addition, multiplication, power, or other functions of a complex quantity, for example, $(z + 1/z)$, $\sin z$, e^z, $\tanh^{-1} z$. An excellent resource for the review of functions of complex numbers is given by Spiegel.[v]

In electrical engineering, it is common for the field vectors $\vec{E}$, $\vec{D}$, $\vec{H}$, $\vec{B}$, and the current density, $\vec{J}$, to be written as complex quantities. Furthermore, real and imaginary parts of the field vectors are likely to be functions of space and time. Such variable fields are called *complex variables*. The real part of the complex field vector (e.g., $\text{Re}\,\vec{E}$) is typically labeled as $u(x, y, z, t)$, and the imaginary part of the complex field vectors (e.g., $\text{Im}\,\vec{E}$) is typically labeled as $v(x, y, z, t)$ in Cartesian coordinates. It is shorthand to just write the electric field vector as $\vec{E}$ without denoting the fact that it is a complex quantity that depends on space and time coordinates. Some texts remind the student of this fact by expressing $\vec{E}$ as $\vec{E}(\vec{x}, t)$ no matter if the coordinate system is Cartesian, cylindrical, spherical, or other. Some texts even put a tilde over the vector field to remind the student that $\vec{E}$ is a complex variable, but we will not choose that complicated notation here. Our advice is to always assume that a quantity in question is a complex variable unless otherwise known or stated (e.g., x, y, z, r, θ, ϕ, and t are always real).

Fortunately, with computers, the tedious manipulation of complex numbers is very easy. Not infrequently, however, we will be required to carry out derivations using complex algebraic expressions. Although even symbolic simplification can be accomplished with computers, it will nonetheless be useful to become adept at doing complex algebra by hand.

1.7 PHASOR NOTATION

In electromagnetic engineering, electric and magnetic fields that vary sinusoidally with time play a large role. In the sense that an arbitrary but otherwise periodic field can be expanded into a Fourier series of sinusoidal components and a transient nonperiodic field can be expressed as a Fourier integral, we can concentrate on analyzing steady and sinusoidal fields with the confidence that our theory can be extended to the more general situation involving nonsinusoidal time dependence.

In this section, we first review the phasor notation and then represent Maxwell's equation with the phasors. Here, we would like to illustrate the uses of phasor notation by looking at some examples. Let us consider the series resistor, inductor, capacitor (RLC) circuit shown in Figure 1.3 with an applied voltage

$$V(t) = V_0 \cos(\omega t), \tag{1.30}$$

where V_0 is the amplitude of the voltage and ω is the angular frequency (rad/s), which is equal to $2\pi f$, with f being the frequency in Hz.

Our objective is to solve for the corresponding current $i(t)$, which, in general, can be expressed as

$$i(t) = I_0 \cos(\omega t + \phi), \tag{1.31}$$

where I_0 is the current amplitude and ϕ designates the current phase.

Using Kirchhoff's voltage law, we have

$$L\frac{di(t)}{dt} + Ri(t) + \frac{1}{C}\int i(t)\,dt = V(t). \tag{1.32}$$

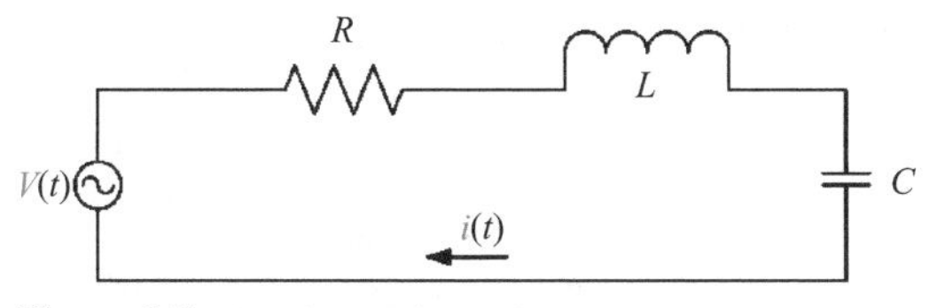

Figure 1.3 A series RLC circuit.

Using the phasor notation, we can express[10]

$$\begin{aligned} V(t) = V_0 \cos(\omega t) &= \mathrm{Re}\left[\left(V_0 e^{j0}\right) e^{j\omega t}\right] \\ &= \mathrm{Re}\left[V_S e^{j\omega t}\right] \end{aligned} \tag{1.33}$$

and

$$\begin{aligned} i(t) = I_0 \cos(\omega t + \phi) &= \mathrm{Re}\left[\left(I_0 e^{j\phi}\right) e^{j\omega t}\right] \\ &= \mathrm{Re}\left[I_S e^{j\omega t}\right] \end{aligned}. \tag{1.34}$$

With the phasor notation, we can deduce that

$$di/dt = \mathrm{Re}\left[j\omega I_S e^{j\omega t}\right] \tag{1.35}$$

$$\int i dt = \mathrm{Re}\left[\frac{1}{j\omega} I_S e^{j\omega t}\right]. \tag{1.36}$$

Substitution of Equations 1.35 and 1.36 into Equation 1.32 leads to

$$\left[R + j\left(\omega L - \frac{1}{\omega C}\right)\right] I_S = V_S \tag{1.37}$$

from which the phasor current, $I_S = V_S / Z_{Series\ RLC}$, can be easily obtained. We note that the phasor current, I_S, includes information about both the magnitude, I_0, and phase, φ, and that the corresponding instantaneous current, $i(t)$, then follows from Equation 1.34.

As seen in Equations 1.35 and 1.36, by using phasor notation, differentiation and integration in the time domain are converted to a simple algebraic operation symbolized by the following time-frequency conversion operations:

$$\frac{d}{dt} \Leftrightarrow j\omega \quad (\text{or } -i\omega \text{ in physics books}) \tag{1.38a}$$

$$\int dt \Leftrightarrow \frac{1}{j\omega} \quad \left(\text{or } \frac{1}{-i\omega} \text{ in physics books}\right). \tag{1.38b}$$

Because algebraic equations are much easier to solve than integral–differential equations, time-harmonic electromagnetic fields are much easier to analyze than time-varying fields.

NOTE Cheng and Hayt and Buck use a subscript S on the complex variable (e.g., I_S or V_S) to remind the reader that the variable is changing only with spatial quantities. Balanis and Inan and Indan use a script $\vec{\mathcal{E}}(x, y, z, t) = \vec{E}(x, y, z)e^{j\omega t}$ convention

[10] Electrical engineers almost always use the convention $e^{j\omega t}$ for the time dependence, while physicists and mathematicians conventionally use $e^{-i\omega t}$. In these cases, $j = \sqrt{-1}$ and $i = \sqrt{-1}$ so that, when we take the real part, $\cos(\omega t)$ results. However, if we take the derivative, $de^{j\omega t}/dt = j\omega e^{j\omega t}$ but $de^{-i\omega t}/dt = -i\omega e^{-i\omega t}$, we see that a negative sign occurs in equations with derivatives. We must take care to know which convention is being used in reading a given text when using Maxwell's equations.

to indicate that $\vec{E}$ varies only with spatial variables. Paul and Edminister use the convention $\vec{E}(x, y, z, t) = \vec{E}(x, y, z)e^{j\omega t}$ to indicate that $\vec{E}$ varies only with spatial variables. Pozar uses the convention that $\vec{E}(x, y, z, t) = \text{Re}[\vec{E}(x, y, z)e^{j\omega t}]$, and Rao use the convention $\vec{E}(x, y, z, t) = \vec{E}(x, y, z)e^{j\omega t}$ to indicate that $\vec{E}$ varies only with spatial variables. We will adopt the Rao convention in the following sections and will color electric fields in red and magnetic fields in blue in equations and figures. When the mathematician's convention for time harmonics, $e^{-i\omega t}$, is used, we will highlight that fact by coloring the imaginary number. When $e^{j\omega t}$ is used, there will be no color for the imaginary number.

Time-Harmonic Electromagnetics

For a general time-varying electromagnetic field, the differential (or point) form of Maxwell's equations can be written as

$$\nabla \times \vec{E} = -\mu \frac{\partial \vec{H}}{\partial t} \tag{1.39a}$$

$$\nabla \times \vec{H} = \vec{J} + \varepsilon \frac{\partial \vec{E}}{\partial t} \tag{1.39b}$$

$$\nabla \cdot \vec{E} = \frac{\rho_v(t)}{\varepsilon} \tag{1.39c}$$

$$\nabla \cdot \vec{B} = 0. \tag{1.39d}$$

Now, considering a time-harmonic electromagnetic field with the time variation of $cos(\omega t)$, we can write the electric and magnetic fields as

$$\vec{E}(x, y, z, t) = \text{Re}\left[\vec{E}(x, y, z)e^{j\omega t}\right] \tag{1.40a}$$

$$\vec{H}(x, y, z, t) = \text{Re}\left[\vec{H}(x, y, z)e^{j\omega t}\right], \tag{1.40b}$$

where $\vec{E}(x, y, z)$ and $\vec{H}(x, y, z)$ are vector phasors that contain information on direction, magnitude, and phase. Using phasor relations in Equation 1.38a, we can simplify Equations 1.39a to 1.39d as

$$\nabla \times \vec{E} = -j\omega\mu\vec{H} \tag{1.41a}$$

$$\nabla \times \vec{H} = \vec{J} + j\omega\varepsilon\vec{E} \tag{1.41b}$$

$$\nabla \cdot \vec{E} = \rho_v/\varepsilon \tag{1.41c}$$

$$\nabla \cdot \vec{B} = 0, \tag{1.41d}$$

where the time variable has been eliminated from the differential form 1.39 of Maxwell's equations. Equations 1.41 are called the *time-harmonic* differential form of Maxwell's equations. Note that all fields in above equations are phasor quantities, and, to convert them to the time domain, we only have to use the relations 1.40a and 1.40b.

PROBLEMS

Using Complex Numbers

1.1 Two complex numbers are given as $z_1 = 5 - j3$ and $z_2 = 4 + j6$.

a. Express z_1 and z_2 in polar form.

b. Determine $z_1 \cdot z_2$ in rectangular and polar forms.

c. Determine z_1/z_2 in rectangular and polar form.

d. Determine $(z_1)^3$ and $(z_2)^5$ in polar form.

1.2 If $z = 3 + j4$, determine following quantities in polar form.

a. z^3

b. $|z^3|$

c. $1/|z^3|$

d. $\mathrm{Re}(|z^3|)$

e. $\mathrm{Im}(1/|z^3|)$

1.3 Complex numbers z_1 and z_2 are given as $z_1 = 10e^{-j\pi/4}$ and $z_2 = 5e^{j30°}$. In polar form, determine the following:

a. product $z_1 \cdot z_2$

b. ratio z_1/z_2

c. ratio z_1^*/z_2^*

d. value $\sqrt{z_1}$

1.4 If two complex number are given as $z_1 = 2 - j3$ and $z_2 = 4 + j5$, find the value of $\ln(z_1) \cdot \ln(z_2)$.

1.5 If two complex number are given as $z_1 = 4 - j3$ and $z_2 = 5 + j4$, find the value of $e^{z_1}e^{z_2}$.

Using Phasor Notation

1.6 A voltage source $V(t) = 100\cos(6\pi 10^9 t - 45°)(V)$ is connected to a series RLC circuit, as shown in Figure 1.3. If $R = 10\,\mathrm{M\Omega}$, $C = 100\,\mathrm{pF}$, and $L = 1\,\mathrm{H}$, use phasor notation to find the following:

a. $i(t)$

b. $V_c(t)$, the voltage cross the capacitor

1.7 Find the phasors for the following field quantities:

a. $E_x(z, t) = E_0\cos(\omega t - \beta z + \phi)(V/m)$

b. $E_y(z, t) = 100e^{-3z}\cos(\omega t - 5z + \pi/4)(V/m)$

c. $H_x(z, t) = H_0\cos(\omega t + \beta z)(A/m)$

d. $H_y(z, t) = 120\pi e^{5z}\cos(\omega t + \beta z + \phi_h)(A/m)$

1.8 Find the instantaneous time domain sinusoidal functions corresponding to the following phasors:

a. $E_x(z) = E_0 e^{j\beta z}$(V/m)

b. $E_y(z) = 100e^{-3z}e^{-j5z}$(V/m)

c. $I_s(z) = 5 + j4$(A)

d. $V_s(z) = j10e^{j\pi/3}$(V)

1.9 Write the phasor expression I for the following current using a cosine reference.

a. $i(t) = I_0 \cos(\omega t - \pi/6)$

b. $i(t) = I_0 \sin(\omega t + \pi/3)$

1.10 Find the instantaneous $V(t)$ for the following phasors using a cosine reference.

a. $V_S = V_0 e^{j\pi/4}$

b. $V_S = [12 - j5]$(V)

1.8 QUATERNIONS

Because of its historical significance as the mathematical language of Maxwell, the subject of quaternions should be briefly known to students of electromagnetics. As mentioned in the "Introduction," Maxwell's equations were in the form of eight field equations that explicitly contained the magnetic vector potential and 12 quaternion equations that contained magnetic mass, magnetic charge, scalar magnetic potential, magnetic charge current, and magnetic conductivity of media. The complete set of equations is given in the next section. However, we must first understand the operations of the four-dimensional (4-D) complex numbers in which the formation exists. This formalization was devised by Sir William Rowan Hamilton in 1843. At that time, vector algebra and matrices had not yet been developed, but the vector dot and cross product were a result of Hamilton's work. It is said that Hamilton was walking across the Royal Canal in Dublin with his wife when the solution to quaternions came to him in the form of an equation, which he inscribed in stone on the bridge now called the Brougham or Broom Bridge. The original inscription has faded but a Quaternion plaque exists there today that reads, "Here as he walked by on the 16th of October 1843, Sir William Rowan Hamilton in a flash of genius discovered the fundamental formula for quaternion multiplication

$$i^2 = j^2 = k^2 = ijk = -1 \tag{1.42}$$

and cut it on a stone of this bridge."[11]

In his formalism, Hamilton devised a four-vector form of a complex number that had the components of a 4-D space just as the two-dimensional (2-D) complex

[11] Equation 1.42 added for this book.

number ($a + ib$), where a and b are real and $i^2 = -1$. In quaternion language, a complex number would be written as

$$Q = a + ib + jc + kd, \tag{1.43}$$

where a, b, c, and d are real. The scalar part of the quaternion is a, and the vector part is $ib + jc + kd$. The appealing characteristics of the quaternions is that they obey the same rules of addition and multiplication as 2-D complex numbers:

$$\begin{aligned}(a_1 + ib_1 + jc_1 + kd_1) &+ (a_2 + ib_2 + jc_2 + kd_2)\\ &= (a_1 + a_2) + (b_1 + b_2)i + (c_1 + c_2)j + (d_1 + d_2)k\end{aligned} \tag{1.44}$$

and

$$\begin{aligned}(a_1 + ib_1 + jc_1 + kd_1)&(a_2 + ib_2 + jc_2 + kd_2)\\ &= (a_1a_2) + (a_1b_2)i + (a_1c_2)j + (a_1d_2)k\\ &\quad + (b_1a_2)i + (b_1b_2)i^2 + (b_1c_2)ij + (b_1d_2)ik\\ &\quad + (c_1a_2)j + (c_1b_2)ji + (c_1c_2)j^2 + (c_1d_2)jk\\ &\quad + (d_1a_2)k + (d_1b_2)ki + (d_1c_2)kj + (d_1d_2)k^2\end{aligned} \tag{1.45}$$

using the additional ring properties:

$$ij = k,\ ji = -k,\ jk = i,\ kj = -i,\ ki = j,\ ik = -j \tag{1.46}$$

from Equation 1.42. Note that, unlike the commutative relations of 2-D complex numbers, Equation 1.46 shows that the 4-D quaternions do not commute (i.e., $ab \neq ba$).

In Appendix B, we have examined the roots of complex number equations like $z^2 + 1 = 0$ in 2-D space and found the roots to be at i and $-i$. Using the analogous equation in 4-D space, we would consider $Q^2 + 1 = 0$ and find an infinite number of solutions. We could draw the locus of these solutions in 3-D space when there was no real part ($a = 0$) for the quaternion with no real part, $Q = ib + jc + kd$ and $b^2 + c^2 + d^2 = 1$. These solutions form a unitary sphere centered on zero in the 3-D pure imaginary subspace of quaternions. We could then say that the locus of the solutions in 3-D space for a fixed real part ($a_1 = c\Delta t$) was a larger sphere with radius squared $b^2 + c^2 + d^2 = 1 + c^2\Delta t^2$ in 3-D space. Thus, the radius of the solution sphere is growing with time at a rate of $c\Delta t$. Sequencing the value of a to successively larger values would correspond to sequential spheres of larger radius. One can see that the appeal for saying the solutions in quaternion space is a movie of solutions with spheres of growing radius like the expansion of a spherical potential at constant velocity, c, in 3-D space (the scalar dimension corresponding to a multiple of c times time).

1.9 ORIGINAL FORM OF MAXWELL'S EQUATIONS

Maxwell originally introduced the following eight equations to represent the components of the electromagnetic field:

$$\vec{J} = \vec{j} + \partial\vec{D}/\partial t \tag{1.47}$$
$$\vec{\nabla} \times \vec{H} = \vec{J} \tag{1.48}$$
$$\mu\vec{H} = \vec{\nabla} \times \vec{A} \tag{1.49}$$
$$\vec{E} = \mu(\vec{v} \times \vec{H}) - \partial\vec{A}/\partial t - \vec{\nabla}\varphi \tag{1.50}$$
$$\vec{D} = \varepsilon\vec{E} \tag{1.51}$$
$$\vec{j} = \sigma\vec{E} \tag{1.52}$$
$$\vec{\nabla} \cdot \vec{D} = -\rho_e \tag{1.53}$$
$$\vec{\nabla} \cdot \vec{j} = -\partial\rho_e/\partial t \tag{1.54}$$

While the original field equations do not exactly correspond to the Heavyside vector formulation, they will be addressed in the coming chapters. For example, the original field equations explicitly contain the magnetic vector potential, $\vec{A}$, which does not appear in the Heavyside vector formulation, but we will define $\mu\vec{H} = \vec{\nabla} \times \vec{A}$ as a mathematical convenience, and $\vec{E}_{Lorenz} = -\partial\vec{A}/\partial t - \vec{\nabla}\varphi$ as part of the Lorenz gauge, in which case the equations look alike. $\vec{E}_{motion} = \mu(\vec{v} \times \vec{H})$ is the one term that appears to be discarded. Hertz interpreted the velocity, v, as the (absolute) motion of charges relative to the luminiferous ether, but, if v is interpreted as relative velocity between charges, then the Maxwell Heavyside equations are defined for the case $v = 0$ (i.e., test charges do not move in the observer's reference frame).

Maxwell also described 12 quaternion equations by employing scalar and vector operators:

$$S \cdot Q = S \cdot (a + ib + jc + kd) = a \tag{1.55}$$
$$V \cdot Q = V \cdot (a + ib + jc + kd) = ib + jc + kd, \tag{1.56}$$

so that when he put S or V in front of a quaternion, he means that S is an operation that yields only the scalar part of the quaternion and V is an operation that yields only the vector part of a quaternion. The original equations are applied to isotropic media, normal letters imply a scalar quantity, and a capital letter implies a quaternion without the scalar:

$$\vec{B} = V \cdot \vec{\nabla}\vec{A} \tag{1.57}$$
$$\vec{E} = V \cdot v\vec{B} - \partial\vec{A}/\partial t - \vec{\nabla}\varphi \tag{1.58}$$
$$\vec{F} = V \cdot v\vec{B} + e\vec{E} - m\vec{\nabla}\Omega \tag{1.59}$$
$$\vec{B} = \vec{H} + 4\pi\vec{M} \tag{1.60}$$
$$4\pi\vec{J}_{tot} = V \cdot \vec{\nabla}\vec{H} \tag{1.61}$$
$$\vec{J} = C\vec{E} \tag{1.62}$$
$$\vec{D} = K\vec{E}/4\pi \tag{1.63}$$
$$\vec{J}_{tot} = \vec{J} + \partial\vec{D}/\partial t \tag{1.64}$$
$$\vec{B} = \mu\vec{H} \tag{1.65}$$
$$e = S \cdot \vec{\nabla}\vec{D} \tag{1.66}$$
$$m = S \cdot \vec{\nabla}\vec{M} \tag{1.67}$$
$$\vec{H} = -\vec{\nabla}\Omega. \tag{1.68}$$

The eight field Equations 1.47–1.54 and the 12 quaternions 1.57–1.68 constitute the original form of Maxwell's equations. The equations include the magnetic scalar potential, Ω, and the magnetic charge, m. The $\vec{\nabla} = id/dx_1 + jd/dx_2 + kd/dx_3$ is a quaternion operator without the scalar part. The factor of 4π came about as a result of using Gaussian or cgs units. As a result of the vector form of Maxwell's equations, we will deduce Equation 1.54, the so-called equation of continuity (or conservation of charge statement). It is also interesting that Maxwell included the equation of Lorentz force, Equation 1.59, as one of his quaternion equations. We will use this force law as the starting point for the development of magnetic field intensity as it pertains to two parallel current-carrying wires through the Biot–Savart formulation, an inverse square law, which is now the National Institute of Standards and Technology standard for measuring the unit of force (the newton).

In Maxwell's original formulation, Faraday's $\vec{A}$ field was central and had physical meaning. The magnetic vector potential was not arbitrary, as defined by boundary conditions and choice of gauge as we will discuss; they were said to be gauge invariant. The original equations are thus often called the Faraday–Maxwell theory. The centrality of the $\vec{A}$ field was abandoned in the later interpretation of Maxwell by Heavyside. In this interpretation, electromagnetic fields $\vec{E}$ and $\vec{D}$, $\vec{H}$ and $\vec{B}$ are the only physical entities, and the magnetic vector potential is considered a mathematical convenience. Some say this perception replaces action-at-a-distance, as defined by Newton; by contact-action, as defined by Descartes; that is, a theory accounting for both local and global effects was replaced by a completely local theory. The local theory can address global effects only with the aid of the Lorenz gauge. These concepts will be more meaningful when we address time-varying fields in Chapter 7.

ENDNOTES

i. James Clerk Maxwell, *A Treatise on Electricity & Magnetism*, Vol. 1, 3rd ed. (New York: Dover, 1954). An unabridged, slightly altered, republication of the third edition, published by the Clarendon Press in 1891.

ii. James Clerk Maxwell, *The Dynamical Theory of the Electromagnetic Field*, ed. Thomas F. Torrance (Eugene, OR: Wipf and Stock Publishers, 1996). A commemorative reprint.

iii. IEEE Spectrum online, www.spectrum.ieee.org, *Gravity Probe B*, P. S. Wesson and M. Anderson, Nov. 3, 2008.

iv. Web sites for the electromagnetic wave spectrum can be found at http://www.ntia.doc.gov/osmhome/allochrt.pdf and http://csep10.phys.utk.edu/astr162/lect/light/waves.html.

v. Murray R. Spiegel, Shaum's Outline series, *Complex Variables: With an Introduction to Conformal Mapping and Its Applications* (McGraw-Hill, 1999).

Chapter 2

Vector Analysis

LEARNING OBJECTIVES

- Use vector algebra to carry out *addition*, *subtraction*, and the *dot product* and *cross product*, of vectors and to understand the results pictorially
- Use fundamental orthogonal coordinate systems—Cartesian, cylindrical, and spherical coordinates—in the description of geometric configurations commonly encountered in the study of fields and convert from one system to another
- Use and interpret the "*del*" or $\vec{\nabla}$ operator in computing spatial derivatives involving vectors, that is, the *gradient*, *divergence*, *curl*, and *Laplacian*
- Derive and understand the *divergence theorem* and *Stokes's theorem*

INTRODUCTION

In electromagnetic engineering, in addition to scalar quantities, there are many quantities defined not only by their amplitudes but also by their directions, for instance, the electric field intensity, $\vec{E}(x, y, z, t)$.

At a given time, t, and position, (x, y, z), a scalar is completely specified by its (possibly signed; i.e., positive or negative) magnitude and with appropriate units. Some representative scalars are mass, and electric charge.

A vector quantity, however, is defined, at a specified location and time, by both a magnitude and a direction. In a three-dimensional (3-D) space, this leads to three spatial components, the projections of the vector onto the three coordinate axes at its location.

2.1 ADDITION AND SUBTRACTION

Definition of a Vector and Unit Vector

In general, an arbitrary vector $\vec{A}$ can be written as

$$\vec{A} = A\,\hat{a}, \tag{2.1}$$

Maxwell's Equations, by Paul G. Huray

where A is the magnitude of the vector that usually has the unit and dimension, and $\hat{a}$ is a dimensionless *unit* vector (with magnitude 1) in the direction of $\vec{A}$. Thus, we have

$$\hat{a} = \vec{A}/|\vec{A}| = \vec{A}/A, \tag{2.2}$$

as shown pictorially in Figure 2.1.

Because of the nature of a vector as a directed quantity, it follows that a parallel displacement of a vector does not alter it materially, or, in other words, two vectors are equal if they have the same magnitude and direction. This is illustrated in Figure 2.2, where we would say $\vec{A} = \vec{A}'$.

Vector Components

It is usually convenient for a vector calculation to express a vector in component form. In the Cartesian coordinate system, the vector $\vec{A}$ can be written as

$$\vec{A} = \vec{A}_x + \vec{A}_y + \vec{A}_z = A_x\hat{a}_x + A_y\hat{a}_y + A_z\hat{a}_z, \tag{2.3}$$

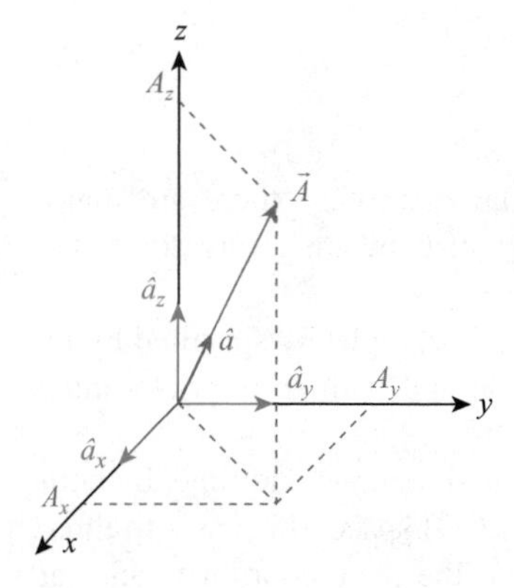

Figure 2.1 Vector $\vec{A}$ and its unit vector $\hat{a}$.

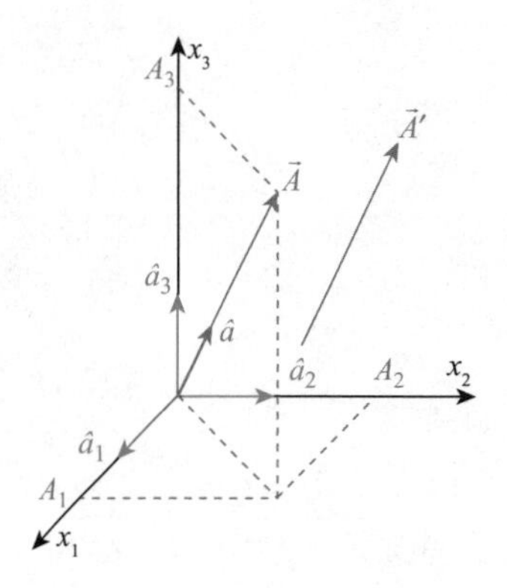

Figure 2.2 Two equal vectors $\vec{A}$ and $\vec{A}'$.

where three scalars, A_x, A_y, A_z (shown in blue in Figure 2.1), are called the components of $\vec{A}$. Thus, the magnitude of a vector can be expressed in terms of components as

$$A = |\vec{A}| = \sqrt{(A_x)^2 + (A_y)^2 + (A_z)^2} \tag{2.4}$$

(the 3-D Pythagorean theorem).

Vector Addition and Subtraction

Two vectors, $\vec{A}$ and $\vec{B}$, can be added graphically by using the parallelogram rule or "head-to-tail" rule. The resultant vector, $\vec{C} = \vec{A} + \vec{B}$, is the diagonal vector of the parallelogram constructed by $\vec{A}$ and $\vec{B}$, as shown in Figure 2.3. This is equivalent to the "head-to-tail" rule, in which the tail of $\vec{A}$ connects to the head of $\vec{B}$, and the vector $\vec{C} = \vec{A} + \vec{B}$ is the vector drawn from the tail of $\vec{A}$ or $\vec{B}$ to their vector sum.

Because both vectors, $\vec{A}$ and $\vec{B}$, can be written in component form, their sum can written as

$$\begin{aligned} \vec{C} &= C_x\hat{a}_x + C_y\hat{a}_y + C_z\hat{a}_z \\ &= \vec{A} + \vec{B} = (A_x\hat{a}_x + A_y\hat{a}_y + A_z\hat{a}_z) + (B_x\hat{a}_x + B_y\hat{a}_y + B_z\hat{a}_z) \\ &= (A_x + B_x)\hat{a}_x + (A_y + B_y)\hat{a}_y + (A_z + B_z)\hat{a}_z \end{aligned} \tag{2.5}$$

Subtraction can be defined in terms of vector addition by a new vector that has the same amplitude as the original vector but in the opposite direction, that is,

$$\vec{D} = \vec{A} - \vec{B} = \vec{A} + (-\vec{B}) \tag{2.6}$$

$$\begin{aligned} \vec{D} &= D_x\hat{a}_x + D_y\hat{a}_y + D_z\hat{a}_z \\ &= \vec{A} - \vec{B} = (A_x\hat{a}_x + A_y\hat{a}_y + A_z\hat{a}_z) - (B_x\hat{a}_x + B_y\hat{a}_y + B_z\hat{a}_z) \\ &= (A_x - B_x)\hat{a}_x + (A_y - B_y)\hat{a}_y + (A_z - B_z)\hat{a}_z \end{aligned} \tag{2.7}$$

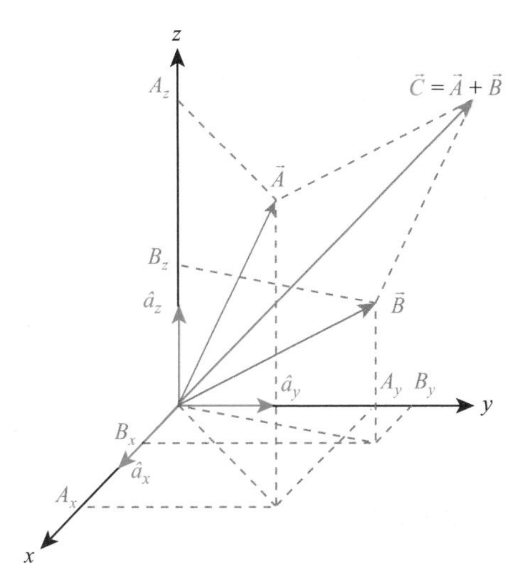

Figure 2.3 Sum of two vectors $\vec{A}$ and $\vec{B}$ to produce a vector $\vec{C} = \vec{A} + \vec{B}$.

2.2 MULTIPLICATION

Multiplication of a Vector by a Scalar

Multiplication of a vector $\vec{A}$ by a positive (*or negative*) scalar, C_0 (or $-C_0$), changes its magnitude by C_0 times in the same (*or opposite*) direction.

$$\vec{C} = C_0\vec{A} = C_0 A\hat{a} = C_0 A_x \hat{a}_x + C_0 A_y \hat{a}_y + C_0 A_z \hat{a}_z \tag{2.8a}$$

and

$$\vec{D} = -C_0\vec{A} = -C_0 A\hat{a} = -C_0 A_x \hat{a}_x - C_0 A_y \hat{a}_y - C_0 A_z \hat{a}_z \tag{2.8b}$$

Scalar Product

We *define* the scalar product (*dot product*) (*inner product*) of two vectors as the *scalar* equal to the product of the magnitudes of the vectors and the cosine of the angle between them, or

$$\vec{A} \cdot \vec{B} \equiv AB\cos\theta_{AB}, \tag{2.9}$$

where θ_{AB} is the angle between the two vectors $\vec{A}$ and $\vec{B}$ in the plane of the two vectors, as shown in Figure 2.4.

In Figure 2.4, we show the *special case* of $\vec{A}$ and $\vec{B}$ in the *x–y* plane in order to make the point that $\theta_{AB} = \theta_B - \theta_A$, where θ_B is the angle between the vector $\vec{B}$ and the x-axis, while θ_A is the angle between the vector $\vec{A}$ and the x-axis.

It is clear from Equation 2.9 that, because the cosine is an even function of θ, the order of terms does not change the scalar product; that is,

$$\vec{A} \cdot \vec{B} = \vec{B} \cdot \vec{A}. \tag{2.10}$$

Hence, the scalar product is commutative.

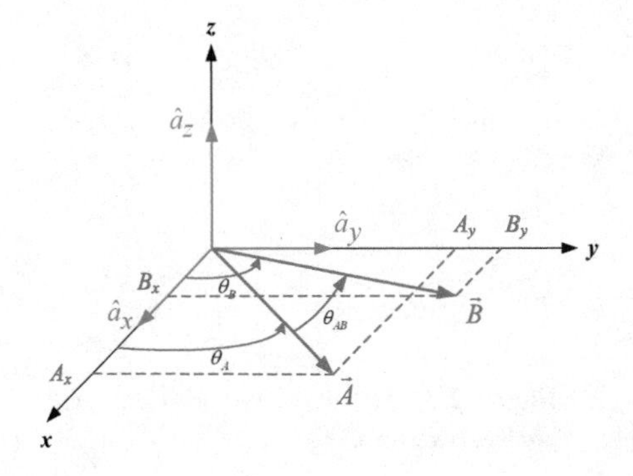

Figure 2.4 Special case in which two vectors, $\vec{A}$ and $\vec{B}$, lie in the *x–y* plane.

We also see

$$\vec{A}\cdot\vec{A} = A^2. \tag{2.11}$$

If we know the Cartesian components of $\vec{A}$ and $\vec{B}$, we can calculate their scalar product directly rather than use Equation 2.9; that is,

$$\begin{aligned}\vec{A}\cdot\vec{B} &= (A_x\hat{a}_x + A_y\hat{a}_y)\cdot(B_x\hat{a}_x + B_y\hat{a}_y) \\ &= A_xB_x + A_yB_y\end{aligned}, \tag{2.12}$$

where we have used the unit vector relations

$$\begin{aligned}\hat{a}_x\cdot\hat{a}_x &= 1 \\ \hat{a}_y\cdot\hat{a}_y &= 1 \\ \hat{a}_x\cdot\hat{a}_y &= 0\end{aligned} \tag{2.13}$$

In the more general case of vectors with three components,

$$\begin{aligned}\vec{A}\cdot\vec{B} &= (A_x\hat{a}_x + A_y\hat{a}_y + A_z\hat{a}_z)\cdot(B_x\hat{a}_x + B_y\hat{a}_y + B_z\hat{a}_z) \\ &= A_xB_x + A_yB_y + A_zB_z\end{aligned} \tag{2.14}$$

In Appendix C, we examine the even more general case of vectors with n-components and even permit the value of n to go to ∞.

Vector Product

The vector product (or *cross product*) is written as $\vec{A}\times\vec{B}$ and is defined as

$$\vec{A}\times\vec{B} = AB\sin\theta_{AB}\hat{a}_n, \tag{2.15}$$

where θ_{AB} is the angle (smaller than π between the vectors $\vec{A}$ and $\vec{B}$ in the plane of $\vec{A}$ and $\vec{B}$, and $\hat{a}_n$ is a unit vector perpendicular to the plane containing $\vec{A}$ and $\vec{B}$.[1] The direction of $\hat{a}_n$ follows the right-hand rule, namely, the thumb of a right hand when the fingers rotate from $\vec{A}$ and $\vec{B}$ through the angle θ_{AB}. This is illustrated in Figure 2.5 for the *special case* of $\vec{A}$ and $\vec{B}$ in the x–y plane.

From the definition of the vector product (Equation 2.15) and the fact that the sin is an odd function, we can see that

$$\vec{A}\times\vec{B} = -\vec{B}\times\vec{A}. \tag{2.16}$$

Thus, the vector product (or cross product) is *not* commutative.

The vector product can also be written in terms of the rectangular components. For the *special case* of the vectors $\vec{A}$ and $\vec{B}$ in the x–y plane as shown in Figure 2.5, we find that

[1] Note that the quantity $AB\sin\theta_{AB}$ is also numerically equal to the area of the parallelogram formed by $\vec{A}+\vec{B}$ in the x–y plane.

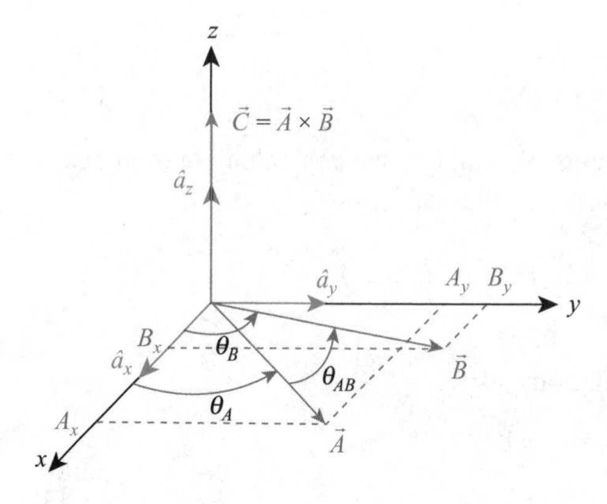

Figure 2.5 Cross product of vectors $\vec{A}$ and $\vec{B}$ for the *special case* of $\vec{A}$ and $\vec{B}$ in the *x*–*y* plane.

$$
\begin{aligned}
\vec{A} \times \vec{B} &= (A_x \hat{a}_x + A_y \hat{a}_y) \times (B_x \hat{a}_x + B_y \hat{a}_y) \\
&= (A_x B_y - A_y B_x)\hat{a}_z
\end{aligned}
\tag{2.17}
$$

For the more general case of vectors $\vec{A}$ and $\vec{B}$ with three components,

$$
\begin{aligned}
\vec{A} \times \vec{B} &= (A_x \hat{a}_x + A_y \hat{a}_y + A_z \hat{a}_z) \times (B_x \hat{a}_x + B_y \hat{a}_y + B_z \hat{a}_z) \\
&= (A_y B_z - A_z B_y)\hat{a}_x + (A_z B_x - A_x B_z)\hat{a}_y + (A_x B_y - A_y B_x)\hat{a}_z
\end{aligned}
\tag{2.18}
$$

Equation 2.18 may be written as a determinant in the form

$$
\vec{A} \times \vec{B} = \begin{vmatrix} \hat{a}_x & \hat{a}_y & \hat{a}_z \\ A_x & A_y & A_z \\ B_x & B_y & B_z \end{vmatrix}. \tag{2.19}
$$

Tensor Product

Finally, we note that a tensor product of the form

$$
\begin{aligned}
\overleftrightarrow{AB} &= \vec{A}\vec{B} = (A_x \hat{a}_x + A_y \hat{a}_y + A_z \hat{a}_z)(B_x \hat{a}_x + B_y \hat{a}_y + B_z \hat{a}_z) \\
\overleftrightarrow{AB} &= A_x B_x \hat{a}_x \hat{a}_x + A_x B_y \hat{a}_x \hat{a}_y + A_x B_z \hat{a}_x \hat{a}_z + A_y B_x \hat{a}_y \hat{a}_x + A_y B_y \hat{a}_y \hat{a}_y \\
&\quad + A_y B_z \hat{a}_y \hat{a}_z + A_z B_x \hat{a}_z \hat{a}_x + A_z B_y \hat{a}_z \hat{a}_y + A_z B_z \hat{a}_z \hat{a}_z
\end{aligned}
\tag{2.20}
$$

can be defined for applications in relativity and for influences brought about by noncubic atomic structures. While this is a special class of important products, it requires an entire formulation of tensors that are beyond the scope of this text. We therefore recommend a study of tensor analysis in classic books.[i]

2.3 TRIPLE PRODUCTS

There are two products involving three vectors, the *scalar triple product* and the *vector triple product*, which will be useful in derivations that follow.

A typical form of the scalar triple product is $\vec{A}\cdot(\vec{B}\times\vec{C})$, which can also be conveniently written in terms of the rectangular components. By using cyclical relations between unit vectors,

$$\hat{a}_x \times \hat{a}_y = \hat{a}_z \tag{2.21a}$$

$$\hat{a}_y \times \hat{a}_z = \hat{a}_x \tag{2.21b}$$

$$\hat{a}_z \times \hat{a}_x = \hat{a}_y \tag{2.21c}$$

and

$$\hat{a}_x \times \hat{a}_x = \hat{a}_y \times \hat{a}_y = \hat{a}_z \times \hat{a}_z = 0. \tag{2.21d}$$

We can verify that

$$\vec{A}\cdot(\vec{B}\times\vec{C}) = \vec{B}\cdot(\vec{C}\times\vec{A}) = \vec{C}\cdot(\vec{A}\times\vec{B}) = \begin{vmatrix} A_x & A_y & A_z \\ B_x & B_y & B_z \\ C_x & C_y & C_z \end{vmatrix}. \tag{2.22}$$

Figure 2.6 gives a graphical interpretation of the scalar triple product $\vec{D}\cdot(\vec{A}\times\vec{B})$ for the *special case* of two vectors $\vec{A}$ and $\vec{B}$ that lie in the *x–y* plane and a vector $\vec{D}$ that has three components.

As noted above, the area of the parallelogram formed by $\vec{A}$ and $\vec{B}$ is $|\vec{A}\times\vec{B}|$. Vector $\vec{D}$ makes angle β with respect to the vector product $\vec{A}\times\vec{B}$, so the volume of the parallelepiped formed by $\vec{A}$, $\vec{B}$, and $\vec{D}$ is $\vec{D}\cdot(\vec{A}\times\vec{B})$.

The *vector triple product*, with form $\vec{A}\times(\vec{B}\times\vec{C})$, is the vector product of one vector, with the result of the vector product of two other vectors. It can be shown that the triple vector product can be expanded as

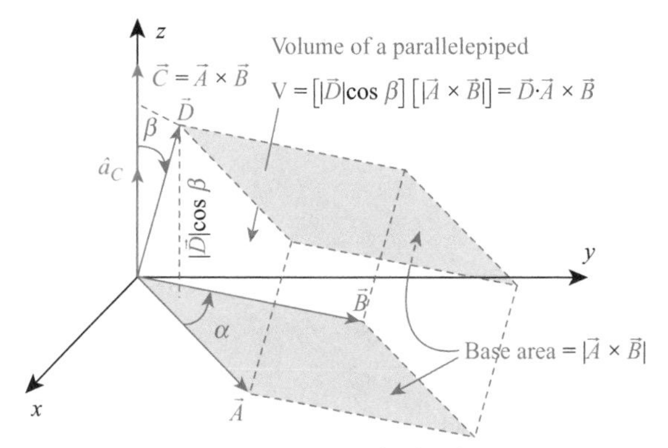

Figure 2.6 Scalar triple product $\vec{D}\cdot(\vec{A}\times\vec{B})$.

$$\vec{A}\times(\vec{B}\times\vec{C})=\vec{B}(\vec{A}\cdot\vec{C})-\vec{C}(\vec{A}\cdot\vec{B}), \tag{2.23}$$

which is known as the "back-cab" rule.

2.4 COORDINATE SYSTEMS

Cartesian Coordinates

In Cartesian coordinates, named after René Descartes (1596–1650), an arbitrary space location $\vec{P}(x_0, y_0, z_0)$ can be specified by three quantities or coordinates, x, y, and z, as shown in Figure 2.7.

The three mutually perpendicular unit vectors, defined in the three coordinate directions, are called the base vectors.

As mentioned earlier, the base vectors satisfy the following relations:

$$\hat{a}_x \times \hat{a}_y = \hat{a}_z \tag{2.24a}$$
$$\hat{a}_y \times \hat{a}_z = \hat{a}_x \tag{2.24b}$$
$$\hat{a}_z \times \hat{a}_x = \hat{a}_y \tag{2.24c}$$

and

$$\hat{a}_x \times \hat{a}_x = \hat{a}_y \times \hat{a}_y = \hat{a}_z \times \hat{a}_z = 0 \tag{2.24d}$$
$$\hat{a}_x \cdot \hat{a}_x = \hat{a}_y \cdot \hat{a}_y = \hat{a}_z \cdot \hat{a}_z = 1. \tag{2.24e}$$

The position vector from the origin to the point $P(x_0, y_0, z_0)$ can then be written as

$$\overrightarrow{OP} = x_0\hat{a}_x + y_0\hat{a}_y + z_0\hat{a}_z. \tag{2.25}$$

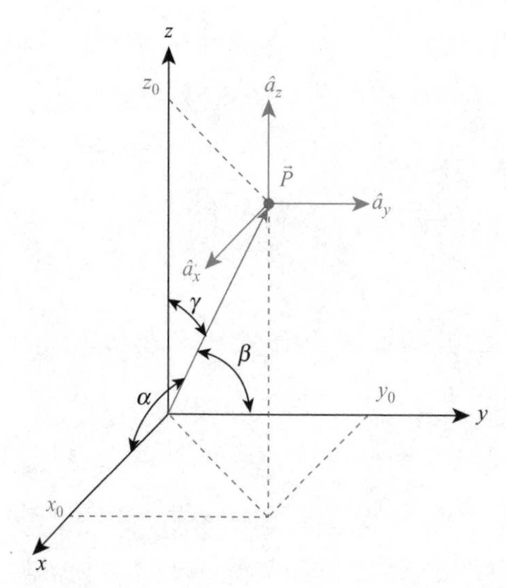

Figure 2.7 Definition of Cartesian coordinates.

Note that the base vectors, $\hat{a}_x$, $\hat{a}_y$, $\hat{a}_z$ in the Cartesian coordinate system can be located anywhere (here shown at the tip of vector $\overrightarrow{OP}$ rather than at the origin). This is to emphasize that the x-, y-, z-directions in Cartesian coordinates are independent of the location of the point P_0 (unlike cylindrical and spherical coordinates in the following discussion).

The vector $\vec{A}$ in Cartesian coordinates can be written in a component form as

$$\vec{A} = A_x\hat{a}_x + A_y\hat{a}_y + A_z\hat{a}_z, \tag{2.26}$$

so its scalar (dot product) and vector (cross product) with $\vec{B}$ are

$$\vec{A}\cdot\vec{B} = A_xB_x + A_yB_y + A_zB_z \tag{2.27}$$

$$\vec{A}\times\vec{B} = \begin{vmatrix} \hat{a}_x & \hat{a}_y & \hat{a}_z \\ A_x & A_y & A_z \\ B_x & B_y & B_z \end{vmatrix}. \tag{2.28}$$

A vector differential length in Cartesian coordinates can be expressed as

$$d\vec{l} = dx\hat{a}_x + dy\hat{a}_y + dz\hat{a}_z \tag{2.29}$$

and a differential volume in Cartesian coordinates is the multiplication of differential lengths in the three coordinate directions:

$$dv = dx\,dy\,dz. \tag{2.30}$$

Note that A_x, A_y, A_z are the components of $\vec{A}$ in the direction of the x, y, z axes, so they are the projection of $\vec{A}$ onto these axes with the angles α, β, γ. Thus, $A_x = A\cos\alpha$, $A_y = A\cos\beta$, $A_z = A\cos\gamma$, and we can write $\vec{A} = A\cos\alpha\hat{a}_x + A\cos\beta\hat{a}_y + A\cos\gamma\hat{a}_z$.

We can use this equation to evaluate

$$A = \sqrt{|\vec{A}|^2} = (\vec{A}, \vec{A})^{1/2} = (A_x^2 + A_y^2 + A_z^2)^{1/2} = (A^2\cos^2\alpha + A^2\cos^2\beta + A^2\cos^2\gamma)^{1/2}$$

or

$$A = A(\cos^2\alpha + \cos^2\beta + \cos^2\gamma)^{1/2}$$

from which we can conclude that

$$1 = \cos^2\alpha + \cos^2\beta + \cos^2\gamma \text{ (law of cosines in 3-D)}. \tag{2.31}$$

Cylindrical Coordinates

In cylindrical coordinates, an arbitrary space location $\vec{P}$ (ρ_0, ϕ_0, z_0) can also be specified by three quantities or coordinates, ρ, ϕ, and z, which is illustrated in Figure 2.8. In cylindrical coordinates, we can now define a set of three mutually perpendicular unit vectors $\hat{a}_\rho$, $\hat{a}_\varphi$, and $\hat{a}_z$.

In Figure 2.8, $\hat{a}_z$ remains in the same direction as in the Cartesian coordinates; $\hat{a}_\rho$ is chosen to be in the direction of increasing ρ and is perpendicular to $\hat{a}_z$ ($\hat{a}_\rho$ is parallel to the x–y plane); and $\hat{a}_\varphi$ is perpendicular to both of these and in the

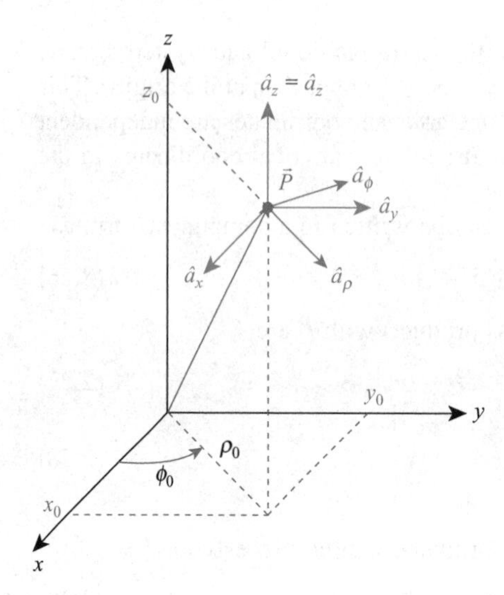

Figure 2.8 Definition of cylindrical coordinates.

direction of increasing ϕ ($\hat{a}_\varphi$ is also parallel to the x–y plane). These unit vectors satisfy the following relations:

$$\hat{a}_\rho \times \hat{a}_\varphi = \hat{a}_z \quad (2.33a)$$
$$\hat{a}_\varphi \times \hat{a}_z = \hat{a}_\rho \quad (2.32b)$$
$$\hat{a}_z \times \hat{a}_\rho = \hat{a}_\varphi \quad (2.32c)$$

and

$$\hat{a}_\rho \cdot \hat{a}_\varphi = \hat{a}_\varphi \cdot \hat{a}_z = \hat{a}_\rho \cdot \hat{a}_z = 0 \quad (2.32d)$$
$$\hat{a}_\rho \cdot \hat{a}_\rho = \hat{a}_\varphi \cdot \hat{a}_\varphi = \hat{a}_z \cdot \hat{a}_z = 1. \quad (2.32e)$$

We can express an arbitrary vector $\vec{A}$ in terms of components in cylindrical coordinates

$$\vec{A} = A_\rho \hat{a}_\rho + A_\varphi \hat{a}_\varphi + A_z \hat{a}_z. \quad (2.33)$$

The resulting scalar or dot product is

$$\vec{A} \cdot \vec{B} = (A_\rho \hat{a}_\rho + A_\varphi \hat{a}_\varphi + A_z \hat{a}_z) \cdot (B_\rho \hat{a}_\rho + B_\varphi \hat{a}_\varphi + B_z \hat{a}_z)$$
$$= A_\rho B_\rho + A_\varphi B_\varphi + A_z B_z \quad (2.34)$$

We can also define the vector differential length in cylindrical coordinates

$$d\vec{l} = du_1\, \hat{a}_\rho + du_2\, \hat{a}_\varphi + du_3\, \hat{a}_z = d\rho\, \hat{a}_\rho + \rho\, d\varphi\, \hat{a}_\varphi + dz\, \hat{a}_z, \quad (2.35)$$

so that its components in the directions of increasing ρ, ϕ, and z are $du_1 = d\rho$, $du_2 = \rho d\phi$, and $du_3 = dz$, respectively. Therefore, the differential volume in cylindrical coordinates is

$$dv = du_1\, du_2\, du_3 = \rho d\rho\, d\varphi\, dz. \quad (2.36)$$

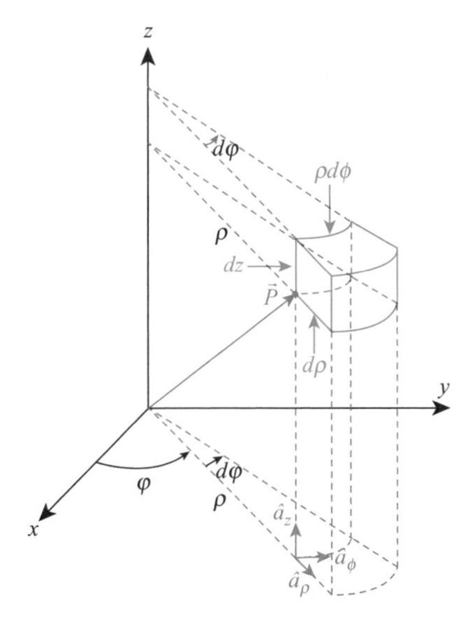

Figure 2.9 Differential volume element in cylindrical coordinates.

The differential volume element in cylindrical coordinates is shown in Figure 2.9.

Spherical Coordinates

In this system, the location of a point $P(R_0, \theta_0, \phi_0)$ is specified by three coordinates, R_0, θ_0, and ϕ_0 shown in Figure 2.10.

It is seen that R_0 is the distance from the origin to observation point, θ_0 is the angle made by R_0 with respect to the positive z axis, while ϕ_0 is again the angle made with the positive x-axis by the projection of R_0 onto the x–y plane.

Similarly as with the Cartesian and cylindrical coordinates, we define a set of three mutually perpendicular unit vectors $\hat{a}_R$, $\hat{a}_\theta$, and $\hat{a}_\varphi$, in the sense of increasing R, θ, and ϕ, respectively, as shown in Figure 2.10. We see that all these unit vectors are functions of the location of P and that they satisfy the relations

$$\hat{a}_R \times \hat{a}_\theta = \hat{a}_\varphi \tag{2.37a}$$

$$\hat{a}_\theta \times \hat{a}_\varphi = \hat{a}_R \tag{2.37b}$$

$$\hat{a}_\varphi \times \hat{a}_R = \hat{a}_\theta \tag{2.37c}$$

and

$$\hat{a}_R \cdot \hat{a}_\theta = \hat{a}_\theta \cdot \hat{a}_\varphi = \hat{a}_\varphi \cdot \hat{a}_R = 0 \tag{2.37d}$$

$$\hat{a}_R \cdot \hat{a}_R = \hat{a}_\theta \cdot \hat{a}_\theta = \hat{a}_\varphi \cdot \hat{a}_\varphi = 1. \tag{2.37e}$$

A vector $\vec{A}$ can be written in terms of components in spherical coordinates:

$$\vec{A} = A_R \hat{a}_R + A_\theta \hat{a}_\theta + A_\varphi \hat{a}_\varphi. \tag{2.38}$$

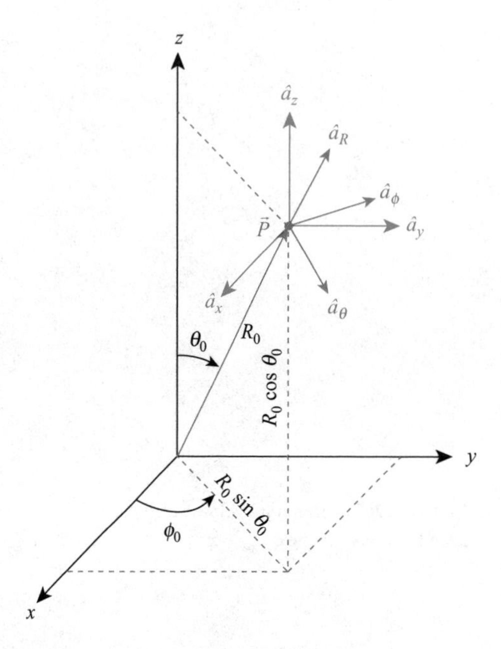

Figure 2.10 Definition of spherical coordinates.

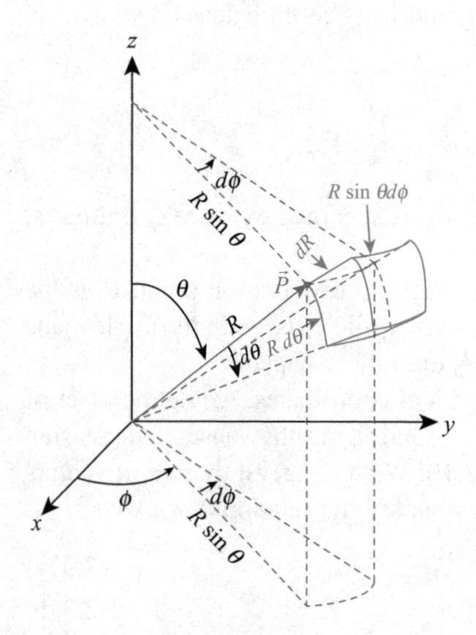

Figure 2.11 Differential volume element in spherical coordinates.

The scalar or dot product is

$$\begin{aligned}\vec{A}\cdot\vec{B} &= \left(A_R\hat{a}_R + A_\theta\hat{a}_\theta + A_\varphi\hat{a}_\varphi\right)\cdot\left(B_R\hat{a}_R + B_\theta\hat{a}_\theta + B_\varphi\hat{a}_\varphi\right)\\ &= A_R B_R + A_\theta B_\theta + A_\varphi B_\varphi\end{aligned} \tag{2.39}$$

The vector differential length in spherical coordinates is

$$d\vec{l} = du_1\,\hat{a}_1 + du_2\,\hat{a}_2 + du_3\,\hat{a}_3 = dR\,\hat{a}_R + R\,d\theta\,\hat{a}_\varphi + R\sin\theta\,d\varphi\,\hat{a}_z, \tag{2.40}$$

so that its components in the directions of increasing R, θ, and φ are $du_1 = dR$, $du_2 = Rd\theta$, and $du_3 = R\sin\theta\, d\varphi$, respectively.

The differential volume in spherical coordinates is

$$dv = du_1\, du_2\, du_3 = R^2 \sin\theta\, dR\, d\theta\, d\varphi, \tag{2.41}$$

as shown in Figure 2.11.

2.5 COORDINATE TRANSFORMATIONS

To solve a practical electromagnetic problem, we often need to concurrently use two coordinates (for instance, to find input impedance and radiation patterns for a rectangular microstrip patch antenna). Here, we need to both use Cartesian and spherical coordinates and go back and forth between them.

In general, we derive the α component in any orthogonal coordinates by computing

$$A_\alpha = \hat{a}_\alpha \cdot \vec{A}, \tag{2.42}$$

where the vector $\vec{A}$ can be defined in any coordinate system. This implies that, as long as we know the relationship between unit vectors, we can find any component of $\vec{A}$ in our preferred coordinates.

Transformation between Cartesian and Other Coordinate Systems—Matrix Representation

In Appendix C, we show how vectors can be expressed in terms of column matrices.

Using a numbered base vector convention (Figure 2.12), we can write $\vec{A} = A_1\hat{a}_1 + A_2\hat{a}_2 + A_3\hat{a}_3$.

Now, let us switch to another set of orthonormal base vectors, $\hat{a}'_1$, $\hat{a}'_2$, $\hat{a}'_3$, that are rotated in some defined way relative to the set $\hat{a}_1$, $\hat{a}_2$, $\hat{a}_3$, as shown in Figure 2.13.

In the primed numbered base vectors, we can write $\vec{A} = A'_1\hat{a}'_1 + A'_2\hat{a}'_2 + A'_3\hat{a}'_3$. It is clear that the vector $\vec{A}$ did not change; only the description of the vector changed.

Thus,

$$A_i\hat{a}_i = A'_j\hat{a}'_j \text{ (Summation convention implied)} \tag{2.43}$$

Transformation of Base Vectors

We just concluded above that each of the $\hat{a}'_j$ could be expressed in terms of the $\hat{a}_i$ by using the direction cosine description

$$\hat{a}'_j = 1\cos\theta_{1j}\,\hat{a}_1 + 1\cos\theta_{2j}\,\hat{a}_2 + 1\cos\theta_{3j}\,\hat{a}_3 = \sum_{i=1}^{3}\cos\theta_{ij}\,\hat{a}_i = \cos\theta_{ij}\,\hat{a}_i. \tag{2.44}$$

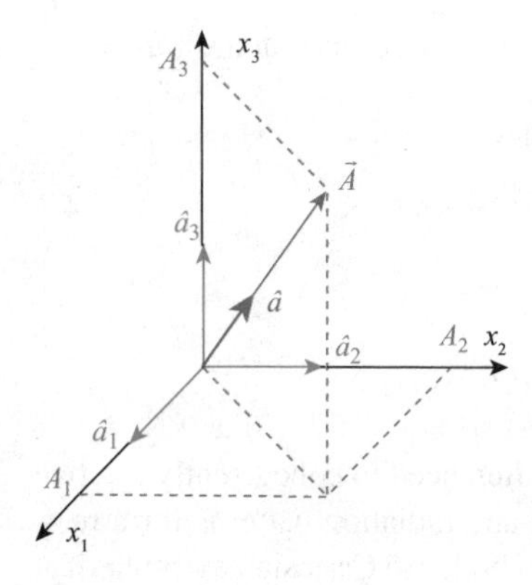

Figure 2.12 Vector $\vec{A}$ in terms of numbered base vectors $\hat{a}_1$, $\hat{a}_2$, $\hat{a}_3$.

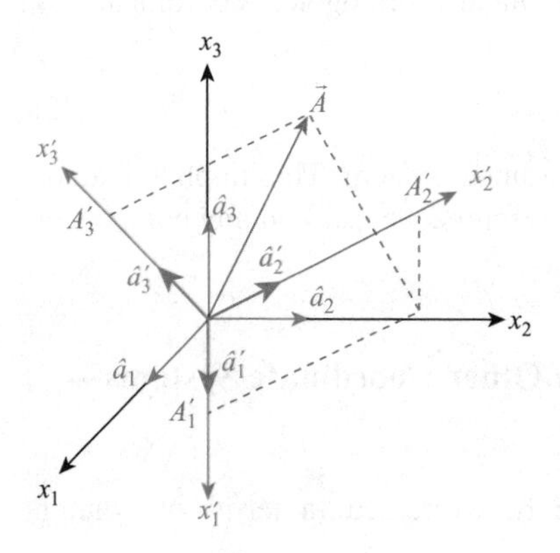

Figure 2.13 Vector $\vec{A}$ in terms of the numbered base vectors $\hat{a}_1'$, $\hat{a}_2'$, $\hat{a}_3'$.

For example, for a_1',

$$\hat{a}_1' = 1\cos\theta_{11}\,\hat{a}_1 + 1\cos\theta_{21}\,\hat{a}_2 + 1\cos\theta_{31}\,\hat{a}_3 = \sum_{i=1}^{3}\cos\theta_{i1}\,\hat{a}_i = \cos\theta_{i1}\,\hat{a}_i.$$

For example, for a_2',

$$\hat{a}_2' = 1\cos\theta_{12}\,\hat{a}_1 + 1\cos\theta_{22}\,\hat{a}_2 + 1\cos\theta_{32}\,\hat{a}_3 = \sum_{i=1}^{3}\cos\theta_{i2}\,\hat{a}_i = \cos\theta_{i2}\,\hat{a}_i.$$

For example, for a_3',

$$\hat{a}_3' = 1\cos\theta_{13}\,\hat{a}_1 + 1\cos\theta_{23}\,\hat{a}_2 + 1\cos\theta_{33}\,\hat{a}_3 = \sum_{i=1}^{3}\cos\theta_{i3}\,\hat{a}_i = \cos\theta_{i3}\,\hat{a}_i.$$

All of these equations can be written as a single matrix equation as

$$(\hat{a}_1' \quad \hat{a}_2' \quad \hat{a}_3') = (\hat{a}_1 \quad \hat{a}_2 \quad \hat{a}_3)\begin{pmatrix}\cos\theta_{11} & \cos\theta_{12} & \cos\theta_{13}\\ \cos\theta_{21} & \cos\theta_{22} & \cos\theta_{23}\\ \cos\theta_{31} & \cos\theta_{32} & \cos\theta_{33}\end{pmatrix} = (\hat{a}_1 \quad \hat{a}_2 \quad \hat{a}_3)T \qquad (2.45)$$

Note that, because $\cos\theta$ is an even function of θ, $\cos\theta_{i1} = \cos\theta_{1i}$. T is a square matrix whose elements are the direction cosines between the primed and the unprimed base vectors. This equation is valid for any rotation of axes. We will next define a rotation of base vectors for a cylindrical coordinate system and then for a spherical coordinate system, and the transformation of base vectors will be written through the transformation matrix T.

Transformation of Components of Vectors

Any vector $\vec{A}$ can be written as $\vec{A} = A_i\hat{a}_i$ or as $\vec{A} = A'_j\hat{a}'_j$ so, from above,

$$A_i\hat{a}_i = A'_j\hat{a}'_j = A'_j(\cos\theta_{ij}\,\hat{a}_i) = (A'_j\cos\theta_{ij})\hat{a}_i.$$

Thus, for any of the coefficients of $\hat{a}_i$,

for example, for $\hat{a}_1$,

$$A_1 = A'_1\cos\theta_{11} + A'_2\cos\theta_{12} + A'_3\cos\theta_{13}$$

for example, for $\hat{a}_2$,

$$A_2 = A'_1\cos\theta_{21} + A'_2\cos\theta_{22} + A'_3\cos\theta_{23}$$

for example, for $\hat{a}_3$,

$$A_3 = A'_1\cos\theta_{31} + A'_2\cos\theta_{32} + A'_3\cos\theta_{33},$$

which may be written in matrix form as

$$\begin{pmatrix} A_1 \\ A_2 \\ A_3 \end{pmatrix} = \begin{pmatrix} \cos\theta_{11} & \cos\theta_{12} & \cos\theta_{13} \\ \cos\theta_{21} & \cos\theta_{22} & \cos\theta_{23} \\ \cos\theta_{31} & \cos\theta_{32} & \cos\theta_{33} \end{pmatrix} \begin{pmatrix} A'_1 \\ A'_2 \\ A'_3 \end{pmatrix} = T \begin{pmatrix} A'_1 \\ A'_2 \\ A'_3 \end{pmatrix}, \tag{2.46}$$

where T is the same square matrix as above. However, the student must recognize that, for the *base vector* row matrix in Equation 2.45, the prime quantities are on the left, while, for the *component* column matrix in Equation 2.46, the prime quantities are on the right.

CONCLUSION Given the components of a vector $\vec{A}$ expressed in any primed coordinate system, we may use Equation 2.46 to find its components in any unprimed coordinate system. Conversely, given the components of a vector $\vec{A}$ expressed in any unprimed coordinate system, we may use

$$\begin{pmatrix} A'_1 \\ A'_2 \\ A'_3 \end{pmatrix} = T^{-1} \begin{pmatrix} A_1 \\ A_2 \\ A_3 \end{pmatrix} \tag{2.47}$$

to find its components in any primed coordinate system.

It is possible to show[ii] that, for any coordinate system that can be expressed as a rotation from another, the transformation is a *Unitary Transformation* for which $T^{-1} = \tilde{T}^*$ (the tilda over T meaning exchange columns for rows). Because all of the components of T are real, the complex conjugate has no effect and we can conclude that

$$T^{-1} = \begin{pmatrix} \cos\theta_{11} & \cos\theta_{21} & \cos\theta_{31} \\ \cos\theta_{12} & \cos\theta_{22} & \cos\theta_{32} \\ \cos\theta_{13} & \cos\theta_{23} & \cos\theta_{33} \end{pmatrix}. \tag{2.48}$$

We will use this transformation matrix for both the cylindrical coordinate system and the spherical coordinate system in the next sections.

Cylindrical Coordinate System Transformations

By careful inspection of Figure 2.8, it can be seen that the unit vectors in Cartesian and cylindrical coordinates can be related by

$$\hat{a}_\rho = \cos\varphi\,\hat{a}_x + \sin\varphi\,\hat{a}_y \tag{2.49a}$$
$$\hat{a}_\varphi = -\sin\varphi\,\hat{a}_x + \cos\varphi\,\hat{a}_y \tag{2.49b}$$
$$\hat{a}_z = \hat{a}_z \tag{2.49c}$$

or, from Equation 2.48,

$$\begin{aligned}(\hat{a}_1' \quad \hat{a}_2' \quad \hat{a}_3') &= (\hat{a}_\rho \quad \hat{a}_\varphi \quad \hat{a}_z) \\ &= (\hat{a}_x \quad \hat{a}_y \quad \hat{a}_z)\begin{pmatrix} \cos\varphi & -\sin\varphi & 0 \\ \sin\varphi & \cos\varphi & 0 \\ 0 & 0 & 1 \end{pmatrix} = (\hat{a}_1 \quad \hat{a}_2 \quad \hat{a}_3)T \end{aligned} \tag{2.50}$$

By Equation 2.46 and by using the Unitary property that $T^{-1} = \tilde{T}^*$, the components of $\vec{A}$ in cylindrical coordinates can be written as

$$\begin{pmatrix} A_1' \\ A_2' \\ A_3' \end{pmatrix} = \begin{pmatrix} A_\rho \\ A_\varphi \\ A_z \end{pmatrix} = \begin{pmatrix} \cos\varphi & \sin\varphi & 0 \\ -\sin\varphi & \cos\varphi & 0 \\ 0 & 0 & 1 \end{pmatrix}\begin{pmatrix} A_x \\ A_y \\ A_z \end{pmatrix} = T^{-1}\begin{pmatrix} A_1 \\ A_2 \\ A_3 \end{pmatrix} \tag{2.51}$$

or inversely

$$\begin{pmatrix} A_x \\ A_y \\ A_z \end{pmatrix} = \begin{pmatrix} \cos\varphi & -\sin\varphi & 0 \\ \sin\varphi & \cos\varphi & 0 \\ 0 & 0 & 1 \end{pmatrix}\begin{pmatrix} A_\rho \\ A_\varphi \\ A_z \end{pmatrix}. \tag{2.52}$$

Spherical Coordinate System Transformations

The unit vectors in spherical coordinates can also be expressed in terms of their components in Cartesian coordinates as

$$\hat{a}_R = \sin\theta\cos\varphi\,\hat{a}_x + \sin\theta\sin\varphi\,\hat{a}_y + \cos\theta\,\hat{a}_z \tag{2.53a}$$
$$\hat{a}_\theta = \cos\theta\cos\varphi\,\hat{a}_x + \cos\theta\sin\varphi\,\hat{a}_y - \sin\theta\,\hat{a}_z \tag{2.53b}$$
$$\hat{a}_\varphi = -\sin\varphi\,\hat{a}_x + \cos\varphi\,\hat{a}_y \tag{2.53c}$$

so, by using Equation 2.46,

$$(\hat{a}_1' \quad \hat{a}_2' \quad \hat{a}_3') = (\hat{a}_R \quad \hat{a}_\theta \quad \hat{a}_\varphi) \\ = (\hat{a}_x \quad \hat{a}_y \quad \hat{a}_z)\begin{pmatrix} \sin\theta\cos\varphi & \cos\theta\cos\varphi & -\sin\varphi \\ \sin\theta\sin\varphi & \cos\theta\sin\varphi & \cos\varphi \\ \cos\theta & -\sin\theta & 0 \end{pmatrix} = (\hat{a}_1 \quad \hat{a}_2 \quad \hat{a}_3)T. \tag{2.54}$$

By Equation 2.48 and by using the Unitary property that $T^{-1} = \tilde{T}^*$, the components of $\vec{A}$ in spherical coordinates can be written as

$$\begin{pmatrix} A_1' \\ A_2' \\ A_3' \end{pmatrix} = \begin{pmatrix} A_R \\ A_\theta \\ A_\varphi \end{pmatrix} = \begin{pmatrix} \sin\theta\cos\varphi & \sin\theta\sin\varphi & \cos\theta \\ \cos\theta\cos\varphi & \cos\theta\sin\varphi & -\sin\theta \\ -\sin\varphi & \cos\varphi & 0 \end{pmatrix}\begin{pmatrix} A_x \\ A_y \\ A_z \end{pmatrix} = T^{-1}\begin{pmatrix} A_1 \\ A_2 \\ A_3 \end{pmatrix} \tag{2.55}$$

or inversely

$$\begin{pmatrix} A_x \\ A_y \\ A_z \end{pmatrix} = \begin{pmatrix} \sin\theta\cos\varphi & \cos\theta\cos\varphi & -\sin\varphi \\ \sin\theta\sin\varphi & \cos\theta\sin\varphi & \cos\varphi \\ \cos\theta & -\sin\theta & 0 \end{pmatrix}\begin{pmatrix} A_R \\ A_\theta \\ A_\varphi \end{pmatrix}. \tag{2.56}$$

2.6 VECTOR DIFFERENTIATION

In this section, we are going discuss several frequently used vector differential operations, gradient and Laplacian operations of a scalar field, and divergence and curl operations of a vector field. Herein, we prefer to explain physical contents of these vector differential operations rather than derive them through strict math approaches.

Gradient of a Scalar Field

Suppose we have a scalar quantity V that is a function of position so that $V = V(x, y, z)$ in 3-D or $V(u_1, u_2, \ldots u_n)$ in n-D space. Such a function is called a scalar field and has only one number associated with each point in space; that is, it has n^0 components.[2] Examples are temperature (K), electric potential (V), density (kg/m^3), and pressure (N/m^2). Such quantities are *invariant* to a transformation of coordinates; that is, at one point in space, they are the same whether expressed in Cartesian, cylindrical, or spherical coordinates.

The vector differential operator $\vec{\nabla}$ (del) (nabla) (grad) is usually defined in 3-D as

$$\vec{\nabla} \equiv \hat{i}\frac{\partial}{\partial x} + \hat{j}\frac{\partial}{\partial y} + \hat{k}\frac{\partial}{\partial z} = \hat{a}_1\frac{\partial}{\partial u_1} + \hat{a}_2\frac{\partial}{\partial u_2} + \hat{a}_3\frac{\partial}{\partial u_3}, \tag{2.57}$$

where u_i indicates the use of curvilinear coordinates (for example cylindrical coordinates or spherical coordinates), as shown in Figure 2.14.

[2] Such quantities are called a tensor of rank 0.

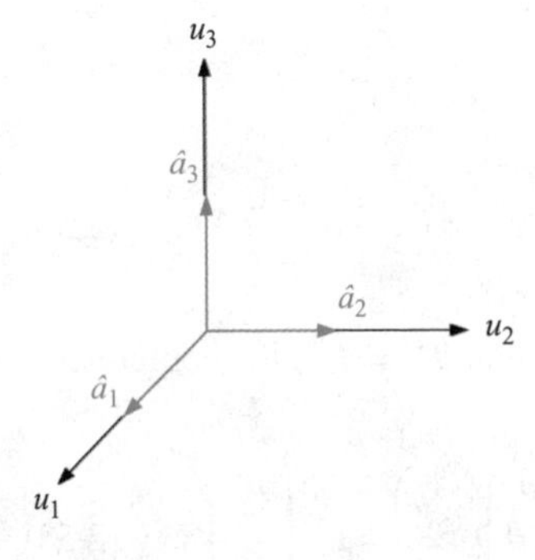

Figure 2.14 Orthonormal curvilinear coordinate system notation.

In cylindrical and spherical coordinates, the gradient of a scalar field is

$$\vec{\nabla} V = \hat{a}_1' \frac{\partial V}{\partial u_1} + \hat{a}_2' \frac{\partial V}{\partial u_2} + \hat{a}_3' \frac{\partial V}{\partial u_3} = \hat{a}_\rho \frac{\partial V}{\partial \rho} + \hat{a}_\varphi \frac{\partial V}{\rho \partial \varphi} + \hat{a}_z \frac{\partial V}{\partial z} \tag{2.58}$$

$$\vec{\nabla} V = \hat{a}_1' \frac{\partial V}{\partial u_1} + \hat{a}_2' \frac{\partial V}{\partial u_2} + \hat{a}_3' \frac{\partial V}{\partial u_3} = \hat{a}_R \frac{\partial V}{\partial R} + \hat{a}_\theta \frac{\partial V}{R \partial \theta} + \hat{a}_\varphi \frac{\partial V}{R \sin\theta \partial \varphi}. \tag{2.59}$$

The gradient of a scalar quantity is a vector quantity that represents both magnitude and direction, so it has three numbers associated with each point in 3-D space and n numbers in n-D space; that is, it has n^1 components.[3] Examples are electric field intensity, $\vec{E}(x, y, z, t)$ (V/m), magnetic field intensity, $\vec{H}(x, y, z, t)$ (A/m), fluid velocity, $\vec{v}(x, y, z, t)v$ (*m/s*), and electric current density, $\vec{J}(x, y, z, t)$ (A/m^2). Such quantities are called a vector field and are typically *variant* to a transformation of coordinates; that is, at one point in space, their components are different when expressed in Cartesian, cylindrical, or spherical coordinates.

We can also write the gradient of a scalar function, $V(u_1, u_2, u_3)$, as

$$\vec{\nabla} V = \hat{a}_n \frac{dV}{dn}, \tag{2.60}$$

where $\hat{a}_n$ is a unit vector in the direction of $\vec{\nabla} V$ and dV/dn is its magnitude. This vector can also be written in matrix notation as

$$\vec{\nabla} V \equiv \hat{a}_1 \frac{\partial V}{\partial u_1} + \hat{a}_2 \frac{\partial V}{\partial u_2} + \hat{a}_3 \frac{\partial V}{\partial u_3} = (\hat{a}_1 \quad \hat{a}_2 \quad \hat{a}_3) \begin{pmatrix} \partial V/\partial u_1 \\ \partial V/\partial u_2 \\ \partial V/\partial u_3 \end{pmatrix} \tag{2.61}$$

and, from Equation 2.47, we can write the components of $\vec{\nabla} V$ in another (primed) coordinate system as

$$\vec{\nabla} V \equiv (\hat{a}_1' \quad \hat{a}_2' \quad \hat{a}_3') \begin{pmatrix} \partial V/\partial u_1' \\ \partial V/\partial u_2' \\ \partial V/\partial u_3' \end{pmatrix} = (\hat{a}_1' \quad \hat{a}_2' \quad \hat{a}_3') T^{-1} \begin{pmatrix} \partial V/\partial u_1 \\ \partial V/\partial u_2 \\ \partial V/\partial u_3 \end{pmatrix}, \tag{2.62}$$

[3] Such quantities are called a tensor of rank 1.

where T^{-1} is given for cylindrical and spherical coordinates by Equations 2.51 and 2.55 respectively.

EXAMPLE

2.1 Suppose f is a function of R (*not of* θ, φ) so that $f(R) = f\left(\sqrt{x^2+y^2+z^2}\right)$.

A specific example would be the electric potential at a point in space for a charge q at the origin of coordinates. Then, $\vec{\nabla} f = \hat{a}_1 \partial f/\partial x + \hat{a}_2 \partial f/\partial y + \hat{a}_3 \partial f/\partial z$, in Cartesian coordinates.

$$\frac{\partial f}{\partial x} = \frac{\partial f}{\partial R}\frac{\partial R}{\partial x} + \frac{\partial f}{\partial \theta}\frac{\partial \theta}{\partial x} + \frac{\partial f}{\partial \phi}\frac{\partial f}{\partial x}$$

by the chain rule and

$$\frac{\partial f}{\partial \theta} = 0 \text{ and } \frac{\partial f}{\partial \phi} = 0$$

and

$$\frac{\partial R}{\partial x} = \frac{\partial}{\partial x}\left(x^2+y^2+z^2\right)^{1/2} = \frac{x}{\left(x^2+y^2+z^2\right)^{1/2}} = \frac{x}{R}.$$

Likewise,

$$\frac{\partial R}{\partial y} = \frac{y}{R} \; and \; \frac{\partial R}{\partial z} = \frac{z}{R},$$

so,

$$\vec{\nabla} f = \hat{a}_1\left(\frac{df}{dR}\frac{x}{R}\right) + \hat{a}_2\left(\frac{df}{dR}\frac{y}{R}\right) + \hat{a}_3\left(\frac{df}{dR}\frac{z}{R}\right) = (\hat{a}_1 x + \hat{a}_2 y + \hat{a}_3 z)\frac{1}{R}\frac{df}{dR} = \frac{\vec{R}}{R}\frac{df}{dR} = \hat{a}_R\frac{df}{dR}.$$

CONCLUSION The gradient of a function of R is a vector ($\hat{a}_R\, df/dR$) in the positive or negative radial direction.

Now, consider a space point $P + \Delta P$, which is displaced by $d\vec{l} = \hat{a}_x dx + \hat{a}_y dy + \hat{a}_z dz$ from the point P. In general, the value of the scalar function $V(P)$ will change to $V(P) + dV$, where

$$dV = (\partial V/\partial x)dx + (\partial V/\partial y)dy + (\partial V/\partial z)dz, \tag{2.63}$$

which can be written as the scalar product of two vectors:

$$dV = (\hat{a}_x\, \partial V/\partial x + \hat{a}_y\, \partial V/\partial y + \hat{a}_z\, \partial V/\partial z)\cdot(\hat{a}_x dx + \hat{a}_y dy + \hat{a}_z dz)$$
$$dV = (\hat{a}_x\, \partial V/\partial x + \hat{a}_y\, \partial V/\partial y + \hat{a}_z\, \partial V/\partial z)\cdot d\vec{l}$$
$$dV = \left(\vec{\nabla} V\right)\cdot d\vec{l} \tag{2.64}$$

Geometrical Representation of the Gradient Operator in 2-D

Suppose $V(x, y)$ describes a surface above the x,y plane. Then, $V(x, y) = C$ defines a curve above the x,y plane, as shown in Figure 2.15.

In what direction is $\vec{\nabla} V$?

$$\vec{\nabla} V \equiv \hat{a}_1 \frac{\partial V}{\partial u_1} + \hat{a}_2 \frac{\partial V}{\partial u_2} + \hat{a}_3 \frac{\partial V}{\partial u_3} = \hat{a}_x \frac{\partial V}{\partial x} + \hat{a}_y \frac{\partial V}{\partial y}$$

Along the curve

$$V = C, dV = \frac{\partial V}{\partial x} dx + \frac{\partial V}{\partial y} dy = 0 \text{ or } \left(\frac{\partial V}{\partial x} \hat{a}_x + \frac{\partial V}{\partial y} \hat{a}_y \right) \cdot \left(dx \hat{a}_x + dy \hat{a}_y \right) = 0.$$

But $(dx\hat{a}_x + dx\hat{a}_y)$ along the curve C is tangent to curve C. Therefore, since $\frac{\partial V}{\partial x} \neq 0$ and $\frac{\partial V}{\partial y} \neq 0$ in general, $\vec{\nabla} V$ must be perpendicular to the curve C.

Geometrical Representation of the Gradient Operator in 3-D

Suppose $V(x, y, z) = C$ defines an iso-surface (isotherm, iso-potential, or equipotential) surface, as shown in Figure 2.16.

On this surface, $dV = 0$, so $(\partial V/\partial x)dx + (\partial V/\partial y)dy + (\partial V/\partial z)dz = 0$. As in Equation 2.63, we can rewrite this expression, as in Equation 2.64, as a dot product,

$$\begin{aligned} dV &= \left(\hat{a}_x \frac{\partial V}{\partial x} + \hat{a}_y \frac{\partial V}{\partial y} + \hat{a}_z \frac{\partial V}{\partial z} \right) \cdot (\hat{a}_x dx + \hat{a}_y dy + \hat{a}_z dz) = 0 \\ &= \left(\hat{a}_x \frac{\partial V}{\partial x} + \hat{a}_y \frac{\partial V}{\partial y} + \hat{a}_z \frac{\partial V}{\partial z} \right) \cdot d\vec{l} = 0 \\ &= \left(\vec{\nabla} V \right) \cdot d\vec{l} = 0 \end{aligned}$$

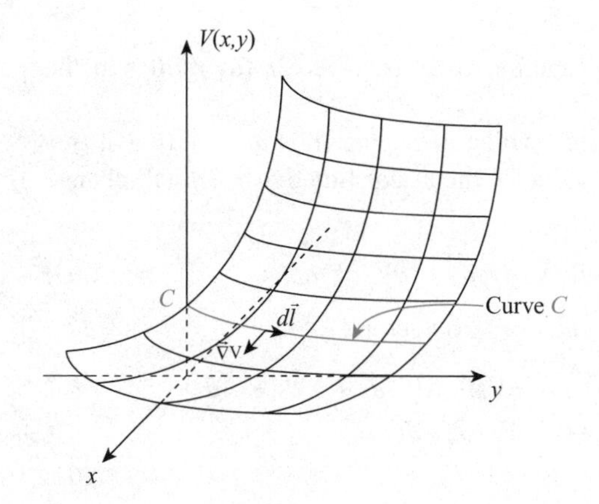

Figure 2.15 Plot of $V(x, y)$ as a function of x and y.

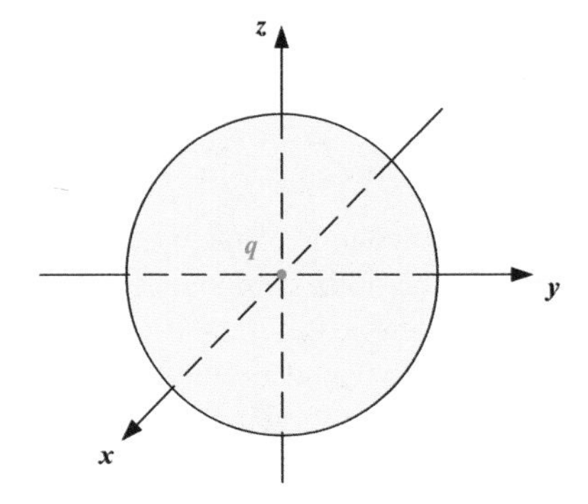

Figure 2.16 An equipotential surface for a charge, q, at the origin.

But we can see that, if we restrict ourselves to the iso-surface (the sphere), then $d\vec{l}$ must be on the surface of the sphere, so we conclude that $\vec{\nabla} V$ must be orthogonal to the sphere.

PROBLEMS

2.1 Find a unit vector parallel to the (x, y) plane and normal to the curve $x^2 - xy + y^2 = 7$ at the following:

a. the point $(-1, 2)$

b. any point (x, y) on the curve

2.2 The electric field intensity $\vec{E} = -\vec{\nabla} V$. Determine $\vec{E}$ at the point (1, 1, 0) if

a. $V = V_0 \, e^{-x} \sin \dfrac{\pi y}{4}$

b. $V = V_0 \, R \cos \theta$.

Divergence of a Vector Field

The *del* operator can be expressed in Cartesian coordinates from Equation 2.57 as

$$\vec{\nabla} = \hat{a}_x \, \partial/\partial x + \hat{a}_y \, \partial/\partial y + \hat{a}_z \, \partial/\partial z \tag{2.65}$$

and, letting $\vec{\nabla}$ operate on a vector field, $\vec{A}(x, y, z)$, we define the divergence of $\vec{A}$ as

$$\begin{aligned} div \, \vec{A} = \vec{\nabla} \cdot \vec{A} &= (\hat{a}_x \, \partial/\partial x + \hat{a}_y \, \partial/\partial y + \hat{a}_z \, \partial/\partial z) \cdot (\hat{a}_x A_x + \hat{a}_y A_y + \hat{a}_z A_z) \\ &= \partial A_x/\partial x + \partial A_y/\partial y + \partial A_z/\partial z \end{aligned}, \tag{2.66}$$

which can be written in matrix notation as

$$\vec{\nabla} \cdot \vec{A} = \begin{pmatrix} \dfrac{\partial}{\partial x} & \dfrac{\partial}{\partial y} & \dfrac{\partial}{\partial z} \end{pmatrix} \begin{pmatrix} A_x \\ A_y \\ A_z \end{pmatrix}. \tag{2.67}$$

Physical Meaning of Divergence

Suppose the current density at the point E at x_0, y_0, z_0 is

$$\vec{J}(x_0, y_0, z_0) = J_x(x_0, y_0, z_0)\hat{a}_x + J_y(x_0, y_0, z_0)\hat{a}_y + J_z(x_0, y_0, z_0)\hat{a}_z, \quad (2.68)$$

as shown in Figure 2.17. Then, the current flowing into the volume element through face *EFGH* is approximately $J_x(x_0, y_0, z_0)\Delta y\Delta z$[4] because the other components of $\vec{J}$ flow parallel to the face. Of course, the current density might be different at various points on this face, so the current flowing through *EFGH* would be exactly this amount only in the limit as $\Delta y \to 0$ and $\Delta z \to 0$. The current flowing out of the volume element through face *ABCD* would be $J_x(x_0 + \Delta x, y_0, z_0)\Delta y\Delta z$ in the limit as $\Delta y \to 0$ *and* $\Delta z \to 0$. The net current flowing out of the volume element in the x-direction is thus

$$I_x = [J_x(x_0 + \Delta x, y_0, z_0) - J_x(x_0, y_0, z_0)]\Delta y\Delta z. \quad (2.69)$$

However, we can see that the difference in the square brackets (in the limit as $\Delta x \to 0$) is just $\left.\dfrac{\partial J_x}{\partial x}\right|_{x_0, y_0, z_0} \Delta x$, so the net current flowing out of the volume element in the x-direction is

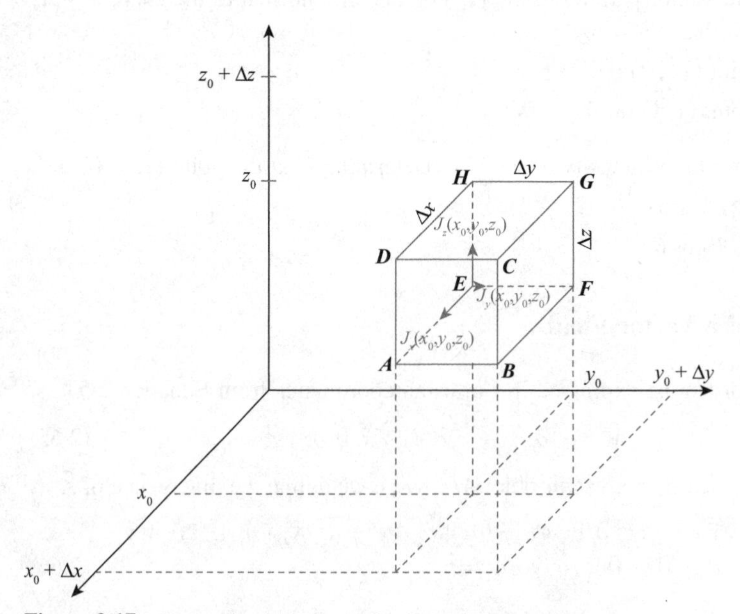

Figure 2.17 Current flow out of a small volume element $\Delta x\Delta y\Delta z$.

[4] We can also write this term as $-\vec{J}(x_0, y_0, z_0) \cdot \Delta\vec{\sigma}_{EFGH}$ if we interpret $\Delta\vec{\sigma}_{EFGH} = -\Delta y\Delta z\hat{a}_x$ as the part of the surface, *EFGH*, of the volume element that points *outward*.

$$I_x = \left.\frac{\partial J_x}{\partial x}\right|_{x_0, y_0, z_0} \Delta x \Delta y \Delta z. \tag{2.70a}$$

Now, we can carry out the same calculation for current flowing into the volume element through face *AEHD* and out of the volume element through face *BFGC* to compute the net current flowing out of the volume element in the *y*-direction as

$$I_y = \left.\frac{\partial J_y}{\partial y}\right|_{x_0, y_0, z_0} \Delta x \Delta y \Delta z. \tag{2.70b}$$

Finally, we can carry out the same calculation for current flowing into the volume element through face *ABFE* and out of the volume element through face *CGHD* to compute the net current flowing out of the volume element in the *z*-direction as

$$I_z = \left.\frac{\partial J_z}{\partial z}\right|_{x_0, y_0, z_0} \Delta x \Delta y \Delta z. \tag{2.70c}$$

Thus, the **net** current flowing out of the volume element is

$$I_x + I_y + I_z = \left[\left.\frac{\partial J_x}{\partial x}\right|_{x_0, y_0, z_0} + \left.\frac{\partial J_y}{\partial y}\right|_{x_0, y_0, z_0} + \left.\frac{\partial J_z}{\partial z}\right|_{x_0, y_0, z_0} \right] \Delta x \Delta y \Delta z. \tag{2.71}$$

However, from Equation 2.66, we recognize the quantity in square brackets as $\vec{\nabla} \cdot \vec{J}\,\Big|_{x_0,\, y_0,\, z_0}$.

CONCLUSION The net current flowing out of a volume element at *x*, *y*, *z* is

$$I_{net} = \vec{\nabla} \cdot \vec{J}\, \Delta x \Delta y \Delta z. \tag{2.72}$$

Because electric charge, q, is conserved in any volume element of space, we can say

$$\frac{\partial q}{\partial t} + \vec{\nabla} \cdot \vec{J}\, \Delta x \Delta y \Delta z = 0,$$

and dividing through by $\Delta x \Delta y \Delta z$,

$$\frac{\partial \rho}{\partial t} + \vec{\nabla} \cdot \vec{J} = 0 \text{ (Continuity equation)}, \tag{2.73}$$

where we have defined $\rho = \lim_{\Delta x, \Delta y, \Delta z \to 0} \frac{q_{in\, \Delta x \Delta y \Delta z}}{\Delta x \Delta y \Delta z}$ as the charge density at *x*, *y*, *z*.

NOTE If $\frac{\partial \rho}{\partial t}$ is a negative quantity (i.e., $\vec{\nabla} \cdot \vec{J}$ is a positive quantity), charge is coming out of the volume element at *x*, *y*, *z*. Thus, we could say that this volume element is a *source* of charge or that charge *diverges* from this volume element (hence the name *divergence*).

We can use a similar outflow of current argument with the differential volume element for cylindrical coordinates (shown in Figure 2.9) to find the divergence of a vector in cylindrical coordinates to be

$$\vec{\nabla}\cdot\vec{A}=\frac{1}{\rho}\frac{\partial}{\partial\rho}(\rho A_\rho)+\frac{1}{\rho}\frac{\partial A_\varphi}{\partial\varphi}+\frac{\partial A_z}{\partial z}. \tag{2.74}$$

We can use a similar out flow of current argument with the differential volume element for spherical coordinates (shown in Figure 2.11) to find the divergence of a vector in spherical coordinates to be

$$\vec{\nabla}\cdot\vec{A}=\frac{1}{R^2}\frac{\partial}{\partial R}(R^2A_R)+\frac{1}{R\sin\theta}\frac{\partial}{\partial\theta}(A_\theta\sin\theta)+\frac{1}{R\sin\theta}\frac{\partial A_\varphi}{\partial\varphi}. \tag{2.75}$$

2.7 DIVERGENCE THEOREM

In developing Equation 2.73, we could have taken note of the footnote 4 to alternatively say that in calculating I_{net} we have taken a sum of $\vec{J}\cdot\Delta\vec{\sigma}_i$ over all six faces of the volume element and the minus signs in the sum account for the fact that $\Delta\vec{\sigma}_i$ points ***out*** of the volume element. We will adopt this convention, and rewrite (2.73) as

$$\sum_{i=1}^{6}\vec{J}\cdot\Delta\vec{\sigma}_i=\vec{\nabla}\cdot\vec{J}\,\Delta x\Delta y\Delta z, \tag{2.76}$$

where it is understood that the six values of $\Delta\vec{\sigma}_i$ are the six faces of the volume element $\Delta x\Delta y\Delta z$ that point ***out*** of the volume. This is the differential form of the divergence theorem. If we now consider a large arbitrary volume V surrounded by a closed surface S, as shown in Figure 2.18:

In this figure, we can consider the volume V to be subdivided into a large number, n, of volume elements ΔV_i. If we take the sum of the differential form of the divergence theorem over all such volume elements in Equation 2.76 in the limit as $\Delta V_i \to 0$, the right-hand side becomes the volume integral over the volume V:

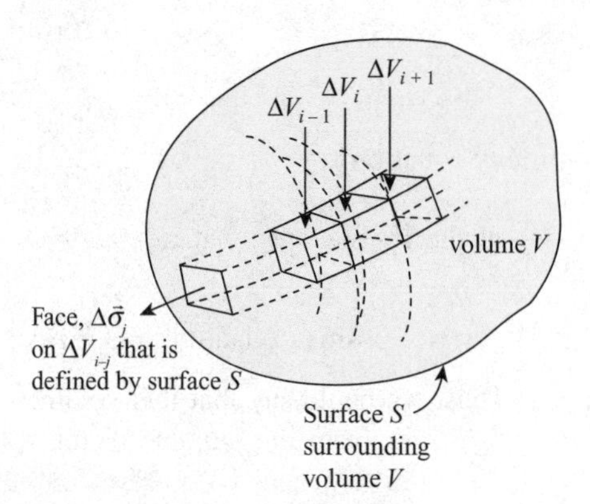

Figure 2.18 Surface S surrounding subdivided volume V.

$$\lim_{\Delta V_i \to 0} \sum_{i=1}^{n} \sum_{i=1}^{6} \vec{J} \cdot \Delta\vec{\sigma}_i = \sum_{i=1}^{n} \vec{\nabla} \cdot \vec{J}\, \Delta V_i = \iiint_V \vec{\nabla} \cdot \vec{J}\, dV. \tag{2.77}$$

On the left-hand side of Equation 2.76, we will take the sum over all six sides, $\vec{J} \cdot \Delta\vec{\sigma}_i$, of every volume element within the volume V, but the $\Delta\vec{\sigma}_j$ of adjacent cells (such as the adjacent elements ΔV_{i-1}, ΔV_i, ΔV_{i+1} shown in Figure 2.18) is in the outward direction from each of those volume elements, so all of the terms cancel one another unless they are the surface element $\Delta\vec{\sigma}_{i-j}$ on the outer surface S (where there is no adjacent cell). Thus, the only remaining terms, in the limit as $\Delta V_i \to 0$, that do not cancel on the left-hand side reduce it to

$$\lim_{\Delta \sigma_i \to 0} \sum_{i=1}^{n} \vec{J} \cdot \Delta\vec{\sigma}_i = \oiint_S \vec{J} \cdot d\vec{\sigma}. \tag{2.78}$$

Equating the right-hand sides of Equations 2.77 and 2.78, we get

$$\oiint_S \vec{J} \cdot d\vec{\sigma} = \iiint_V \vec{\nabla} \cdot \vec{J}\, dV \quad \text{(The divergence theorem)}. \tag{2.79}$$

We have considered a special case of the vector field of current density $\vec{J}$ to get a physical meaning for the divergence, but, in general, this could have been **any** vector field, $\vec{A}$. Thus, the divergence theorem applies to **all** vector fields and is most commonly written as

$$\oiint_S \vec{A} \cdot d\vec{\sigma} = \iiint_V \vec{\nabla} \cdot \vec{A}\, dV \quad \textbf{(The divergence theorem)} \tag{2.80}$$

The quantity on the left hand side of (2.80) is conventionally called the flux, Φ, of a vector field $\vec{A}(x, y, z)$ written as

$$\Phi = \oiint_S \vec{A} \cdot d\vec{s}. \tag{2.81}$$

The divergence theorem is one of the most powerful theorems of vector mathematics and is used in the manipulation of Maxwell's equations when we consider Gauss's laws of electrical and magnetic charge. In the electric field intensity case, the electric flux is defined as $\Phi_E = \oiint_S \vec{E} \cdot d\vec{s}$, and, in the case of magnetic field intensity, it is defined as $\Phi_B = \oiint_S \vec{B} \cdot d\vec{s}$.

Curl of a Vector Field

While the divergence gives a measure of the strength of a radiating source, another type of source, called a vortex source, is described by a circulation of the vector field. Examples of such a source would be fluid mass in a sink that spirals as it goes down a drain, the air molecules in a hurricane or passing over an airfoil at high velocity, and the supercurrent (from Cooper pairs of electrons) that surrounds normal regions in a Type II superconductor. The curl of a vector field gives a measure of the strength of the vortex source.

Physical Meaning

Because most of us have experienced the first of these examples, and this was the historical first use of the measure, we will consider the physical meaning of the curl by considering an incompressible fluid whose mass density, ρ (kg/m^3), is constant (the fluid is incompressible) and whose velocity is $\vec{v}(x, y, z)$. Figure 2.19 shows a rotating fluid around a square path, C, in the x, y plane.

The angular momentum (per unit volume) of the mass flowing (circulating) around the small area $\Delta\sigma = \Delta x \Delta y$ (path C) in 2-D is defined as $\int_{ABCDA}(\rho\vec{v})\cdot d\vec{l} = \oint_C(\rho\vec{v})\cdot d\vec{l}$, where C is the contour bounding $\Delta\sigma = \Delta x\Delta y$ and, by convention, its direction or rotation is in the $\hat{a}_z$ direction (defined by the right-hand rule for counterclockwise flow). $\oint_C \vec{v}\cdot d\vec{l}$ in the x, y plane is defined as the $\hat{a}_z$ component of the *circulation* around $\Delta\sigma = \Delta x\Delta y$ is $\int_1 v_x dx + \int_2 v_y dy + \int_3 v_x(-dx) + \int_4 v_y(-dy)$ and the *circulation per unit area* (*rotation*) of the fluid in the x, y plane (the $\hat{a}_z$ component) is defined as

$$rot\ v_z \equiv \lim_{\Delta\sigma\to 0}\frac{\oint_C \vec{v}\cdot d\vec{l}}{\Delta\sigma}. \tag{2.82}$$

Looking at path segments 1 and 3 in Figure 2.19, we see the net "circulation" along these two paths is

$$\int_1 v_x dx + \int_3 v_x(-dx) = \lim_{\Delta s\to 0}\{[v_x(y_0)]\Delta x + [v_x(y_0+\Delta y)](-\Delta x)\}$$

or

$$\lim_{\Delta\sigma\to 0}\frac{[v_x(y_0+\Delta y) - v_x(y_0)]}{\Delta y}(-\Delta x\Delta y) = -\left(\frac{\partial v_x}{\partial y}\right)\Delta x\Delta y. \tag{2.83a}$$

And the net circulation along the two paths 2 and 4 in Figure 2.19 is

$$\int_2 v_y dy + \int_4 v_y(-dy) = \lim_{\Delta s\to 0}\{[v_y(x_0+\Delta x)]\Delta y + [v_y(x_0)](-\Delta y)\}$$

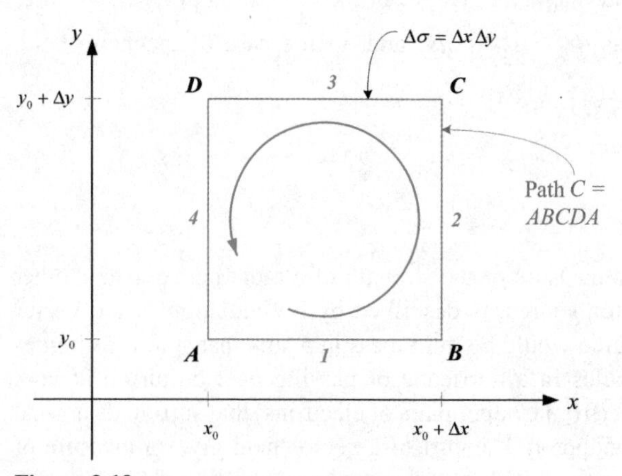

Figure 2.19 Circulation around a path C.

or

$$\lim_{\Delta\sigma\to 0} \frac{[v_y(x_0+\Delta x)-v_y(x_0)]}{\Delta x}(\Delta x\Delta y) = \left(\frac{\partial v_y}{\partial x}\right)\Delta x\Delta y. \tag{2.83b}$$

Thus, the net "rotation" in the $\hat{a}_z$-direction is

$$\left(\frac{\partial v_y}{\partial x} - \frac{\partial v_x}{\partial y}\right) \tag{2.84a}$$

and, similarly, by looking at a small area in the y, z plane, we find the "rotation" in the $\hat{a}_x$-direction is

$$\left(\frac{\partial v_z}{\partial y} - \frac{\partial v_y}{\partial z}\right) \tag{2.84b}$$

and, similarly, by looking at a small area in the z, x plane, we find the "rotation" in the $\hat{a}_y$-direction is

$$\left(\frac{\partial v_x}{\partial z} - \frac{\partial v_z}{\partial x}\right). \tag{2.84c}$$

Adding all of the components of the rotations, we get the net vector rotation as

$$rot\ \vec{v} = \left(\frac{\partial v_y}{\partial x} - \frac{\partial v_x}{\partial y}\right)\hat{a}_z + \left(\frac{\partial v_z}{\partial y} - \frac{\partial v_y}{\partial z}\right)\hat{a}_x + \left(\frac{\partial v_x}{\partial z} - \frac{\partial v_z}{\partial x}\right)\hat{a}_y. \tag{2.85}$$

In general, we can use Equation 2.82 to define the net circulation (circulation) of **any** vector field, $\vec{A}$, around a closed path as the scalar line integral over the path:

$$\text{Circulation of } \vec{A} \text{ around contour C} = \oint_C \vec{A}\cdot d\vec{l}. \tag{2.86}$$

Curl

We can define the curl of a vector field $\vec{A}$ at a point x, y, z to be a vector whose magnitude is the maximum net circulation of $\vec{A}$ per unit area (as the area tends to zero) and whose direction is the normal direction of the area when the area is oriented to make the net circulation maximum:

$$curl\ \vec{A} = \vec{\nabla}\times\vec{A} = \lim_{\Delta\sigma\to 0}\frac{1}{\Delta\sigma}\left[a_n \oint_C \vec{A}\cdot d\vec{l}\right]_{\max} \tag{2.87}$$

But, as we have seen, we can also treat the *del* operator, $\vec{\nabla}$, as a vector operator, which is taken as the cross product with $\vec{A}$ to give the curl. In Cartesian coordinates, we have

$$\begin{aligned} curl\ \vec{A} = \vec{\nabla}\times\vec{A} &= \begin{vmatrix} \hat{a}_x & \hat{a}_y & \hat{a}_z \\ \partial/\partial x & \partial/\partial y & \partial/\partial z \\ A_x & A_y & A_z \end{vmatrix} \\ &= \hat{a}_x(\partial A_z/\partial y - \partial A_y/\partial z) + \hat{a}_y(\partial A_x/\partial z - \partial A_z/\partial x) + \hat{a}_z(\partial A_y/\partial x - \partial A_x/\partial y) \end{aligned} \tag{2.88}$$

In *cylindrical* coordinates, we have

$$\vec{\nabla}\times\vec{A}=\frac{1}{\rho}\begin{vmatrix}\hat{a}_\rho & \hat{a}_\varphi\rho & \hat{a}_z\\ \partial/\partial\rho & \partial/\partial\varphi & \partial/\partial z\\ A_\rho & \rho A_\varphi & A_z\end{vmatrix}. \tag{2.89}$$

In *spherical* coordinates, we have

$$\vec{\nabla}\times\vec{A}=\frac{1}{R^2\sin\theta}\begin{vmatrix}\hat{a}_R & \hat{a}_\theta\rho & \hat{a}_\varphi R\sin\theta\\ \partial/\partial R & \partial/\partial\theta & \partial/\partial\varphi\\ A_R & RA_\theta & (R\sin\theta)A_\varphi\end{vmatrix}. \tag{2.90}$$

CONCLUSION $\vec{\nabla}\times\vec{A}$ is a measure of the rotation of a field. If $\vec{\nabla}\times\vec{A}=0$, we say the field $\vec{A}$ is "irrotational."

2.8 STOKES'S THEOREM

We have seen that $\vec{\nabla}\times\vec{A}$ as defined by Equation 2.88 gives us the rotation of $\vec{A}$ and we can see that, for an infinitesimal area $\Delta\vec{\sigma}_j$,

$$\vec{\nabla}\times\vec{A}\cdot\Delta\vec{\sigma}_j=\oint_{C_j}\vec{A}\cdot d\vec{l}. \tag{2.91}$$

If area $\Delta\vec{\sigma}_j$ is a part of a larger 2-D surface, as shown in Figure 2.20, it has direction $\hat{a}_z$.

On the right-hand side of Equation 2.91, we can see for this surface that the line integrals around adjacent areas also cancel one another, except for those small areas that have no adjacent elements (i.e., on curve C).

Suppose $\Delta\vec{\sigma}_j$ is a part of a complex *open* surface S, as shown in Figure 2.21.

On the right-hand side of Equation 2.91, we can see for the surface in Equation 2.21 that the line integrals around adjacent areas also cancel one another, except for

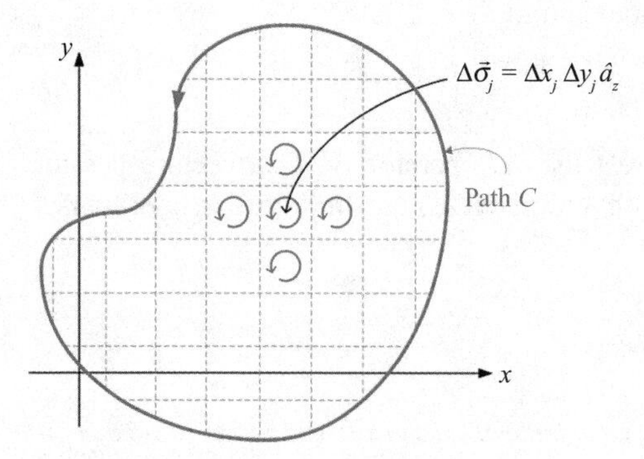

Figure 2.20 Subdivided surface enclosed by C.

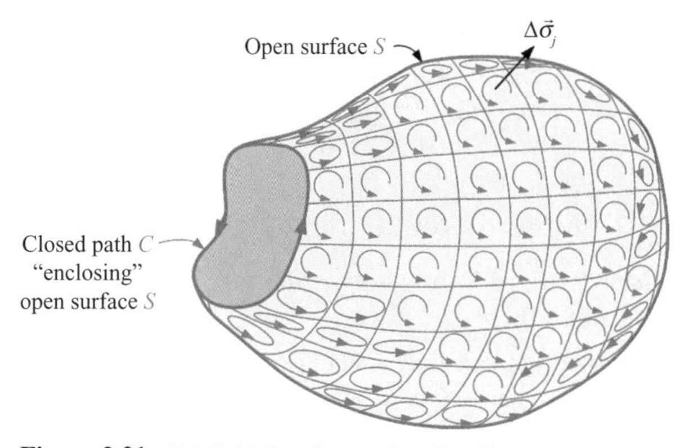

Figure 2.21 Subdivided surface *enclosed* by C.

those small areas that have no adjacent elements (i.e., on curve C). Thus, taking the sum of all such small areas,

$$\lim_{\Delta\sigma_j \to 0} \sum_j \left(\vec{\nabla} \times \vec{A}\right)_j \cdot \Delta\vec{\sigma}_j = \lim_{\Delta\sigma_j \to 0} \sum_j \oint_{C_j} \vec{A} \cdot d\vec{l} \tag{2.92}$$

or

$$\int_S (\vec{\nabla} \times \vec{A}) \cdot d\vec{\sigma} = \oint_C \vec{A} \cdot d\vec{l} \quad \text{(Stokes's theorem).} \tag{2.93}$$

SPECIAL CASE If the surface S is a closed surface (i.e., if we shrink the path C to a point), there is no integral over C, so

$$\oint_S \left(\vec{\nabla} \times \vec{A}\right) \cdot d\vec{\sigma} = 0.$$

2.9 LAPLACIAN OF A VECTOR FIELD

If $V(x, y, z)$ is a scalar function of x, y, z, the operation of ∇^2 in Cartesian coordinates is defined as

$$\nabla^2 V \equiv \vec{\nabla} \cdot \vec{\nabla} V = (\hat{a}_x\, \partial/\partial x + \hat{a}_y\, \partial/\partial y + \hat{a}_z\, \partial/\partial z) \cdot (\hat{a}_x\, \partial V/\partial x + \hat{a}_y\, \partial V/\partial y + \hat{a}_z\, \partial V/\partial z)$$
$$\vec{\nabla}^2 V = \partial^2 V/\partial x^2 + \partial^2 V/\partial y^2 + \partial^2 V/\partial z^2 \tag{2.94}$$

$\nabla^2 = \partial^2/\partial x^2 + \partial^2/\partial y^2 + \partial^2/\partial z^2$ is called the Laplacian operator.

NOTE If $V(x, y, z)$ satisfies $\nabla^2 V = 0$, then $\vec{\nabla} V$ is a vector field that is both (a) solenoidal (divergenceless) and (b) irrotational (curl free).

PROOF

a. $\nabla^2 V = \vec{\nabla}\cdot\vec{\nabla} V = 0$ so if $\vec{E} = -\vec{\nabla} V$, $\vec{\nabla}\cdot\vec{E} = 0$.

b. $\vec{\nabla}\times\vec{E} = -\vec{\nabla}\times\vec{\nabla} V = -\vec{\nabla}\times(\hat{a}_x\partial V/\partial x + \hat{a}_y\partial V/\partial y + \hat{a}_z\partial V/\partial z)$ so

$$\vec{\nabla}\times\vec{E} = -\begin{vmatrix} \hat{a}_x & \hat{a}_y & \hat{a}_z \\ \partial/\partial x & \partial/\partial y & \partial/\partial z \\ \partial V/\partial x & \partial V/\partial y & \partial V/\partial z \end{vmatrix}$$

$$= -\hat{a}_x\,(\partial^2 V/\partial y\partial z - \partial^2 V/\partial z\partial y) + \hat{a}_y\,(\partial^2 V/\partial x\partial z - \partial^2 V/\partial z\partial x) - \hat{a}_z\,(\partial^2 V/\partial y\partial x - \partial^2 V/\partial y\partial x) = 0$$

Conversely, if $\vec{\nabla}\times\vec{E} = 0$, then we can find a $V(x, y, z)$, so that $\vec{E} = -\vec{\nabla} V$ (called a conservative vector field), where V is a scalar potential that satisfies $\nabla^2 V = 0$.

This is a very powerful proof that we shall use many times in the following chapters.

Laplacian in Curvilinear Coordinates

We have found that

$$\vec{\nabla} V = \hat{a}_\rho\frac{\partial V}{\partial\rho} + \hat{a}_\varphi\frac{\partial V}{\rho\partial\varphi} + \hat{a}_z\frac{\partial V}{\partial z} \quad \text{cylindrical} \tag{2.95}$$

$$\vec{\nabla} V = \hat{a}_R\frac{\partial V}{\partial R} + \hat{a}_\theta\frac{\partial V}{R\partial\theta} + \hat{a}_\varphi\frac{\partial V}{R\sin\theta\partial\varphi} \quad \text{spherical} \tag{2.96}$$

and that

$$\vec{\nabla}\cdot\vec{E} = \frac{1}{\rho}\frac{\partial}{\partial\rho}(\rho E_\rho) + \frac{1}{\rho}\frac{\partial E_\varphi}{\partial\varphi} + \frac{\partial E_z}{\partial z} \quad \text{cylindrical} \tag{2.97}$$

$$\vec{\nabla}\cdot\vec{E} = \frac{1}{R^2}\frac{\partial}{\partial R}(R^2 E_R) + \frac{1}{R\sin\theta}\frac{\partial}{\partial\theta}(E_\theta\sin\theta) + \frac{1}{R\sin\theta}\frac{\partial E_\varphi}{\partial\varphi} \quad \text{spherical} \tag{2.98}$$

Therefore,

$$\vec{\nabla}\cdot\vec{\nabla} V = \frac{1}{\rho}\frac{\partial}{\partial\rho}\left(\rho\frac{\partial V}{\partial\rho}\right) + \frac{1}{\rho}\frac{\partial}{\partial\varphi}\left(\frac{1}{\rho}\frac{\partial V}{\partial\varphi}\right) + \frac{\partial}{\partial z}\left(\frac{\partial V}{\partial z}\right) \quad \text{cylindrical}$$

$$\vec{\nabla}\cdot\vec{\nabla} V = \frac{1}{R^2}\frac{\partial}{\partial R}\left(R^2\frac{\partial V}{\partial R}\right) + \frac{1}{R\sin\theta}\frac{\partial}{\partial\theta}\left(\sin\theta\frac{1}{R}\frac{\partial V}{\partial\theta}\right) + \frac{1}{R\sin\theta}\frac{\partial}{\partial\varphi}\left(\frac{1}{R\sin\theta}\frac{\partial V}{\partial\varphi}\right) \quad \text{spherical}$$

or

$$\nabla^2 V = \frac{1}{\rho}\frac{\partial}{\partial\rho}\left(\rho\frac{\partial V}{\partial\rho}\right) + \frac{1}{\rho^2}\frac{\partial^2 V}{\partial\varphi^2} + \frac{\partial^2 V}{\partial z^2} \quad \text{cylindrical} \tag{2.99}$$

$$\nabla^2 V = \frac{1}{R^2}\frac{\partial}{\partial R}\left(R^2\frac{\partial V}{\partial R}\right) + \frac{1}{R^2\sin\theta}\frac{\partial}{\partial\theta}\left(\sin\theta\frac{\partial V}{\partial\theta}\right) + \frac{1}{R^2\sin^2\theta}\frac{\partial^2 V}{\partial\varphi^2} \quad \text{spherical} \tag{2.100}$$

EXAMPLE

2.2 If $f(R) = R^n$, evaluate $\vec{\nabla} f(R)$, $\vec{\nabla}\cdot[\hat{a}_R f(R)]$, $\nabla^2 f(R)$ and $\vec{\nabla} \times [\hat{a}_R f(R)]$.

a. $\vec{\nabla} f(R) = \vec{\nabla}(R^n) = \hat{a}_R \dfrac{dR^n}{dR} = \hat{a}_R n R^{n-1}$ (see example following Equation 2.63)

b. $\vec{\nabla}\cdot[\hat{a}_R f(R)] = \dfrac{1}{R^2}\dfrac{\partial}{\partial R}[R^2 f(R)] = \dfrac{2}{R} f(R) + \dfrac{\partial f(R)}{\partial R} = \dfrac{2}{R} + nR^{n-1} = (n+2)R^{n-1}$

c. $\nabla^2 f(R) = \dfrac{1}{R^2}\dfrac{\partial}{\partial R}\left[R^2 \dfrac{\partial}{\partial R} R^n\right] = \dfrac{1}{R^2}\dfrac{\partial}{\partial R}\left[nR^{n+1}\right] = n(n+1)R^{n-2}$

d. $\vec{\nabla}\times[\hat{a}_R f(R)] = \dfrac{1}{R^2 \sin\theta}\begin{vmatrix} \hat{a}_R & R\hat{a}_\theta & R\sin\theta\hat{a}_\varphi \\ \partial/\partial R & \partial/\partial\theta & \partial/\partial\varphi \\ f(R) & 0 & 0 \end{vmatrix}$

$\vec{\nabla} \times [\hat{a}_R f(R)] = (1/R^2 \sin\theta)\,[\hat{a}_\theta R \partial f/\partial\varphi - \hat{a}_\varphi R \sin\theta(\partial f/\partial\theta)] = 0$

Divergence and Curl

The divergence and the curl of a vector field are basic operations used in the solutions of Maxwell's equations. Two of the most important applications of their use are stated by the **Helmholtz's theorems**:

1. A vector field is uniquely specified if its divergence and its curl are known within a region and its boundary conditions[5] are specified over a surface (perhaps at infinity).
2. Any vector field with both source and circulation densities vanishing at infinity may be written as the sum of two parts, one of which is irrotational (curl free) and the other is solenoidal (divergenceless).

Boundary conditions (BCs) may be simple Dirichlet or Neumann BC or may be mixed on different parts of a closed surface but cannot both be specified on the same part of a surface without being too restrictive; that is, there is no solution. Often, the BCs are *implied* (e.g., the vector field goes to zero on the surface at $R = \infty$), so it is important for the analyst to ask, "Where is the surface that encloses the region where a solution is desired?"

Sometimes, the solutions to Maxwell's equations require an arbitrary (convenient) choice of the form of the divergence of a field to uncouple the resulting differential equations. For example, Lorenz chose a restriction on the magnetic vector potential $\vec{A}$ to be $\vec{\nabla}\cdot\vec{A} + \mu_0\varepsilon_0\partial V/\partial t = 0$ (V being the scalar electric potential), while Coulomb chose $\vec{\nabla}\cdot\vec{A} = 0$. As we will see, these choices are convenient in different circumstances; they change the value of the magnetic vector potential $\vec{A}$ but retain unique values for the measurable quantities of scalar potential (e.g., V) and vector

[5] If the vector field has a specified value on a boundary, we say the BCs are Dirichlet BC if the function (e.g., the electric potential) is specified and Neumann BC if the normal derivative (e.g., the electric field intensity) is specified.

fields (e.g., $\vec{E}$ or $\vec{H}$). These choices are called the Lorentz gauge and the Coulomb gauge, respectively.

From the physical meaning discussions above, we have seen that the strength of the source flowing out of an infinitesimal volume is $\vec{\nabla}\cdot\vec{A}$, and the rotation of a vortex source at a point in space is $\vec{\nabla}\times\vec{A}$, so it is not surprising that a complete description of a source field that vanishes at infinity would include a sum of the two, as stated by Helmholtz's second theorem. Mathematically, we will see the time harmonic version of this statement written as

$$\vec{E}(x, y, z) = -\vec{\nabla}V(x, y, z) + j\omega\vec{\nabla}\times\vec{A}(x, y, z), \tag{2.101}$$

where $\vec{E}$ is the electric vector field, V is the electric scalar potential field, and $\vec{A}$ is the magnetic vector potential field.

The first term in Helmholtz's second theorem is based on the fact that the curl of a vector field integrated over a closed surface $\iint_s[\vec{\nabla}\times(\vec{\nabla}V)]\cdot d\vec{\sigma}$ is zero. We can see this mathematically from Stokes's theorem:

$$\iint_S\left[\vec{\nabla}\times\left(\vec{\nabla}V\right)\right]\cdot d\vec{\sigma} = \oint_C\left(\vec{\nabla}V\right)\cdot d\vec{l} = \oint_C dV = 0.$$

In electrical engineering, we say the change in electric potential around a closed path that returns to its starting point is zero (a conservative field). Because this is true for any surface (even an infinitesimal surface), we can see the integrand $\vec{\nabla}\times(\vec{\nabla}V) = 0$ at any point in space. In other words, **the curl of the gradient of any scalar field is zero** or conversely, **if a vector field is curl free, then it can be expressed as the gradient of a scalar field**. We thus see that a curl-free vector field is a conservative field, so **an irrotational (conservative) vector field can always be expressed as the gradient of a scalar field.**

The second term in Helmholtz's second theorem is based on the volume integral of $\vec{\nabla}\cdot(\vec{\nabla}\times\vec{A})$ and the use of the divergence theorem followed by Stokes's theorem, where the surface S is a closed surface (see Figure 2.21; there is no path C that encloses S):

$$\iiint_V \vec{\nabla}\cdot\left(\vec{\nabla}\times\vec{A}\right)dV = \oiint_S (\vec{\nabla}\times\vec{A})\cdot d\vec{\sigma} = 0 \tag{2.102}$$

Point Conclusion

Because the volume V is arbitrary, we may consider it to be an infinitesimal volume in which case, we see the integrand must itself be zero, that is, $\vec{\nabla}\cdot(\vec{\nabla}\times\vec{A}) = 0$ at any point in space. We conclude that

1. the divergence of the curl of a vector field identically zero, and
2. conversely, if a vector field is divergenceless, then it is solenoidal and can be expressed as the curl of another vector field.

We will use this fact in the construction of magnetic flux density by asserting that, because $\vec{\nabla}\cdot\vec{B} = 0$ in Maxwell's equations, we can define a magnetic vector potential field $\vec{A}$ such that $\vec{B} = \vec{\nabla}\times\vec{A}$.

In Chapter 3, we apply the vector field conclusions of Chapter 2 to electromagnetic fields.

ENDNOTES

i. A. P. Wills, *Vector Analysis with an Introduction to Tensor Analysis* (New York: Dover, 1958).
ii. Jon Mathews and R. L. Walker, *Mathematical Methods of Physics* (New York: W. A. Benjamin, 1965), 145.

Chapter 3

Static Electric Fields

LEARNING OBJECTIVES

- Understand the differential and integral forms of two fundamental properties of electrostatics that specify the divergence and curl of the electric field intensity, $\vec{E}$
- Understand the concepts of the electric field intensity, $\vec{E}$, scalar electric potential, V, electric flux density or electric displacement, $\vec{D}$, and dielectric constant
- Understand Coulomb's and Gauss's laws
- Know how to derive tangential and normal boundary conditions for $\vec{E}$ and $\vec{D}$ for static fields
- Understand the definitions of capacitance and capacitors

INTRODUCTION

Electrostatics is the study of the electric phenomena at rest and the electric fields, generated by discrete point charge or continuous distribution of charge that does not change with time. It is the basis of time-varying electromagnetic fields, and many concepts developed in electrostatics can be extended in future chapters for discussion of time-varying electromagnetic fields.

3.1 PROPERTIES OF ELECTROSTATIC FIELDS

Point Charge

It is convenient to begin with the case of a point charge in which it is assumed that all of the charge is located at a geometric point in space. This is obviously an idealization but can be approximated in the laboratory by using distances of separation that are large compared with the dimensions of charged objects.

We can compare the magnitudes of point charges q_1 and q_2 by introducing another arbitrary test charge q_t, putting it at a fixed distance R from q_1 and measuring the resultant force $\vec{F}_1$ on q_t, as illustrated in Figure 3.1. Then, we can do the same measurement of $\vec{F}_2$ by replacing q_1 by q_2 at the same distance R from q_t, as also shown in Figure 3.1. Because both q_t and R are the same in the two cases, the

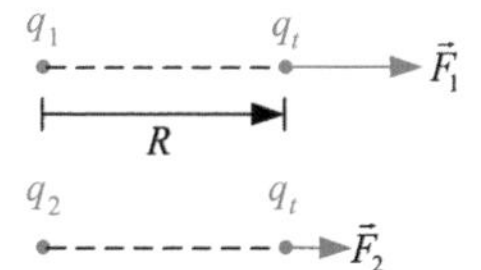

Figure 3.1 Electric charges and their exerted forces.

difference in forces can only be a result of the difference in the numerical values of the charges q_1 and q_2, and it is natural to ascribe the magnitudes of the forces as being directly proportional to the magnitudes of q_1 and q_2, namely,

$$\frac{|\vec{F}_1|}{|\vec{F}_2|} = \frac{|q_1|}{|q_2|}. \tag{3.1}$$

This result can be further extended to the Coulomb's law that describes the relation between two charges q_1 and q_2 and force.

Coulomb's Law

As mentioned in Chapter 1, Coulomb's law is an experimental law.[1] It states that force between two point charges or charge bodies, q_1 and q_2, which are very small in comparison with their distance of separation, R_{12}, is proportional to the product of the two charges and inversely proportional to the square of the distance between the two charges, with the direction of the force being along the line connecting the charges, as illustrated in Figure 3.2; that is,

$$\vec{F}_{12} = \hat{a}_{12} k_e \frac{q_1 q_2}{R_{12}^2} \tag{3.2}$$

with

$$R_{12} = |\vec{R}_{12}| = |\vec{R}_2 - \vec{R}_1|, \tag{3.3}$$

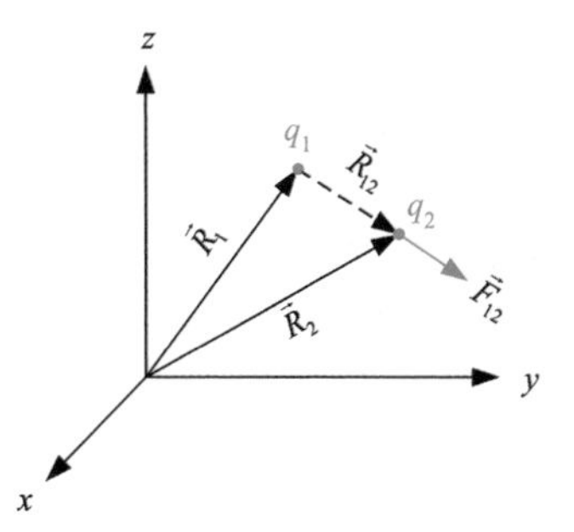

Figure 3.2 Two charges and position vectors in Coulomb's law.

[1] Some researchers have tried to measure a deviation of force from the inverse square relationship without definite success to date. Many of the relationships that follow are a consequence of the inverse square law and would need to be modified if any such deviation were ever discovered.

where $\vec{F}_{12}$ is the vector force exerted by q_1 on q_2, $\hat{a}_{12}$ is a unit vector in the direction from q_1 to q_2, and k_e is a proportionality constant depending on the medium and system units.

If q_1 and q_2 are of the same sign (both positive or both negative), $\vec{F}_{12}$ is positive (repulsive); and if q_1 and q_2 are of opposite signs, $\vec{F}_{12}$ is negative (attractive). In the later discussion, we will write the constant coefficient, k_e, that appears in Equation 3.2 as

$$k_e = (1/4\pi\varepsilon_0), \tag{3.4}$$

where ε_0 ($\varepsilon_0 \equiv 1/c^2\mu_0 \approx 10^{-9}/36\pi$) is called the permittivity of free space, a universal constant and from Equation 3.2, its units must be $C^2/N\ m^2$.

Electric Field Intensity

It is convenient to introduce a physical parameter to describe the strength of the electric force per unit charge generated by arbitrary electric charges. The electric field intensity, $\vec{E}$, is defined as the force per unit charge when a very small stationary test charge, q_t, is placed in the neighborhood of another charged particle, namely,

$$\vec{E} = \lim_{q_t \to 0} \vec{F}/q_t \quad (\text{V/m}) \tag{3.5}$$

It can be seen that the electric field intensity, $\vec{E}$, is proportional to and in the direction of the force $\vec{F}$, where $\vec{F}$ is the measured electric force on q_t in Newtons (*N*), q_t is in Coulombs (*C*), and $\vec{E}$ is in volts per meter (V/m) in the International System of Units. From the analysis above, V/m = N/C.

Electrostatic Fields in Free Space

Based on Coulomb's force law (Equation 3.2) and the electric field intensity definition for static point charges (Equation 3.5), two properties of *electrostatic* fields in free space can be deduced:

$$\nabla \cdot \vec{E} = \rho_V / \varepsilon_0 \tag{3.6}$$

$$\nabla \times \vec{E} = 0, \tag{3.7}$$

where ρ_V is the volume charge density of free charge (C/m^3). Equation 3.7 asserts that *static* electric field intensity is irrotational, and Equation 3.6 implies that a static electric field represents a charge source and is solenoidal only if $\rho_V = 0$. These properties are deduced in the following sections.

3.2 GAUSS'S LAW

Let us calculate the electric flux, Φ_e, for the case of a sphere of radius R that surrounds a point charge Q located at the origin of coordinates, as shown in Figure 3.3.

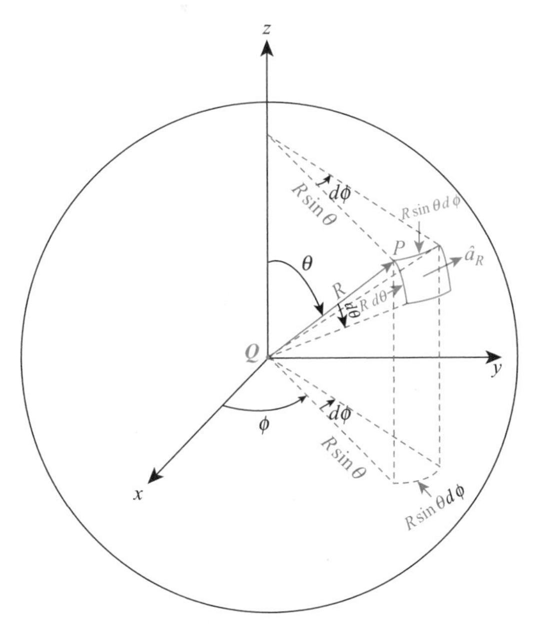

Figure 3.3 A sphere of radius R surrounding a point charge Q at the origin.

On the surface of the sphere at radius R, the electric field intensity is

$$\vec{E} = \vec{F}/q_t = k_e\left(Q/R^2\right)\hat{a}_R.$$

We can find the integrated electric flux on the surface of the sphere to be

$$\begin{aligned} \Phi_e &= \oiint_S \vec{E}\cdot d\vec{\sigma} = \oiint_S \left[k_e\left(Q/R^2\right)\hat{a}_R\right]\cdot(R\sin\theta d\phi)(Rd\theta)\,\hat{a}_R \\ &= k_e Q\int_0^{2\pi}\left(\int_0^{\pi}\sin\theta\, d\theta\right)d\phi = 2\pi k_e Q\int_0^{\pi}\sin\theta\, d\theta = Q/\varepsilon_0 \end{aligned} \tag{3.8}$$

CONCLUSION The total electric flux, Φ_e, passing through a sphere that surrounds a point charge Q in free space is $\Phi_e = Q/\varepsilon_0$. This answer is independent of the radius R of the sphere. Thus, the sphere can be of arbitrary radius. We can even take the radius of the sphere to be R in the limit as $R \to 0$.

We can also use the divergence theorem in Equation 3.8 to see that

$$\Phi_e = \oiint_S \vec{E}\cdot d\vec{\sigma} = \iiint_V \vec{\nabla}\cdot\vec{E}\,dv = Q/\varepsilon_0, \tag{3.9}$$

and, if we take the limit as $R \to 0$, the sphere of volume V is of infinitesimal size, and $\vec{\nabla}\cdot\vec{E}$ is evaluated at the origin. Thus,

$$\vec{\nabla}\cdot\vec{E} = (Q/V)/\varepsilon_0 = \rho_V/\varepsilon_0. \tag{3.10}$$

Equation 3.10 is a point differential relation that is valid for charges at the origin and it is one of Maxwell's equations commonly called the point form of Gauss's law.

Note that, while we used the surface S to be a sphere above, we can reuse the divergence theorem for any surface S' that surrounds the origin. That is,

$$\iiint_V \vec{\nabla} \cdot \vec{E}\, dv = \oint\!\!\oint_{S'} \vec{E} \cdot d\vec{\sigma} = Q/\varepsilon_0. \tag{3.11}$$

The surface S' could be the surface shown in Figure 3.4.

But, even if Q is not at the origin of coordinates, we can make a transformation of coordinates to show that Equation 3.11 is valid for the surface S', as shown in Figure 3.5.

We can see that, if there is only one point charge Q inside the surface S', we can shrink it down arbitrarily to infinitesimal size. Thus, we conclude that

$$\vec{\nabla} \cdot \vec{E} = (Q/V)/\varepsilon_0 = \rho_V/\varepsilon_0 \text{ at any point in space.} \tag{3.12}$$

SPECIAL CASE In the absence of charge, $\vec{\nabla} \cdot \vec{E} = 0$ and $\vec{E}$ is a *solenoidal* field.

We can see that Equation 3.12 is valid no matter what the size of Q is or even if there are two, three, or n charges, q_i, within the surface S'; it only matters that the charges are *inside* S'. Thus,

$$\oint\!\!\oint_{S'} \vec{E} \cdot d\vec{\sigma} = Q/\varepsilon_0 = \sum_{i \text{ inside } S} q_i/\varepsilon_0 \quad \text{Gauss's Law.} \tag{3.13}$$

The integral form of Gauss's law given by Equation 3.13 is one of the most important relations in electrostatics. It states that the outward electric flux of the

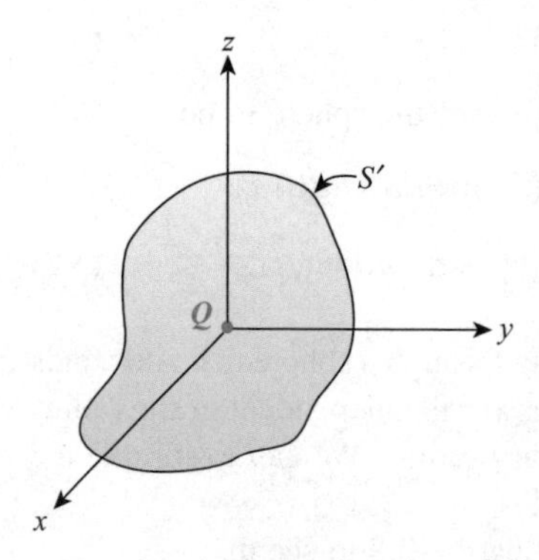

Figure 3.4 Arbitrary surface S' surrounding Q at the origin.

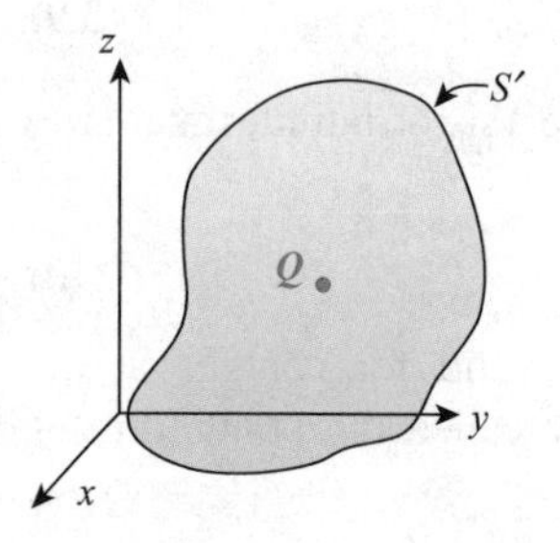

Figure 3.5 Arbitrary surface S' surrounding Q at any point.

electric field intensity, $\Phi_e = \oiint_{S'} \vec{E} \cdot d\vec{\sigma}$ integrated over any closed surface S' in free space is equal to the total charge inside that surface divided by ε_0 and may be used to calculate the electric field intensity generated by charges in specific symmetrical charged configurations. We give examples of its use in the following sections.

EXAMPLE

3.1 **a.** Use Gauss's law to determine the electric field intensity for an infinitely long, straight line in free space with a uniform line charge density ρ_l.

SOLUTION Because the line charge density ρ_L is infinitely long and straight, we can observe that the resulting electric field intensity must be along the radial direction and perpendicular to the line; that is, $\vec{E} = \hat{a}_\rho E_\rho$. Any other components will be canceled by each other because of symmetry of the structure. By mentally constructing a Gaussian surface, as shown in Figure 3.6,

$$\oint_S \vec{E} \cdot d\vec{s} = \int_0^{2\pi} \int_0^{\rho} \hat{a}_\rho E_\rho \cdot \hat{a}_z \rho d\rho d\phi - \int_0^{2\pi} \int_0^{\rho} \hat{a}_\rho E_\rho \cdot \hat{a}_z \rho d\rho d\phi + \int_0^{2\pi} \int_0^{L} \hat{a}_\rho E_\rho \cdot \hat{a}_\rho \rho d\phi dz$$

$$\oint_S \vec{E} \cdot d\vec{s} = 2\pi\rho L E_\rho, \tag{3.14}$$

where the first and second terms integrated on top and bottom surfaces, respectively, are zero because the unit vectors of the differential surfaces are perpendicular to electric field intensity on that surface. Thus, the surface integration includes only the integral over the side-wall surface.

Using Gauss's law, we have

$$2\pi\rho L E_\rho = L\rho_l / \varepsilon_0 \tag{3.15}$$

or

$$\vec{E} = \hat{a}_\rho E_\rho = \frac{\rho_l}{2\pi\varepsilon_0} \frac{1}{\rho} \hat{a}_\rho, \tag{3.16}$$

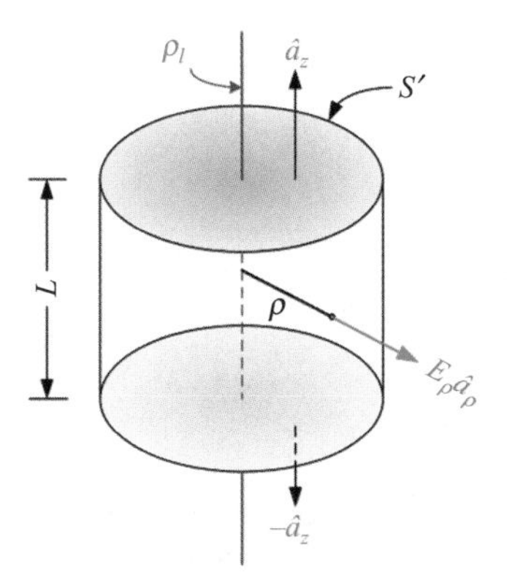

Figure 3.6 Gaussian surface used to calculate the electric field intensity produced by a uniform infinite line charge density, ρ_l, using Gauss's law.

Notice that the length L of the cylindrical Gaussian surface does not contribute to electric field intensity, so we could choose a cylinder with arbitrary height.

b. Use direct integration to determine the electric field intensity for an infinitely long, straight line in free space with a uniform line charge density ρ_l, as shown in Figure 3.7.

SOLUTION Because the line charge is uniform, infinitely long, and straight, the electric field intensity at a point P caused by a charge element Δz_i *at* $z = z_i$ must produce a z component that cancels the z component of the charge element Δz_i *at* $z = -z_i$.

The component of the electric field intensity in the radial direction at point P caused by the charge element Δz_i at $z = z_i$ must add to the component of the electric field intensity in the radial direction caused by the charge element Δz_i at $z = -z_i$.

Thus, the net vector electric field intensity at point P can be written as

$$\Delta\vec{E}_i = \hat{a}_\rho E_{i,\rho} = \hat{a}_\rho k_e\left(\Delta q_{at\, z_i} / R^2\right) 2\cos\theta, \tag{3.17}$$

where

$$\cos\theta = \rho \big/ \sqrt{\rho^2 + z_i^2} \quad \text{and} \quad R^2 = (\rho^2 + z_i^2).$$

Any other components of the electric field intensity at P will be canceled by each other because of symmetry of the structure. Using $\Delta q_{at\ zi} = \rho_l \Delta z_i$, we find

$$\Delta\vec{E}_i = \hat{a}_\rho k_e \frac{\rho_l \Delta z_i}{(\rho^2 + z_i^2)} 2 \frac{\rho}{\sqrt{\rho^2 + z_i^2}} = 2k_e\rho_l \frac{\rho}{(\rho^2 + z_i^2)^{3/2}} \hat{a}_\rho \tag{3.18}$$

Now, if we take the sum of all such elements Δz_i from $z = 0$ to $l/2$ (we have already taken into account the charge elements at negative values of z_i),

$$\vec{E} = \frac{\rho_l}{2\pi\varepsilon_0} \lim_{\Delta z_i \to 0} \sum_{i=1}^{\infty} \frac{\rho}{(\rho^2 + z_i^2)^{3/2}} \Delta z_i \hat{a}_\rho = \frac{\rho_l}{2\pi\varepsilon_0} \int_0^{l/2} \frac{\rho}{(\rho^2 + z^2)^{3/2}} dz \hat{a}_\rho \tag{3.19}$$

and, if we look up the integral in tables,

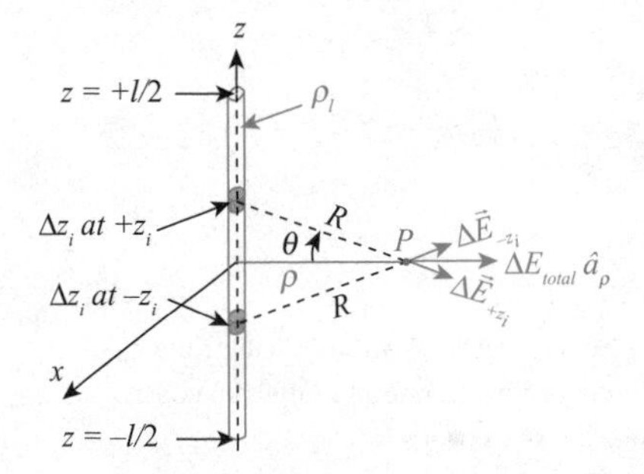

Figure 3.7 Electric field components at point P caused by infinitesimal charge elements at z_i and $-z_i$.

$$\int_0^{l/2} \frac{\rho}{(\rho^2+z^2)^{3/2}}dz = \rho \frac{z}{\rho^2(\rho^2+z^2)^{1/2}}\bigg|_0^{l/2} = \frac{l/2}{\rho[\rho^2+(l/2)^2]^{1/2}} \tag{3.20}$$

and in the limit for $l/2 \gg \rho$

$$\vec{E} = \hat{a}_\rho E_\rho = \frac{\rho_l}{2\pi\varepsilon_0}\frac{1}{\rho}\hat{a}_\rho, \tag{3.21}$$

which is the same as Equation 3.16.

CONCLUSION Gauss's law, with the appropriate Gaussian surface, produces the electric field intensity in a much easier manner than a direct integration technique.

3.2 **a.** Use Gauss's law to determine the electric field intensity for an infinitely large, charged surface with a uniform surface charge density ρ_S (alternately labeled Σ_S).

SOLUTION Because the surface charge is infinite in the plane of the surface, the electric field intensity must be perpendicular to the surface. Thus, if we let the surface be in the x–y plane, $\vec{E} = \hat{a}_z E_z$. Any other components will be canceled by each other because of symmetry of the structure. By mentally constructing a Gaussian surface, as shown in Figure 3.8,

Here, the Gaussian surface is in the form of a pillbox of radius b and height h that penetrates the x–y plane at its geometric center.

By symmetry, we can see that the electric field intensity points in the $\boldsymbol{a}_z$-direction above the x–y plane and in the $-\boldsymbol{a}_z$-direction below the x–y

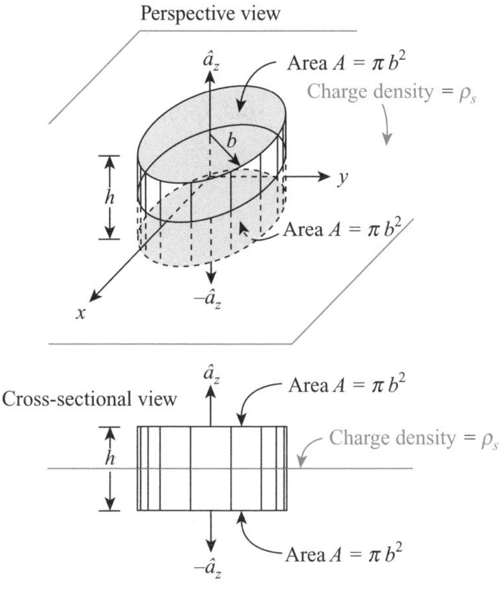

Figure 3.8 Electric field intensity caused by a uniformly charged, infinitely large surface.

plane. Thus, the dot product of the electric field intensity and the differential area on the sides of the pillbox are zero, while the dot product of the electric field intensity on the upper circular surface gives an electric flux on the upper surface of $\Phi_e = E_z\pi b^2$ and an electric flux on the bottom surface of $\Phi_e = E_z\pi b^2$. Because the charge inside the pillbox is $\rho_S\pi b^2$, we can use Gauss's law:

$$2E_z \pi b^2 = \rho_S \pi b^2/\varepsilon_0 \tag{3.22}$$

to find the electric field intensity above a uniformly charged plane as

$$\vec{E} = (\rho_S/2\varepsilon_0)\hat{a}_Z. \tag{3.23}$$

b. Use direct integration to determine the electric field intensity for an infinitely large, charged surface with a uniform surface charge density ρ_S.

SOLUTION We will first consider the electric field intensity at a point P along the z-axis due to an infinitesimally small charged area, $\rho\Delta\phi\Delta\rho$, as shown in Figure 3.9.

If charge Q is spread uniformly on a circular disk of radius a in the x–y plane, its charge density per square meter is $\rho_S = Q/\pi a^2$. From Coulomb's law 3.2 and 3.5,

$$\Delta\vec{E}_R = k_e \frac{(Q/\pi a^2)\rho\Delta\phi\Delta\rho}{(z^2+\rho^2)}\hat{a}_R \tag{3.24}$$

and the z-component of this electric field intensity is

$$\Delta E_z = k_e \frac{(Q/\pi a^2)\rho\Delta\phi\Delta\rho}{(z^2+\rho^2)}\cos\theta = \frac{k_e Q}{\pi a^2}\frac{\rho z}{(z^2+\rho^2)^{3/2}}\Delta\phi\Delta\rho. \tag{3.25}$$

So if we take the limit as $\Delta\varphi \to 0$ and $\Delta\rho \to 0$ and add the electric field intensity components in the z-direction due to all such infinitesimal areas,

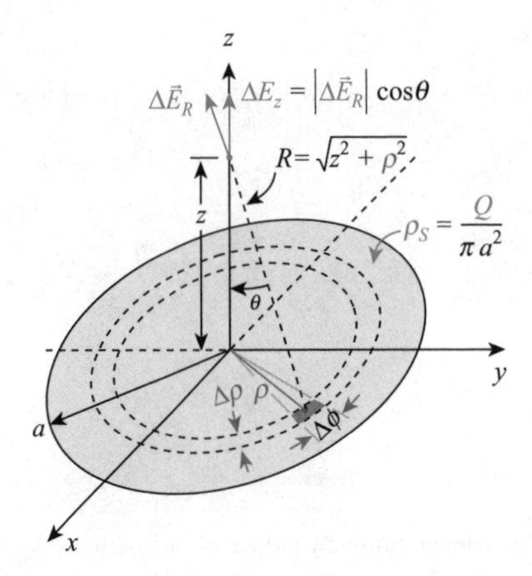

Figure 3.9 Electric field intensity at a point on the z axis due to an infinitesimal element of charge in the x–y plane.

$$E_z = \frac{k_e Q}{\pi a^2} \int_0^a \frac{\rho z}{\left(z^2+\rho^2\right)^{3/2}} \left[\int_0^{2\pi} d\phi\right] d\rho = \frac{2k_e Q z}{a^2} \int_0^a \frac{\rho}{\left(z^2+\rho^2\right)^{3/2}} d\rho \tag{3.26}$$

and from integral tables

$$E_z = -\left(2k_e Q z/a^2\right)\left(z^2+\rho^2\right)^{-1/2}\Big|_0^a = \left(2k_e Q/a^2\right)\left[1 - z\left(z^2+a^2\right)^{-1/2}\right]. \tag{3.27}$$

Equation 3.27 is the exact answer for the electric field intensity caused by a uniformly charged disk of radius a. Now, if we let $Q = \rho_S \pi a^2$ and take the limit for $z \ll a$, we can write

$$E_z = \left(2k_e \rho_S \pi a^2/a^2\right)\left[1 - z\left(z^2+a^2\right)^{-1/2}\right] = (\rho_S/2\varepsilon_0)\left[1 - (z/a)\left(1+(z/a)^2\right)^{-1/2}\right] \tag{3.28}$$

and expanding

$$\left(1+(z/a)^2\right)^{-1/2} = 1 - 1/2(z/a)^2 + (1/2)(3/2)/2!(z/a)^4 - (1/2)(3/2)(5/2)/3!(z/a)^6 + \ldots$$

For small values of (z/a), we see that

$$\vec{E} \approx (\rho_S/2\varepsilon_0)\hat{a}_z, \tag{3.29}$$

which is consistent with Equation 3.23.

CONCLUSION Direct integration and the use of Gauss's law yield the same result, but it is much easier to use Gauss's law.

3.3 Use Gauss's law to determine the electric field intensity for a large, uniformly charged, spherical shell of radius a.

SOLUTION If the charge, Q, on the sphere is uniformly distributed, $\rho_S = Q/4\pi a^2$, and we can find the electric field intensity using a Gaussian surface of radius R, as shown in Figure 3.10.

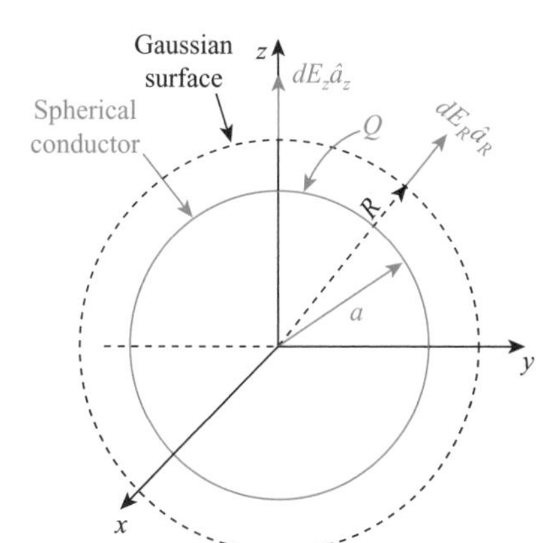

Figure 3.10 Gaussian sphere of radius R surrounding a uniformly charged spherical shell of radius a.

Using symmetry arguments for the electric field intensity, $d\vec{E}_R$, we can see that it will have only a radial component and that this component will be parallel to a differential surface element at every point on the Gaussian surface, so

$$\Phi_e = Q_{enclosed}/\varepsilon_0 = Q/\varepsilon_0 = \oint_S \vec{E}\cdot d\vec{\sigma} = E_R 4\pi R^2 \tag{3.30}$$

and, thus,

$$\vec{E} = \left(Q/4\pi\varepsilon_0 R^2\right)\hat{a}_R \tag{3.31}$$

PROBLEM

3.1 Use direct integration to determine the electric field intensity for a large, uniformly charged spherical shell. ***HINT*** Consider the charged shell, as shown in Figure 3.11.

Parallel Plate Capacitor

Let us apply Gauss's law to a large parallel plate capacitor, as shown in Figure 3.12.

In the upper-left corner of Figure 3.12a, Gaussian (black) surface (like that shown in Figure 3.8) encloses an amount of positive charge, $Q = +\rho_{S,1}\pi b^2$, so there is an equal electric flux, $\Phi_e = \int_A \vec{E}\cdot d\vec{\sigma} = E_{z,1}\pi b^2$, as a result of this charge on the top surface of the pillbox and on the lower surface (there is no electric flux on the cylindrical side surface). Using Gauss's law, we thus conclude that

$$2E_{z,1}\pi b^2 = Q/\varepsilon_0 = \rho_{S,1}\pi b^2/\varepsilon_0 \tag{3.32}$$

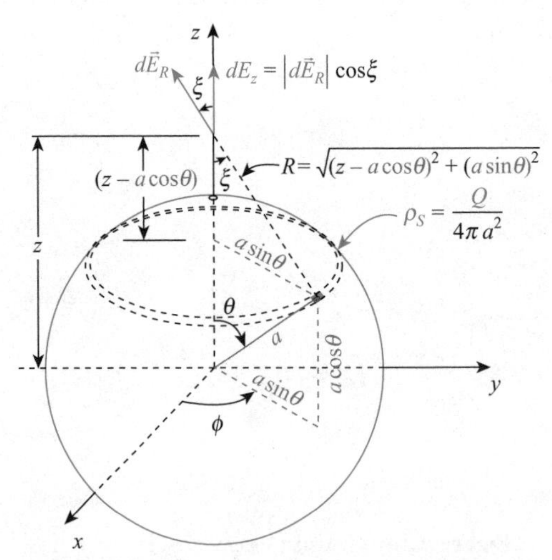

Figure 3.11 Electric field intensity at a point on the z-axis due to an infinitesimal element of charge on the surface of a sphere of radius a.

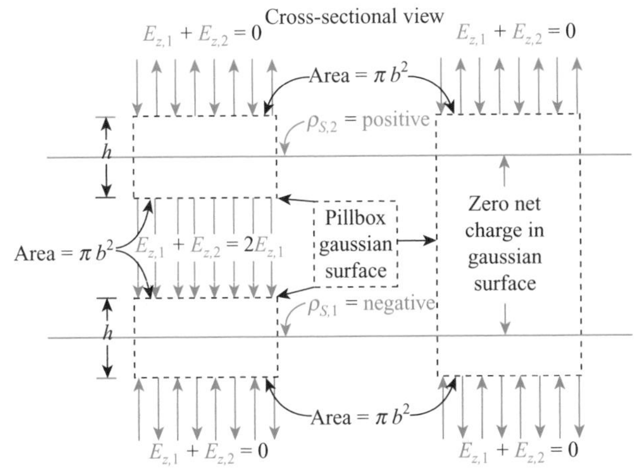

Figure 3.12 Cross-sectional view of the charges and electric field intensity produced by a large parallel plate capacitor.

$$\vec{E}_1 = (\rho_{S,1}/2\varepsilon_0)\hat{a}_z \text{ on the top and } \vec{E}_1 = (\rho_{S,1}/2\varepsilon_0)(-\hat{a}_z) \text{ on the bottom.} \tag{3.33}$$

In the lower-left corner of Figure 3.12, another Gaussian surface encloses an amount of negative charge so the electric field intensity on its two surfaces is exactly opposite to those of Equation 3.23.

We can conclude that the electric field intensity in the region between the two parallel plates due to the two charged surfaces reinforces to produce a *net* electric field intensity

$$\vec{E} = (-\rho_S/\varepsilon_0)\hat{a}_z \text{ in the region between the plates.} \tag{3.34}$$

We can also use a larger Gaussian cylindrical surface like that shown on the right-hand side of Figure 3.12 to see that there is no *net* charge enclosed, so the electric field intensity on the upper surface due to the positive charge is exactly canceled by the electric field intensity on the upper surface due to the negative charge.

CONCLUSION The electric field intensity in the neighborhood of a large parallel plate capacitor is zero outside the capacitor and is $\vec{E} = (-\rho_S/\varepsilon_0)\hat{a}_z$ in the region between them.

Spherical Charged Cloud

1. Suppose a uniformly charged spherical cloud inside a sphere of radius b, as shown in Figure 3.13 with a Gaussian surface at radius $R > b$.

   From symmetry, we can see that the electric field intensity is of constant magnitude anywhere on the Gaussian surface and points in the outward radial direction parallel to a surface element on the Gaussian surface, so

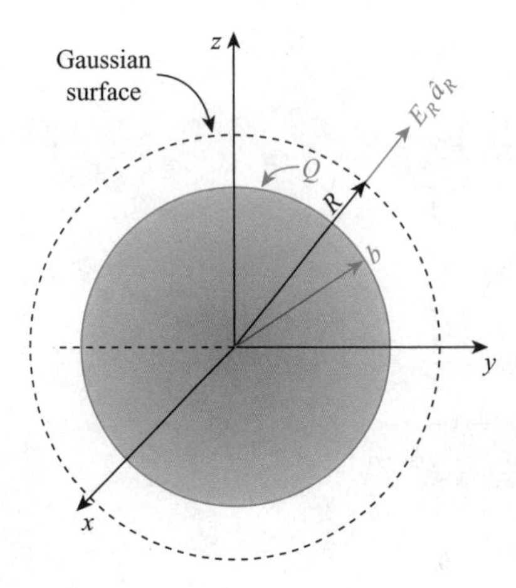

Figure 3.13 Uniform charged cloud inside a sphere of radius b with a Gaussian surface at R.

$$\Phi_e = \oiint_S \vec{E} \cdot d\vec{\sigma} = E_R 4\pi R^2 = Q_{inside}/\varepsilon_0 \tag{3.35}$$

and because $Q_{inside} = Q$,

$$E_R = (1/4\pi\varepsilon_0)(Q/R^2) \tag{3.36}$$

as if **all** the charge were located at the origin.

2. Now, suppose the Gaussian surface is of radius $R < b$, as shown in Figure 3.14.

   Again, from symmetry, we can see that the electric field intensity is of constant magnitude anywhere on the Gaussian surface and points in the outward radial direction parallel to a surface element on the Gaussian surface, so that

$$\Phi_e = \oiint_S \vec{E} \cdot d\vec{\sigma} = E_R 4\pi R^2 = Q_{inside}/\varepsilon_0. \tag{3.37}$$

   But, in this case,

$$Q_{inside} = (4\pi R^3/3)[Q/(4\pi b^3/3)]$$

   so that

$$E_R = (1/4\pi\varepsilon_0)(Q/b^2)(R/b). \tag{3.38}$$

   Note that, at $R = b$, the two values of E_R (Equations 3.36 and 3.38) are the same. Thus, if we plot the magnitude of the electric field intensity as a function of R, we get the dependence shown in Figure 3.15.

GOOD QUESTIONS Given that the electric field intensity inside a perfectly conducting material is zero, use Gauss's law arguments to answer the following questions:

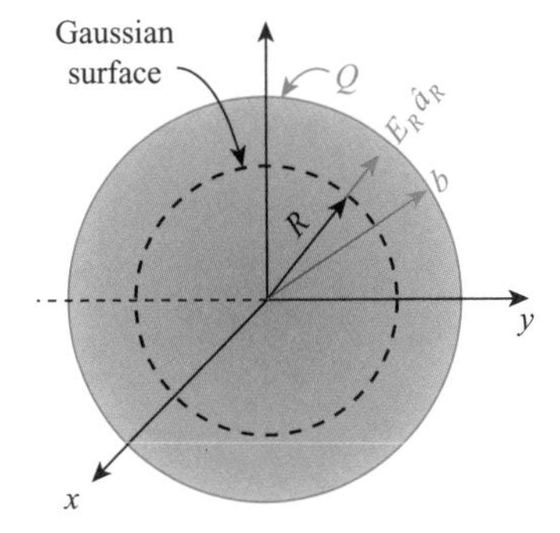

Figure 3.14 Uniform charged cloud inside a sphere of radius b with a Gaussian surface at R.

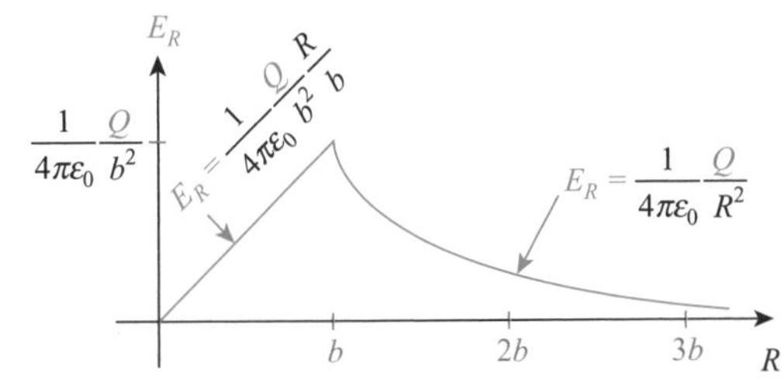

Figure 3.15 Electric field intensity as a function of R for a uniformly charged spherical cloud.

a. How will a charge Q distribute itself on the perfectly conducting sphere shown in Figure 3.16a?

b. How will a charge Q distribute itself on the perfectly conducting spherical shell shown in Figure 3.16b?

c. If a charge Q is located at the center of a conducting spherical shell, as shown in Figure 3.16c, how does the magnitude of the electric field intensity behave with radius R?

d. If a charge Q is located at the center of a conducting spherical shell and another charge $-Q$ is added to the shell, how does the charge distribute itself and how does the electric field intensity vary with radius R?

HINT Mentally construct a spherical Gaussian surface concentric with the origin with subsequently larger radius. If the radius of this surface is in a conductor the field **must** be zero at this point.

Using this classical picture of charge distributions how do you suppose the charge is distributed within a spherically shaped electron, proton or neutron? Is it a uniform cloud, a point, a shell? Could there be a mixture of positive and negative charges for these fundamental particles? Could you tell without probing inside them? Do the answers change if you ground the sphere?

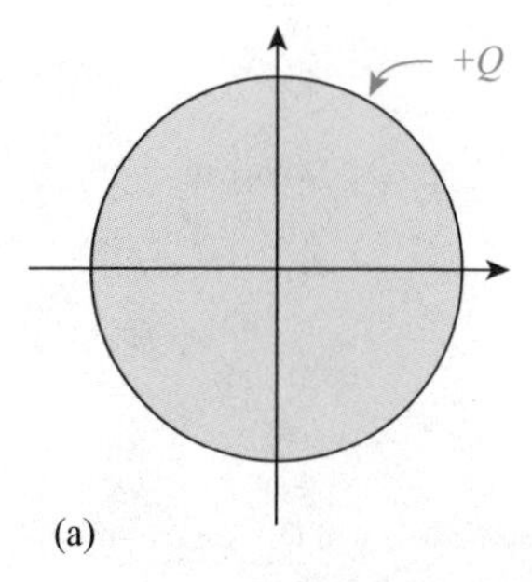

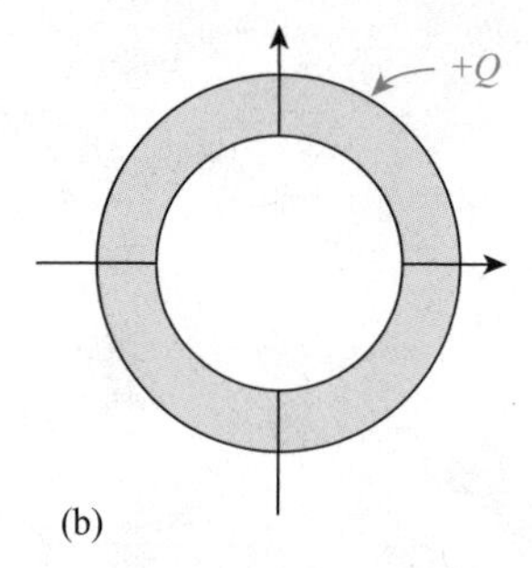

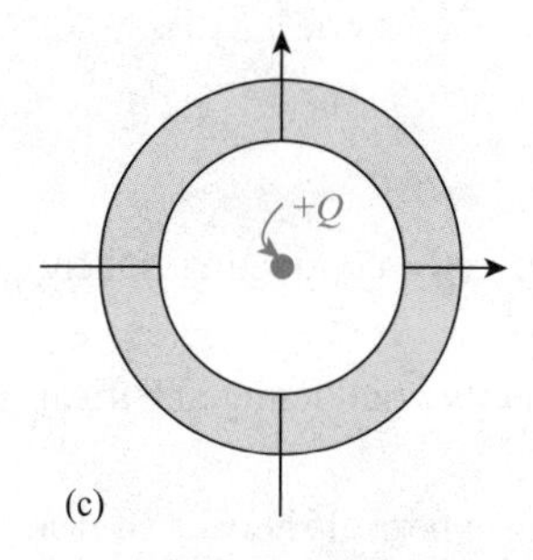

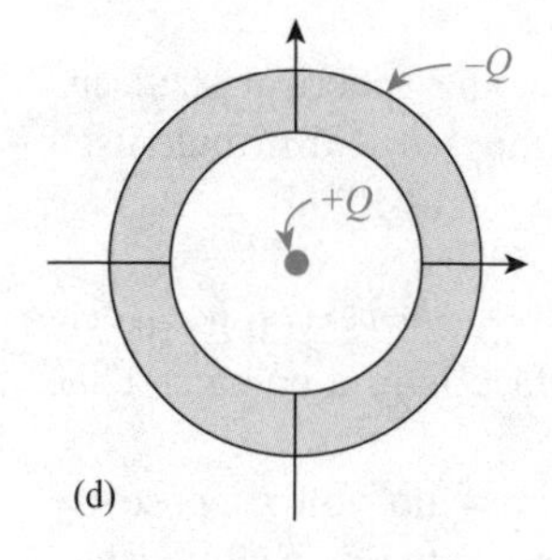

Figure 3.16 Perfectly conducting spherical conductors with (a) charge Q on a solid sphere, (b) charge Q on a sphere with an interior spherical cavity, (c) charge Q in the interior of a sphere with a spherical cavity, (d) charge Q in the interior and charge $-Q$ on the exterior of a sphere with a spherical cavity.

If the cross sections of Figure 3.16 represent a cut through a long cylindrical conductor or cylindrical shell with charge per unit length, $\rho_l = Q/l$, how would the electric field intensity distributions differ from the above answers?

3.3 CONSERVATION LAW

The curl of the electric field intensity (Equation 3.7), $\vec{\nabla} \times \vec{E}$, may be evaluated by using Stokes's theorem for *any* arbitrary vector field, $\vec{A}$, as we deduced in Chapter 2 (Figure 2.21). Here, we recall Stokes theorem

$$\int_S (\nabla \times \vec{A}) \cdot d\vec{s} = \oint_C \vec{A} \cdot d\vec{l} \tag{2.93}$$

for *any* open surface S enclosed by the curve C.

However, when we consider the particular case of an integral of *static* electric field intensity, $\vec{A} \Rightarrow \vec{E}$ and $\vec{\nabla} \times \vec{E} = 0$ everywhere on the surface S, so the line integral around the closed path C in Equation 3.38 vanishes:

$$\int_S (\nabla \times \vec{E}) \cdot d\vec{s} = \oint_C \vec{E} \cdot d\vec{l} = 0. \tag{3.39}$$

Because scalar line integral of the *static* electric field intensity around any closed path is adding up the differential electric potential elements, $dV = \vec{E} \cdot d\vec{l}$, around a path that returns to the point where it started. Equation 3.39 is an expression of Kirchhoff's voltage law in circuit theory that states that the algebraic sum of *voltage drop around any closed circuit is zero.*

Conservation Law for Electric Field Intensity

We may also see that Equation 3.39 implies that, in electrostatics, the voltage between any two points depends only on the starting and end points and is not relevant to voltage integral path by considering the curve C in Figure 3.17a to be composed of two parts, C_1 and C_2, as shown in figure Figure 3.17b. Mathematically, we would write this as

$$\oint_C \vec{E} \cdot d\vec{l} = \int_{C_1} \vec{E} \cdot d\vec{l} + \int_{C_2} \vec{E} \cdot d\vec{l}. \tag{3.40}$$

In Figure 3.17c, we consider the integral between P_2 and P_1 to be the negative of the integral between P_1 and P_2. Thus, Equation 3.40 may be rewritten as

$$\begin{aligned}
&\oint_C \vec{E} \cdot d\vec{l} = \int_{P_1(\text{Path } C_1)}^{P_2} \vec{E} \cdot d\vec{l} + \int_{P_2(\text{Path } C_2)}^{P_1} \vec{E} \cdot d\vec{l} = 0 \quad \text{or} \\
&\oint_C \vec{E} \cdot d\vec{l} = \int_{P_1(\text{Path } C_1)}^{P_2} \vec{E} \cdot d\vec{l} - \int_{P_1(\text{Path } C_2)}^{P_2} \vec{E} \cdot d\vec{l} = 0 \quad \text{or} \\
&\int_{P_1(\text{Path } C_1)}^{P_2} \vec{E} \cdot d\vec{l} = \int_{P_1(\text{Path } C_2)}^{P_2} \vec{E} \cdot d\vec{l} \quad \text{or finally} \\
&\int_{P_1(\text{Path } C_1')}^{P_2} \vec{E} \cdot d\vec{l} = \int_{P_1(\text{Path } C_2')}^{P_2} \vec{E} \cdot d\vec{l},
\end{aligned} \tag{3.41}$$

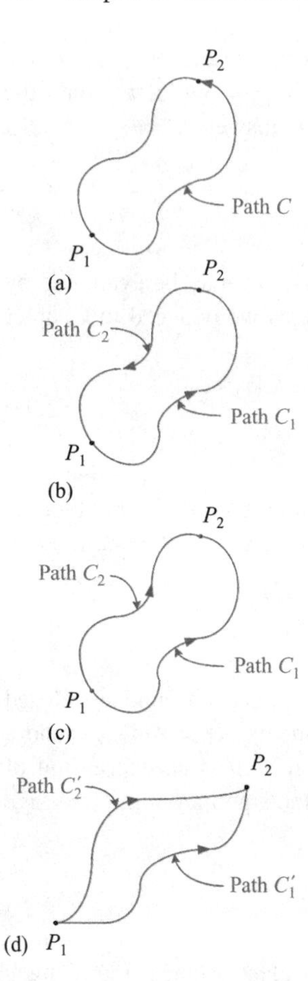

Figure 3.17 (a) Arbitrary closed path C considered as (b) the sum of two parts (C_1 between P_1 and P_2 and C_2 between P_2 and P_1) (c) (C_1 between P_1 and P_2 and $-C_2$ between P_1 and P_2) and (d) (C_1' between P_1 and P_2 and $-C_2'$ between P_1 and P_2).

which states that the scalar line integral of the *irrotational* electric field intensity is *independent of the integral path*; it depends only on the starting and end points. If the value of a line integral is independent of the path taken, we say that such fields are *conservative*.

CONCLUSION Static electric fields are irrotational ($\vec{\nabla} \times \vec{E} = 0$) and are thus conservative; that is, the line integral $\int_{P_1}^{P_2} \vec{E} \cdot d\vec{l}$ is *independent of the path* taken between P_1 and P_2.

3.4 ELECTRIC POTENTIAL

If an electric charge q_t is located at a point (x, y, A) in a uniform electric field intensity (like that produced by a parallel plate capacitor), as shown in Figure 3.18, and the charge is displaced to $z = B$, a constant force $\vec{F} = q_t E_z \hat{a}_z$ will be required. Thus, the work carried out by the force F in displacing the charge will be

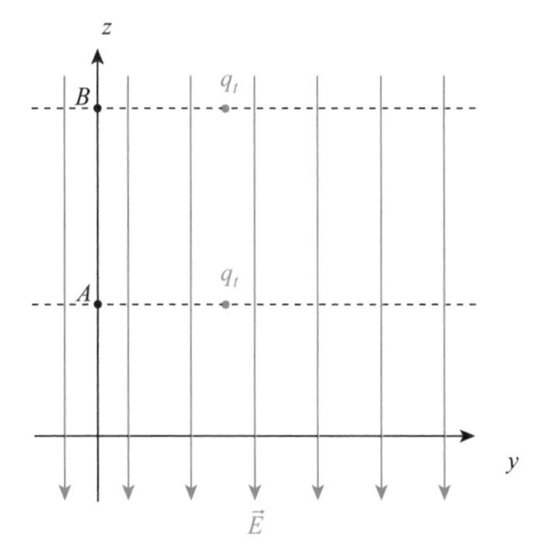

Figure 3.18 A charge q_t in a uniform electric field intensity $\vec{E} = -E_z\hat{a}_z$.

$$\Delta W_{AB} = \int_A^B \vec{F} \cdot d\vec{l} = \int_A^B q_t\, E_z\, dz = q_t\, E_z(B - A). \tag{3.42}$$

Note that ΔW_{AB} is a positive quantity as long as q_t is positive. We also note that a charge q_t at point $z = B$ has the potential to give up that work if we allow it to "fall" to $z = A$; that is, it has electric potential energy, ΔU_{AB}, at point B relative to point A of

$$\Delta U_{AB} = U_B - U_A = q_t\, E_z\, B - q_t\, E_z\, A. \tag{3.43}$$

We *could* define the electric potential energy at B to be $U_B = q_tE_zB$, but that would be merely a statement of its energy relative to $z = 0$.

Electric Potential

We will define the "electric potential" difference between two points, ΔV, to be the electric potential energy per unit charge:

$$\Delta V_{AB} \equiv \Delta U_{ab}/q_t = E_z\,(B - A). \tag{3.44}$$

Because the electric field intensity, $\vec{E} = -E_z\hat{a}_z$, and the external force needed to move the charge, $\vec{F} = q_tE_z\hat{a}_z = -q_t\vec{E}$, are in opposite directions,

$$\Delta V_{AB} = \int_A^B -\vec{E} \cdot d\vec{l} \tag{3.45}$$

Note that, if B is infinitesimally close to A, $dV = -\vec{E} \cdot d\vec{l} = -E_z dz$, so

$$E_z = -dV/dz. \tag{3.46}$$

We could make a similar argument for electric field intensity in the $-x$-direction to show that $E_x = -dV/dx$ and likewise for electric field intensity in the $-y$-direction to show that $E_y = -dV/dy$. Thus, if a field has components in the $-x$-, $-y$-, and $-z$-directions,

$$\vec{E} = -\partial V/\partial x\,\hat{a}_x - \partial V/\partial y\,\hat{a}_y - \partial V/\partial z\,\hat{a}_z = -\vec{\nabla} V \tag{3.47}$$

Note that the units of $\Delta V_{AB} \equiv \Delta U_{ab}/q_t$ are volts, so 1 V = 1 J/C = 1 N m/C. A good conversion to remember is that V/m = N/C.

Electric Potential of a Point Charge at the Origin

Suppose a charge Q at the origin and another charge q_t at radius R_A, as shown in Figure 3.19.

From Coulomb's law, we can find the electric field intensity at R as $\vec{E} = k_e Q/R^2 \hat{a}_R$ and the force required to push q_t toward the origin as $\vec{F} = -k_e q_t Q/R^2 \hat{a}_R$, so the work carried out in pushing q_t from R_A to R_B will be

$$\Delta W_{AB} = \int_{R_A}^{R_B} -k_e \frac{q_t Q}{R^2} \hat{a}_R \cdot dR\hat{a}_R = k_e q_t Q \frac{1}{R}\bigg|_{R_A}^{R_B} = -k_e q_t Q\left(\frac{1}{R_A} - \frac{1}{R_B}\right) \tag{3.48}$$

This work is a negative quantity for $R_B > R_A$ so the charge q_t at R_A has the potential to do work as it "falls" to R_B; that is, it has positive "electric potential energy" $\Delta U_{AB} = k_e q_t Q(1/R_A - 1/R_B)$ at point A relative to point B or it has electric potential:

$$\Delta V_{AB} = k_e Q(1/R_A - 1/R_B) \tag{3.49}$$

If we agree that $R_B = \infty$ is a common reference radius, we can say that the *absolute potential* at R_A is $V_A = k_e Q/R_A$. Unless it is otherwise stated, we will assume a reference point at $R_B = \infty$ and write

$$V(R) = k_e Q/R \tag{3.50}$$

as the *absolute potential* around a point charge Q.

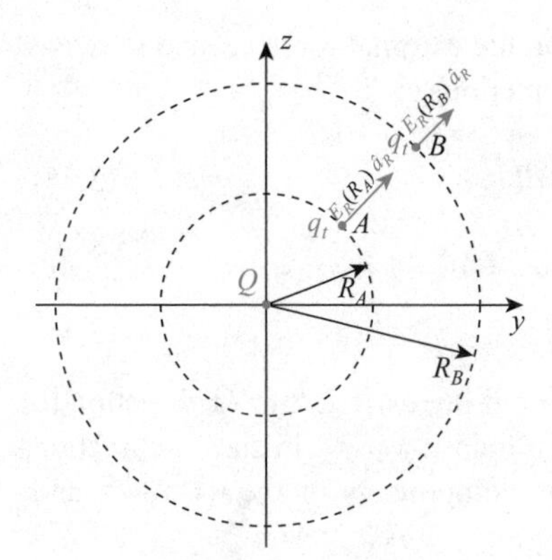

Figure 3.19 A charge q_t in an electric field intensity caused by a point charge Q at the origin.

Electric Potential of a Point Charge at R′

The electric potential at a point P due to a point charge, q at the origin, is a scalar function of R, where R is the radius of a sphere from the point charge. If the charge is not at the origin but at $\vec{R}'$, as shown in Figure 3.20, then the electric potential at point P will be

$$V(R) = k_e q / |\vec{R} - \vec{R}'|, \tag{3.51}$$

where the observation position vector, source position vector, and distance between the source point and observation point in Cartesian coordinates are specified by

$$\vec{R} = \hat{a}_x x + \hat{a}_y y + \hat{a}_z z \tag{3.52a}$$

$$\vec{R}' = \hat{a}_x x' + \hat{a}_y y' + \hat{a}_z z' \tag{3.52b}$$

$$|\vec{R} - \vec{R}'| = \sqrt{(x - x')^2 + (y - y')^2 + (z - z')^2}. \tag{3.52c}$$

In this book, we will generally use the convention that any quantities with a prime sign are source parameters and that unprimed quantities will be points of observation. Therefore, using Equation 3.47,

$$\vec{E} = \hat{a}_n E_P = \hat{a}_n q / 4\pi\varepsilon_0 |\vec{R} - \vec{R}'|^2, \tag{3.53}$$

where the normal unit vector, $\hat{a}_n$, on the Gaussian surface can be written as

$$\hat{a}_n = (\vec{R} - \vec{R}')/|\vec{R} - \vec{R}'| \tag{3.54}$$

then Equation 3.53 becomes

$$\vec{E}_P = q(\vec{R} - \vec{R}')/4\pi\varepsilon_0 |\vec{R} - \vec{R}'|^3 \ (V/m). \tag{3.55}$$

Equation 3.55 indicates that the electric field intensity of a positive point charge, q at $\vec{R}'$, is in the outward radial direction of the Gaussian surface and has a magnitude proportional to the charge and inversely proportional to the square of the distance from the charge. In particular, if the charge q is located at the origin of the coordinates, the electric field intensity can be simplified as

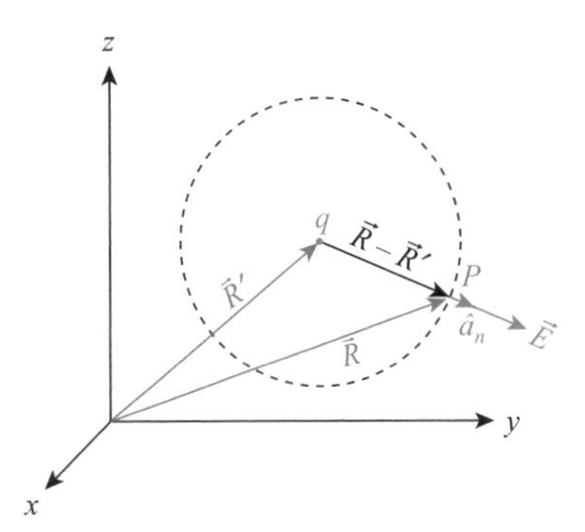

Figure 3.20 Electric field intensity due to a point charge, q, at an arbitrary position, $\vec{R}'$.

$$\vec{E} = \hat{a}_R E_R = \hat{a}_R \frac{q}{4\pi\varepsilon_0 R^2}, \tag{3.56}$$

consistent with Coulomb's law.

3.5 ELECTRIC FIELD FOR A SYSTEM OF CHARGES

Now, we solve the electric field intensity created by a group of n-discrete charges, $q_1, q_2, \ldots, q_n$, which are located at different positions. According to the principle of superposition, the total field at point P is simply the vector summation of all the electric field intensity generated by individual charges:

$$\vec{E}_P = (1/4\pi\varepsilon_0)\sum_{k=1}^{n} q_k \left(\vec{R} - \vec{R}'_k\right) \Big/ \left|\vec{R} - \vec{R}'_k\right|^3. \tag{3.57}$$

EXAMPLE

3.4 Starting with Equation 3.57, find the electric field intensity for an electric dipole, as shown in Figure 3.21.

SOLUTION Here, the position vectors for two point charges are

$$\vec{R}'_{+q} = (d/2)\hat{a}_z \quad \text{and} \quad \vec{R}'_{-q} = -(d/2)\hat{a}_z,$$

so we have

$$\vec{E} = \frac{q}{4\pi\varepsilon_0} \frac{\vec{R} - \vec{R}'_{+q}}{\left|\vec{R} - \vec{R}'_{+q}\right|^3} - \frac{q}{4\pi\varepsilon_0} \frac{\vec{R} - \vec{R}'_{-q}}{\left|\vec{R} - \vec{R}'_{-q}\right|^3} = \frac{q}{4\pi\varepsilon_0}\left(\frac{\vec{R} - (d/2)\hat{a}_z}{\left|\vec{R} - (d/2)\hat{a}_z\right|^3} - \frac{\vec{R} + (d/2)\hat{a}_z}{\left|\vec{R} + (d/2)\hat{a}_z\right|^3}\right). \tag{3.58}$$

If the observation point is far enough, then $d \ll R$ (the "far field" region). We can write

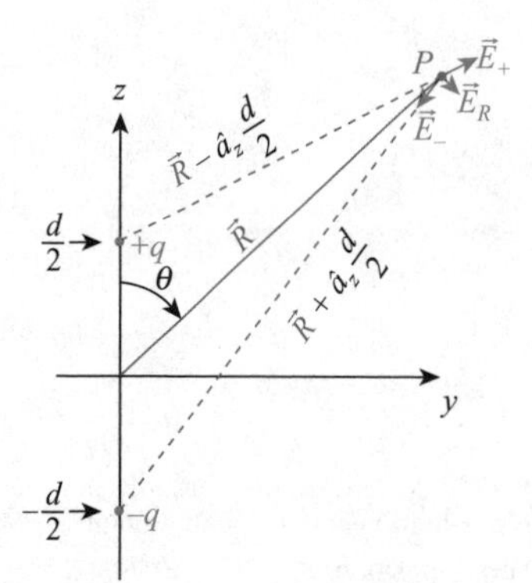

Figure 3.21 Configuration of an electric dipole.

$$|\vec{R}-(d/2)\hat{a}_z|^{-3} = [(\vec{R}-(d/2)\hat{a}_z)\cdot(\vec{R}-(d/2)\hat{a}_z)]^{-3/2} = [R^2 - \vec{R}\cdot\hat{a}_z d + (d/2)^2]^{-3/2}$$
$$\cong R^{-3}[1-\vec{R}\cdot\hat{a}_z d/R^2]^{-3/2} \cong R^{-3}[1+3\vec{R}\cdot\hat{a}_z d/2R^2] \quad (3.59)$$

where we have expended the above equation by using the series replacement

$$(a-x)^{-m} = a^{-m} + ma^{-m-1}x + m(m+1)/2!a^{-m-2}x^2 + \cdots (m>0) \quad (3.60)$$

and taken first two terms in the approximation. Similarly, we have

$$|\vec{R}+(d/2)\hat{a}_z|^{-3} \cong R^{-3}[1-(3/2)(\vec{R}\cdot\hat{a}_z d/R^2)]. \quad (3.61)$$

In Equation 3.58 we may now substitute the approximate value of $|\vec{R}-(d/2)\hat{a}_z|^{-3}$ by the value given in Equation 3.59 and the approximate value of $|\vec{R}+(d/2)\hat{a}_z)|^{-3}$ by the value given in Equation 3.61 to obtain

$$\vec{E} \cong (q/4\pi\varepsilon_0 R^3)[3(\vec{R}\cdot\hat{a}_z d/R^2)\vec{R}-\hat{a}_z d]. \quad (3.62)$$

We can define the electric dipole moment as

$$\vec{p} = \hat{a}_z qd \quad (3.63)$$

in which case Equation 3.62 can then be written as

$$\vec{E} \cong (1/4\pi\varepsilon_0 R^3)[3(\vec{R}\cdot\vec{p}/R^2)\vec{R}-\vec{p}]. \quad (3.64)$$

By converting the electric dipole vector into spherical coordinates, we have

$$\vec{p} = \hat{a}_z p = p(\hat{a}_R \cos\theta - \hat{a}_\theta \sin\theta) \quad (3.65)$$
$$\vec{R}\cdot\vec{p} = Rp\cos\theta \quad (3.66)$$
$$\vec{E}_{dipole} \cong (p/4\pi\varepsilon_0 R^3)[\hat{a}_R 2\cos\theta + \hat{a}_\theta \sin\theta]. \quad (3.67)$$

It is seen that the electric field intensity of an electric dipole decays much faster in space (i.e., $|\vec{E}_{dipole}| \propto 1/R^3$) than $1/R^2$ in Coulomb's law for the electric field intensity produced by a point charge (also called an electric monopole). This is because the electric field intensities generated by $+q$ and $-q$ tend to cancel each other as R increases.

3.6 ELECTRIC POTENTIAL FOR A SYSTEM OF CHARGES

The *electric potential* created by a group of n-discrete charges, $q_1, q_2, \ldots, q_n$, which are located at different positions are also found by the principle of superposition. The total electric potential at point P is simply the algebraic summation of all the electric potentials generated by individual charges:

$$V_P = (1/4\pi\varepsilon_0)\sum_{k=1}^{n} q_k/|\vec{R}-\vec{R}_k'|. \quad (3.68)$$

EXAMPLE

3.5 Starting with Equation 3.68, find the electric potential for the electric dipole shown in Figure 3.21.

SOLUTION Using the notation in Example 3.4, for the position vectors for two point charges $\vec{R}'_{+q} = (d/2)\hat{a}_z$ and $\vec{R}'_{-q} = -(d/2)\hat{a}_z$, we have

$$V_P = (q/4\pi\varepsilon_0)(1/|\vec{R}-(d/2)\hat{a}_z| - 1/|\vec{R}+(d/2)\hat{a}_z|) \tag{3.69}$$

If the observation point is in the "far field" region, then $d \ll R$ and we can write

$$\begin{aligned} |\vec{R}-(d/2)\hat{a}_z|^{-1} &= [(\vec{R}-(d/2)\hat{a}_z)\cdot(\vec{R}-(d/2)\hat{a}_z)]^{-1/2} \\ &\cong R^{-1}[1-\vec{R}\cdot\hat{a}_z d/R^2]^{-1/2} \cong R^{-1}[1+(1/2)\vec{R}\cdot\hat{a}_z d/R^2] \end{aligned} \tag{3.70}$$

and, substituting Equation 3.70 into Equation 3.69,

$$V_P \cong (q/4\pi\varepsilon_0 R)(\vec{R}\cdot\hat{a}_z d/R^2) = \vec{p}\cdot\hat{a}_R/4\pi\varepsilon_0 R^2. \tag{3.71}$$

PROBLEM

3.2 Use $\vec{E} = -\vec{\nabla}V$ in spherical coordinates to show that Equation 3.71 gives the same electric field intensity as Equation 3.67.

EXAMPLE

3.6 Starting with Equation 3.57, find the electric field intensity for a linear electric quadrupole, as shown in Figure 3.22.

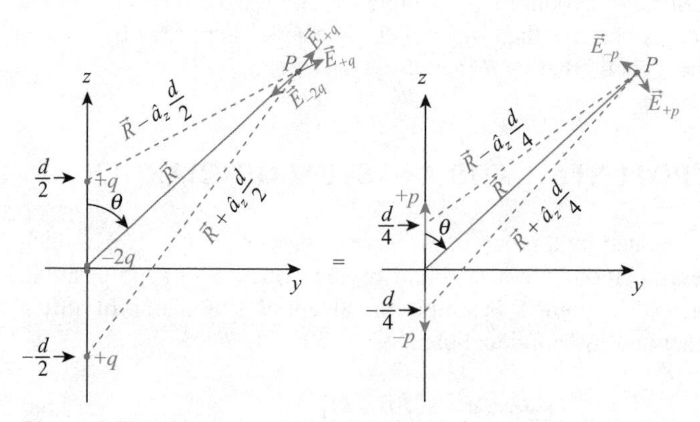

Figure 3.22 Configuration of a linear electric quadrupole.

Here, the position vectors for four point charges are $\vec{R}'_{+q} = (d/2)\hat{a}_z$, $\vec{R}'_{-2q} = 0$, and $\vec{R}'_{+q} = -(d/2)\hat{a}_z$ so we have

$$\vec{E}_{quadrupole} = (q/4\pi\varepsilon_0)(\vec{R} - \vec{R}'_{+q})/|\vec{R} - \vec{R}'_{+q}|^3 - 2q\vec{R}/4\pi\varepsilon_0|\vec{R}|^3 + q(\vec{R} - \vec{R}'_{+q})/4\pi\varepsilon_0|\vec{R} - \vec{R}'_{+q}|^3 \tag{3.72}$$

PROBLEM

3.3 In the "far field" region, show that the electric field intensity of a linear electric quadrupole decays much faster in space (i.e., $|\vec{E}_{quadrupole}| \propto 1/R^4$) than $1/R^3$ for the electric field intensity produced by an electric dipole and $1/R^2$ for an electric monopole.

3.4 In spherical coordinates, show that the electric potential of an electric quadrupole behaves as $V_{quadrupole}(R) \propto 1/R^3$.

A *multipole* charge distribution requires 2^{n-1} charges and its potential behaves as follows (Table 3.1):

3.7 ELECTRIC FIELD FOR A CONTINUOUS DISTRIBUTION

We often encounter situations in which charges are so close together compared with a discrete state that we can regard them as being continuously distributed. We can deal with such cases by considering a region of the charge distribution that is so small that the charge with it can be written as dq' and treated as a point charge, as illustrated in Figure 3.23.

By defining that ρ (C/m^3) is volume charge density, a function of coordinates; then $dq = \rho\, dv'$ is a differential charge; according to Equation 3.57, we have

$$d\vec{E} = \hat{a}_{R-R'}\left(dq/4\pi\varepsilon_0|\vec{R} - \vec{R}'|^2\right) = \hat{a}_{R-R'}\left(\rho dv'/4\pi\varepsilon_0|\vec{R} - \vec{R}'|^2\right), \tag{3.73}$$

Table 3.1 Electric field intensity and electric potential radial behavior of mulltipole charge configurations

Name	n	Number of charges	$\lvert\vec{E}\rvert \propto$	$V(R) \propto$
Monopole	1	$2^0 = 1$	k_e/R^2	k_e/R
Dipole	2	$2^1 = 2$	k_e/R^3	k_e/R^2
Quadrupole	3	$2^2 = 4$	k_e/R^4	k_e/R^3
Octapole	4	$2^3 = 8$	k_e/R^5	k_e/R^4
Hexadecapole	5	$2^4 = 16$	k_e/R^6	k_e/R^5
:	:	:	:	:
Multipole	n'	$2^{n'-1}$	$k_e/R^{n'+1}$	$k_e/R^{n'}$

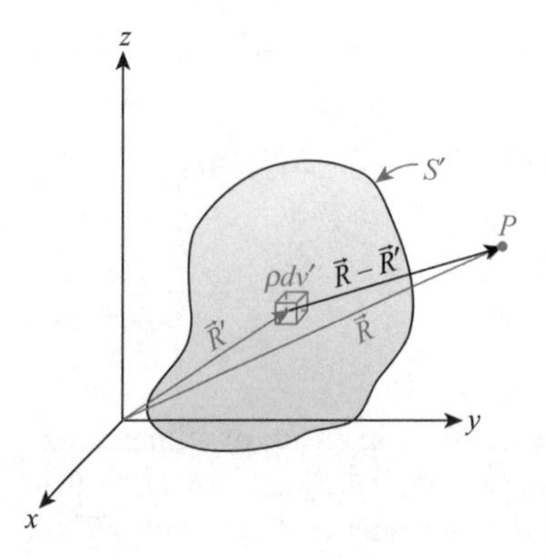

Figure 3.23 Electric field intensity due to a charge distribution.

where $\hat{a}_{R-R'}$ is the unit vector in the direction of $\vec{R} - \vec{R}'$, which is a vector from the source pointing toward the observation point P. It should be noted that $\hat{a}_{R-R'}$ varies with different space locations, $\vec{R} - \vec{R}'$. In Cartesian coordinates,

$$|\vec{R}-\vec{R}'| = \sqrt{(x-x')^2+(y-y')^2+(z-z')^2}. \tag{3.74}$$

Then, the total electric field intensity generated by the entire volume of charge distribution is

$$\vec{E} = \iiint_{V'} d\vec{E} = \iiint_{V'} \hat{a}_{R-R'}\, \rho dv' \big/ 4\pi\varepsilon_0 |\vec{R}-\vec{R}'|^2. \tag{3.75a}$$

Because $\hat{a}_{R-R'} = \vec{R} - \vec{R}'/|\vec{R} - \vec{R}'|$, Equation 3.75a can be rewritten as

$$\vec{E} = (1/4\pi\varepsilon_0) \iiint_{V'} \rho\left[(\vec{R}-\vec{R}')/|\vec{R}-\vec{R}'|^3\right] dv' \tag{3.75b}$$

If the charge is distributed on a surface with a surface charge density ρ_s (C/m^2), thus, the electric field intensity becomes

$$\vec{E} = (1/4\pi\varepsilon_0) \iint_{S'} \hat{a}_{R-R'}\left[\rho_S/|\vec{R}-\vec{R}'|^2\right] ds', \tag{3.76}$$

where the integration is on the charge distributed surface. Similarly, if the charge is distributed along a line with a line charge density ρ_l (C/m), thus,

$$\vec{E} = (1/4\pi\varepsilon_0) \int_{L'} \hat{a}_{R-R'}\left[\rho_l/|\vec{R}-\vec{R}'|^2\right] dl', \tag{3.77}$$

where the integration only needs to be taken along the charge distributed line.

3.8 CONDUCTOR IN A STATIC ELECTRIC FIELD

Classification of Materials

We usually classify materials into three types: conductors, semiconductors, and insulators (or dielectrics). In general terms, a conductor can be defined as a region in which charges are free to move under the influence of electric field intensity. A conductor has many free charges (often one free electron per atom) and is an excellent material for electric conducting. A conductor can be crudely thought of by using the periodic table and thinking of a metal as a periodic structure of atoms with positively charged nuclei and orbiting electrons. The inner (tightly bound) electrons remain associated with a specific nucleus even though there are nearby neighbors, but the electrons in the outermost shells of the atoms (loosely bound) of a conductor atoms are very loosely held and associate with their neighbors just as easily as their parent nucleus. These electrons are more or less "free" to migrate from one atom to another. Most metals belong to this group.

An insulator or dielectric is a material that is unable to conduct electricity. In the periodic atomic model, the electrons in the atoms of insulators or dielectrics are bound more tightly to their parent nuclei and are thus more or less confined to their atomic orbits; they are not "free" to move from atom to atom in normal circumstances, even with the application of external electric field intensity. The electrical properties of semiconductors fall between those of conductors and insulators in that they possess a relatively small number of "free" charges.

Electric Fields inside Conductors

If electric field intensity were present in a perfect conductor, the "free" electrons would move under its influence, and we would not have the static situation we are assuming in this chapter. Hence, inside a perfect electric conductor (PEC) under static conditions,

$$\rho_{volume} = 0 \qquad (\textit{inside} \text{ a PEC}) \tag{3.78}$$

$$\vec{E} = 0 \qquad (\textit{inside} \text{ a PEC}). \tag{3.79}$$

It then follows directly from Equations 3.68 and 3.75b that the electric potential V is constant with the conductor; that is, the conductor is an equipotential volume, or

$$V = \text{constant} \quad (\textit{inside} \text{ a PEC}) \tag{3.80}$$

and that the electric field intensity *inside* a perfect conductor is zero.[2] This can be explained with the following argument. Assume that that some positive (or negative) charges are introduced in the interior of a PEC. These charges will set up electric field intensity in the conductor, and the field will exert a force on the "free" electrons to make them move. This movement will continue until all charges reach the conductor surface and redistribute themselves in such a way that both the net charge and electric field intensity inside vanish. In a PEC, we assume that the redistribution will happen instantaneously. In a good conductor, we will have to permit fields to exist for a short period of time (about 10^{-17} s for Cu) until the charges can redistribute themselves. This more complicated case will be considered when we take up the subject of external electric field intensity that changes with time.

[2] Note it is possible to have a free charge density, ρ, *on the surface* of a PEC.

Boundary Conditions at a Conductor Interface

Now, let us consider the situation at the surface of a conductor that is adjacent to free space. As we have seen, the electric field intensity in the nearby free-space region outside of a conductor certainly can be different from zero.

Suppose that $\vec{E}_{outside}$ could make an angle with the surface, as shown in Figure 3.24. For this *hypothetical* example, we have resolved $\vec{E}$ into a perpendicular (normal) component to the surface $\vec{E}_n$ and a parallel (tangential) component to the surface $\vec{E}_t$.

From our previous arguments, if $\vec{E}_t$ were different from zero *inside* the conductor, there would be a tangential force on the "free" charges inside the conductor and they would move parallel to the surface, in which case we would no longer have the static situation we are assuming in this chapter. Therefore, for static fields, $\vec{E}_{t\ inside}$ = 0. The same argument can be made about $\vec{E}_{n\ inside}$ because the surface of a conductor may be charged, but the charges are not "free" to move into the free-space region outside the conductor. Thus, we can conclude that, *inside* the surface of a conductor,

$$\vec{E}_t = 0 \qquad (\textit{inside} \text{ the surface of a PEC}) \tag{3.81}$$

$$\vec{E}_n = 0 \qquad (\textit{inside} \text{ the surface of a PEC}) \tag{3.82}$$

$$V = \text{constant} \qquad (\textit{inside} \text{ the surface of a PEC}) \tag{3.83}$$

CONCLUSION Under static conditions, the electric field intensity *inside* a PEC surface is zero, so the surface and every point interior to the PEC are an equipotential. The same conclusion may be drawn for a *good* conductor if the "free" charges have come to equilibrium. We say that, for copper, after a period of 10^{-17} s, there is no flow of "free" electrons, so the conductor is "quasi-static."

Now, we can derive boundary conditions at a point just *outside* a conductor in free space from the line integral form of Maxwell's equation.

By choosing an enclosed line integral contour C, with thickness Δt, as shown in Figure 3.25 in blue, and taking the limit as $\Delta t \to 0$,

$$\oint_C \vec{E} \cdot d\vec{l} = E_{t\,inside}\Delta w - E_{t\,outside}\Delta w = 0,$$

where Δw is the *finite* width of the integral contour (i.e., $\Delta w \neq 0$). But, because $E_{t\ inside} = 0$, then $E_{t\ outside}$ must also be zero. Thus,

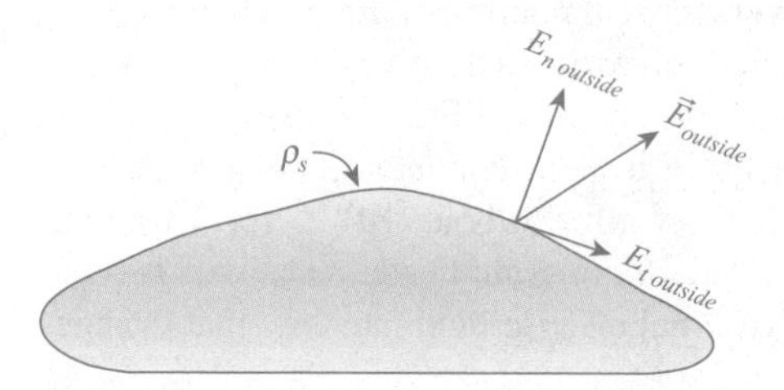

Figure 3.24 *Hypothetical* direction of $\vec{E}$ just outside a conducting surface.

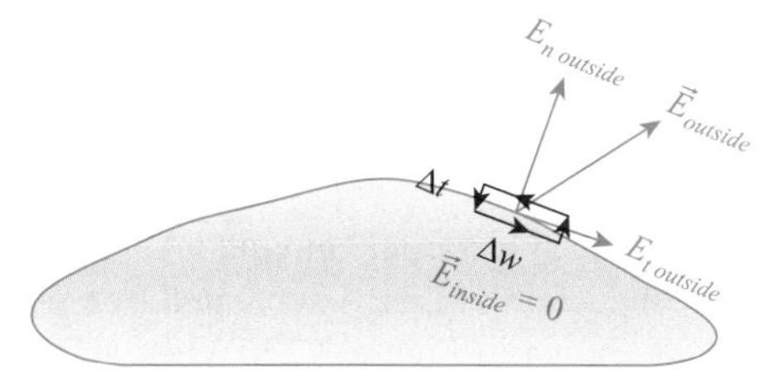

Figure 3.25 An infinitesimal path C for calculating the boundary condition for the electric field intensity in a direction tangential to the surface just outside of a conductor.

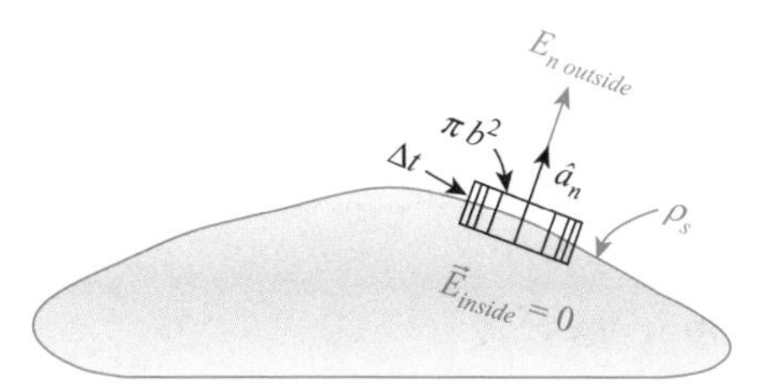

Figure 3.26 A Gaussian (pillbox) surface for calculating the boundary condition for the electric field intensity in a direction normal to the surface of a conductor.

$$E_{t\ inside} = E_{t\ outside} = 0 \tag{3.84}$$

CONCLUSION The tangential component of the electric field intensity just *outside* a good conductor surface under quasi-static static conditions must be zero.

Next, we derive the boundary condition of the electric field intensity along a normal direction to the surface of a conductor.

By constructing a Gaussian closed surface, as shown in Figure 3.26, in black and letting the radius of the "pillbox" be small, we can see that the direction of $\vec{E}_n$ is orthogonal to a surface element on the cylindrical sides of the pillbox and is parallel to the surface element, $\hat{a}_n$, on the top of the pillbox. The charge inside the Gaussian surface is the surface charge density, ρ_s, times the area of the pillbox. Using Gauss's law, we have

$$\oiint_S \vec{E} \cdot d\vec{\sigma} = E_{n\ inside}\hat{a}_n \cdot \pi b^2 \hat{a}_{n\ bottom} + E_{n\,outside}\hat{a}_n \cdot \pi b^2 \hat{a}_{n\ top} = \rho_s \pi b^2 / \varepsilon_0$$

and, because $E_{n\ inside} = 0$,

$$E_{n\ outside} = \rho_s / \varepsilon_0 \tag{3.85}$$

CONCLUSION The normal component of the electric field intensity just outside a conductor-free space interface is equal to the surface charge density, ρ_s, on the conductor divided by the free-space permittivity.

3.9 CAPACITANCE

Definition of a Capacitor

A capacitor is a frequently used circuit component in electronic circuits. A typical parallel plate capacitor consists of two large conductors of area A separated by free space or a dielectric medium by distance d, as shown in Figure 3.27. When a DC voltage source, V, is connected between the conductors, a charge transfer occurs, resulting in a charge $+Q$ on one conductor and $-Q$ on the other.

We define the capacitance of a capacitor to be

$$C \equiv Q/V \tag{3.86}$$

But, because $E = \rho_s/\varepsilon_0 = (Q/A)/\varepsilon_0 = V/d$, as was shown in Equation 3.33, we can see that $(Q/V) = \varepsilon_0 A/d$, so the quantity C depends only on the *geometry* of the capacitor and the permittivity of the medium between conductors of the capacitor.

Series and Parallel Connections of Capacitors

Capacitors are frequently connected in series and parallel connections, as shown in Figure 3.28.

a. In series connection, when a potential difference (electrostatic voltage V) is applied on the capacitor circuit, charges $+Q$ and $-Q$ will be induced on the internally connected conductors, so that the same amount of charge, $+Q$ and $-Q$, will appear across each capacitor independently of its capacitance. Hence, we have

$$V = \sum_i V_i = Q/C_{series} = Q/C_1 + Q/C_2 + \cdots Q/C_n, \tag{3.87}$$

which implies that the equivalent capacitance, C_{series}, of series-connected capacitors is

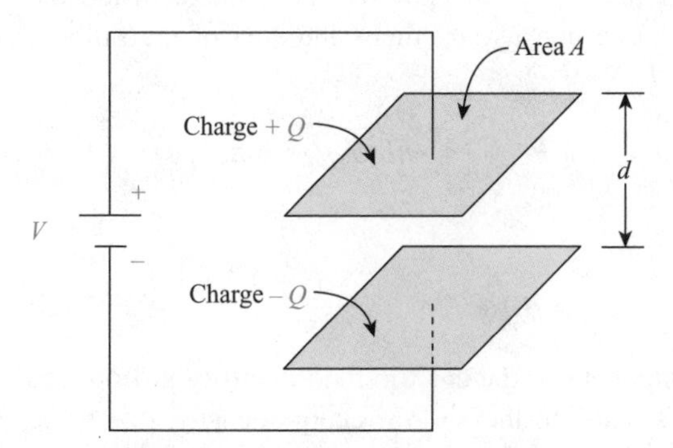

Figure 3.27 A typical parallel-plate capacitor.

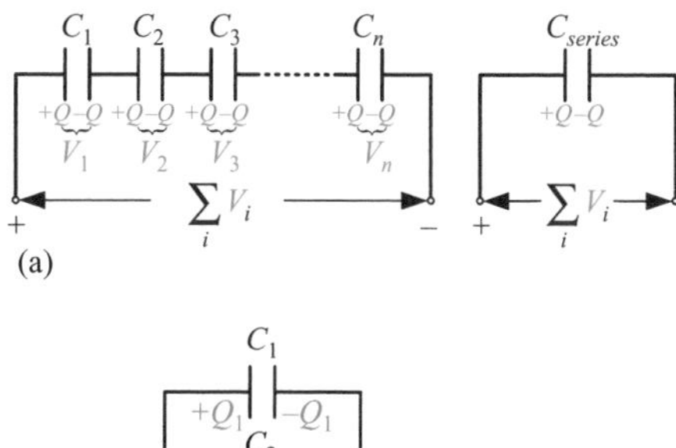

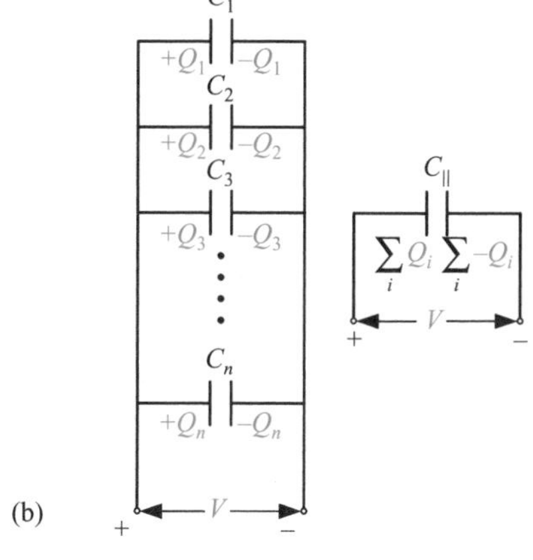

Figure 3.28 (a) Series connection of capacitors; (b) parallel connection of capacitors.

$$1/C_{series} = 1/C_1 + 1/C_2 + \cdots + 1/C_n \tag{3.88}$$

b. In parallel connection of capacitors, the applied potential difference (electrostatic voltage V) will be the same for all capacitors, and the total charge across all capacitors is the summation of all charges.

$$Q = \sum_i Q_i = Q_1 + Q_2 + \cdots + Q_n = C_1V + C_2V + \cdots C_nV = C_{parallel}V \tag{3.89}$$

Thus, equivalent capacitance, $C_{||}$, of parallel-connected capacitors:

$$C_{parallel} = C_1 + C_2 + \cdots + C_n \tag{3.90}$$

Capacitance in Multiconductor Systems

Multiconductor systems have many applications in printed circuit boards. We now consider the situation of more than two conducting bodies, as shown in Figure 3.29 in an isolated system. The positions and geometric shapes of conductors are arbitrary, and one of conductors may represent ground. Obviously, the presence of charges in the system will affect the electric potential distribution of the entire systems.

Because electric potentials and charges are linearly related, we may write n equations relating the potentials $V_1, V_2, \ldots, V_n$, of the n conductors to the n charges $Q_1, Q_2, \ldots, Q_n$:

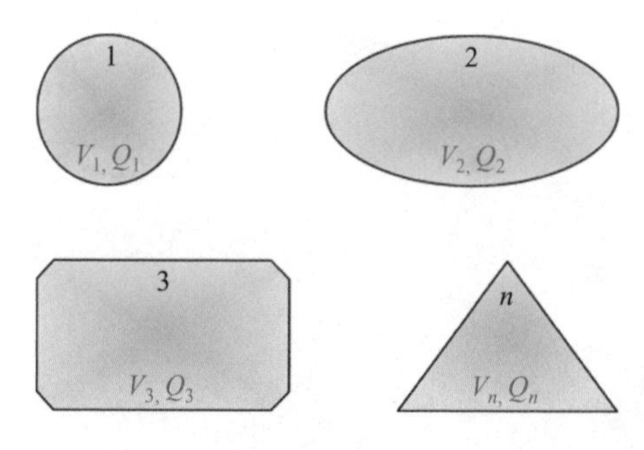

Figure 3.29 An isolated multiconductor system.

$$\begin{aligned} V_1 &= p_{11}Q_1 + p_{12}Q_2 + \cdots + p_{1n}Q_n \\ V_2 &= p_{21}Q_1 + p_{22}Q_2 + \cdots + p_{2n}Q_n \\ &\vdots \\ V_n &= p_{n1}Q_1 + p_{n2}Q_2 + \cdots + p_{nn}Q_n, \end{aligned} \tag{3.91}$$

where the p_{ij}'s are called the coefficients of potential, which are constants whose values depend on the *shape* and relative position of the conductors as well as the *permittivity* of the surrounding medium. This set of linear equations may be written in matrix format as

$$\begin{pmatrix} V_1 \\ V_2 \\ \vdots \\ V_n \end{pmatrix} = \begin{pmatrix} p_{11} & p_{12} & \cdots & p_{1n} \\ p_{21} & p_{22} & \cdots & p_{2n} \\ \vdots & \vdots & \vdots & \vdots \\ p_{n1} & p_{n2} & \cdots & p_{nn} \end{pmatrix} \begin{pmatrix} Q_1 \\ Q_2 \\ \vdots \\ Q_n \end{pmatrix} \tag{3.92}$$

In an isolated system, we note

$$Q_1 + Q_2 + ... + Q_n = 0 \tag{3.93}$$

By inverting Equation 3.92, we can express the charges as functions of potentials as follows:

$$\begin{aligned} Q_1 &= c_{11}V_1 + c_{12}V_2 + \cdots + c_{1n}V_n \\ Q_2 &= c_{21}V_1 + c_{22}V_2 + \cdots + c_{2n}V_n \\ &\vdots \\ Q_n &= c_{n1}V_1 + c_{n2}V_2 + \cdots + c_{nn}V_n \end{aligned}, \tag{3.94}$$

where the c_{ij} are constants, whose values depend only on the values of the inverse matrix elements, p_{ij}^{-1}. The coefficients, c_{ii}, are called the coefficients of self-capacitance, which are equal to the ratios of the charge Q_i to the electric potential V_i of the ith conductor (i = 1, 2,..., n) with all other conductors grounded. The c_{ij} ($i \neq j$) are called the capacitive coefficients of mutual induction. The equations may be written in matrix format as

$$\begin{pmatrix} Q_1 \\ Q_2 \\ \vdots \\ Q_n \end{pmatrix} = \begin{pmatrix} c_{11} & c_{12} & \cdots & c_{1n} \\ c_{21} & c_{22} & \cdots & c_{2n} \\ \vdots & \vdots & \vdots & \vdots \\ c_{n1} & c_{n2} & \cdots & c_{nn} \end{pmatrix} \begin{pmatrix} V_1 \\ V_2 \\ \vdots \\ V_n \end{pmatrix} \tag{3.95}$$

If a positive Q_i exists on the ith conductor, V_i will be positive relative to ground, but, for an isolated system, the Q_j induced on the jth $(i \neq j)$ conductor will be negative; the c_{ii} are positive, and the c_{ij} are negative. Reciprocity guarantees $p_{ij} = p_{ji}$ and $c_{ij} = c_{ji}$. Numerical codes such as Synopsys' of Mountain View, CA, H-Spice yield negative values for the c_{ij} $(i \neq j)$ terms.

The diagonal elements of the matrix (c) are called the self-capacitance of the ith object relative to everything else. Thus, if one of the objects is ground, then the c_{ij} term will be the capacitance of an object relative to ground plus the capacitance elements relative to all of the other objects:

$$c_{ii} = c_{i,\,ground} + \sum_{j(i \neq j)} c_{ij} \tag{3.96}$$

3.10 DIELECTRICS

Concept of the Induced Electric Dipole in Dielectric

As opposed to conductors, ideal dielectrics do not contain "free" or mobile charges. When a dielectric is placed in external electric field intensity, there are no induced free charges that can move to surface and make the interior charge density and electric field intensity vanish, as with conductors. However, dielectrics contain bound charges. The presence of external electric field intensity can cause a torque to be exerted on each charged particle pair that results in small displacements of positive and negative charge in opposite directions. In some atoms or molecules, the presence of external electric field intensity will cause an internal shift of charge in an otherwise neutral atom or molecule so we say that electric dipoles have been induced in the material. Whether the dipoles are permanent or induced, the applied electric field intensity both inside and outside the dielectric material will be modified as a result of such electric dipoles.

In the absence of applied electric field intensity, permanent dipoles are randomly oriented, but in the presence of applied electric field intensity, these permanent electric dipoles tend to align with the applied field, as is illustrated in Figure 3.30.

A solid dielectric possessing a persistent polarization is called an "electret," which is an analog of a permanent magnet. Depending on the temperature of the sample, the partially aligned electric dipoles interact with one another to cause a general orientation in the direction of the applied electric field intensity.

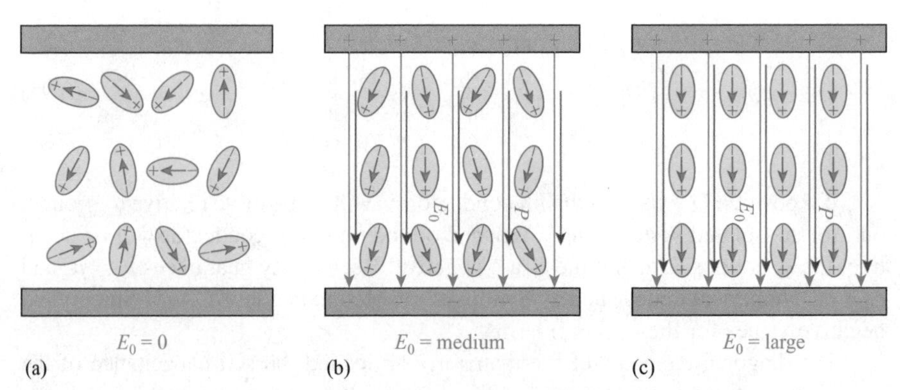

Figure 3.30 Orientation of permanent electric dipoles in external electric field intensity: (a) no applied field; (b) a medium applied field; (c) a large applied field.

Equivalent Charge Distributions of Polarized Dielectrics

To analyze the microscopic effect of induced electric dipoles, we consider the permanent electric dipole moment of a molecule (e.g., water) as shown in Figure 3.31:

We can define the polarization vector, $\vec{P}$, for a collection of such molecules as

$$\vec{P} = \lim_{\Delta v \to 0} \sum_{k=1}^{n\Delta v} \vec{p}_k / \Delta v = N_i \vec{p}_i \ (\mathrm{C/m^3}), \tag{3.97}$$

where N_i is the number of molecules per unit volume, and $\vec{p}_k$ is the electric dipole moment of a single molecular dipole. The numerator represents the vector sum of the induced electric dipole moments contained in a very small volume, Δv. $\vec{P}$ is the average vector volume density of the electric dipole moments.

If the infinitesimal dipole moment $dp = Pdv$, it produces an electric potential

$$dV = \frac{\vec{P} \cdot \hat{a}_R}{4\pi\varepsilon_0 R^2} dv' \tag{3.98}$$

as described in Equation 3.71), and, if we define $\vec{R} - \vec{R}'$ as the distance from the source point dv' to an observation point, $|\vec{R} - \vec{R}'| = \sqrt{(x-x')^2 + (y-y')^2 + (z-z')^2}$ and the electric potential due to the volume V' of the dielectric is

$$V = (1/4\pi\varepsilon_0) \int_{V'} \left(\vec{P} \cdot \hat{a}_R / |\vec{R} - \vec{R}'|^2 \right) dv'. \tag{3.99}$$

By direct differentiation of Equation 3.99, it can be shown that

$$\vec{\nabla}' \left(|\vec{R} - \vec{R}'|^{-1} \right) = \hat{a}_{R-R'} / |\vec{R} - \vec{R}'|^2. \tag{3.100}$$

Substitution of Equation 3.100 into Equation 3.99 leads to

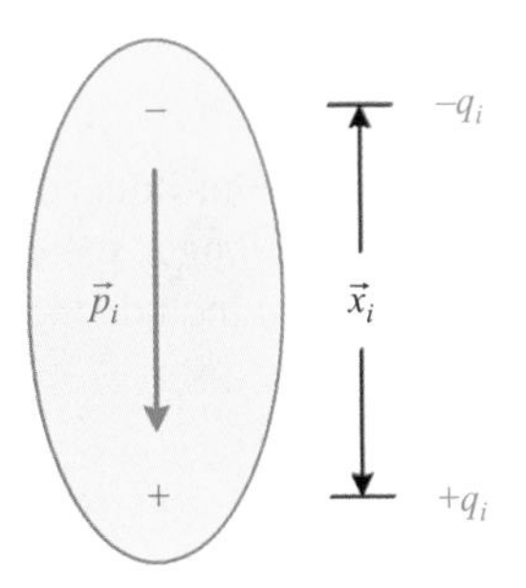

Figure 3.31 Electric dipole moment formed by the separation of electric charge, q_i, in a molecule by distance, x_i.

$$V = (1/4\pi\varepsilon_0)\int_{V'} \vec{P}\cdot\vec{\nabla}'\left(\left|\vec{R}-\vec{R}'\right|^{-1}\right)dv', \tag{3.101}$$

and, by recalling the vector identity

$$\vec{\nabla}\cdot\left(a\vec{B}\right) = a\vec{\nabla}\cdot\vec{B} + \vec{B}\cdot\vec{\nabla}a \tag{3.102}$$

and applying this in Equation 3.101, we can rewrite the equation as

$$V = (1/4\pi\varepsilon_0)\left[\int_{V'} \nabla'\cdot\left(\vec{P}/\left|\vec{R}-\vec{R}'\right|\right)dv' - \int_{V'}\left(\vec{\nabla}'\cdot\vec{P}/\left|\vec{R}-\vec{R}'\right|\right)dv'\right], \tag{3.103}$$

where the first term on the right side of Equation 3.103 can be converted to an enclosed surface integral by the divergence theorem to produce

$$V = (1/4\pi\varepsilon_0)\oint_{S'}\left(\vec{P}\cdot\hat{a}_n'/\left|\vec{R}-\vec{R}'\right|\right)ds' + (1/4\pi\varepsilon_0)\int_{V'}\left[\left(-\vec{\nabla}'\cdot\vec{P}\right)/\left|\vec{R}-\vec{R}'\right|\right]dv', \tag{3.104}$$

where the primed *del* ($\vec{\nabla}'$) operator implies that the differential operation being exerted on source region point and $\hat{a}_n'$ is the outward normal from the surface element ds' of the dielectric. Hence, the electric potential due to a polarized dielectric may be calculated from the contributions of surface and volume charge distributions having, respectively, densities

$$\rho_{ps} = \vec{P}\cdot\hat{a}_n \tag{3.105}$$

$$\rho_V = -\vec{\nabla}\cdot\vec{P}. \tag{3.106}$$

We may infer that these are polarization-charge densities or bound-charge densities. In other words, a polarized dielectric may be replaced by an equivalent polarization surface charge density ρ_{ps} and an equivalent polarization volume charge density ρ_V for field calculations to produce

$$V = \frac{1}{4\pi\varepsilon_0}\oint_{S'}\frac{\rho_{ps}}{\left|\vec{R}-\vec{R}'\right|}ds' + \frac{1}{4\pi\varepsilon_0}\int_{V'}\frac{\rho_V}{\left|\vec{R}-\vec{R}'\right|}dv'. \tag{3.107}$$

3.11 ELECTRIC FLUX DENSITY

In a polarized dielectric, an equivalent volume charge density ρ_V is thus introduced. In particular, the divergence relation 3.6 postulated in free space is no longer valid for a given source distribution *in a dielectric*, and it must be modified to include the bound-charge density ρ_V:

$$\vec{\nabla}\cdot\vec{E}=\frac{\rho_{total}}{\varepsilon_0}=\frac{1}{\varepsilon_0}(\rho_{free}+\rho_V) \tag{3.108}$$

Because $\rho_V = -\vec{\nabla}\cdot\vec{P}$, we obtain

$$\vec{\nabla}\cdot\left(\varepsilon_0\vec{E}+\vec{P}\right)=\rho_{free} \tag{3.109}$$

We can now define a new fundamental quantity to describe the electric field intensity associated with a dielectric, the electric flux density, or electric displacement, $\vec{D}$, such that

$$\vec{D}=\varepsilon_0\vec{E}+\vec{P}\ \left(\mathrm{C/m^2}\right), \tag{3.110}$$

where

$$\nabla\cdot\vec{D}=\rho_{free}, \tag{3.111}$$

and where ρ_{free} is the volume density of free charges. Its volume integral leads to

$$\int_V \vec{\nabla}\cdot\vec{D}dv=\int_V \rho_{free}dv. \tag{3.112}$$

By using divergence theorem, we have

$$\oint_S \vec{D}\cdot d\vec{s}=Q_{free}\,(C). \tag{3.113}$$

Equation 3.108 is the form of Gauss's law in a dielectric, which states that the total outward electric flux of the electric displacement, or, simply, the total outward electric flux, over any closed surface is equal to the total free charge enclosed by the surface.

When the dielectric properties of the medium are linear and isotropic, the polarization vector is directly proportional to the electric field intensity, and the proportionality constant is independent of the direction of the electric field intensity. Hence, we can write

$$\vec{P}=\varepsilon_0\chi_e\vec{E}, \tag{3.114}$$

where χ_e is a dimensionless quantity defined as electric susceptibility. A dielectric is said to be linear if χ_e is independent of the electric field and is homogeneous if χ_e is independent of space coordinates.

Substitution of Equation 3.114 into 3.110 yields

$$\vec{D} = \varepsilon_0 \vec{E} + \vec{P} = \varepsilon_0 (1 + \chi_e) \vec{E} = \varepsilon_0 \varepsilon_r \vec{E} = \varepsilon \vec{E}, \tag{3.115}$$

where ε_r is the relative permittivity or the dielectric constant of the medium and ε is the absolute permittivity of the medium.

Note that the dielectric constant ε_r can be a function of space coordinates. If ε_r is independent of position, the medium is said to be homogeneous. A linear, homogeneous, isotropic medium is called a simple medium.

In general, $\vec{D} = \bar{\bar{\varepsilon}} \cdot \vec{E}$ for an anisotropic medium can be expanded as

$$\begin{bmatrix} D_x \\ D_y \\ D_z \end{bmatrix} = \begin{bmatrix} \varepsilon_{11} & \varepsilon_{12} & \varepsilon_{13} \\ \varepsilon_{21} & \varepsilon_{22} & \varepsilon_{23} \\ \varepsilon_{31} & \varepsilon_{32} & \varepsilon_{33} \end{bmatrix} \begin{bmatrix} E_x \\ E_y \\ E_z \end{bmatrix}, \tag{3.116}$$

where $\bar{\bar{\varepsilon}}$ is called the permittivity *tensor*.

For many crystals

$$\begin{bmatrix} D_x \\ D_y \\ D_z \end{bmatrix} = \begin{bmatrix} \varepsilon_{11} & 0 & 0 \\ 0 & \varepsilon_{22} & 0 \\ 0 & 0 & \varepsilon_{33} \end{bmatrix} \begin{bmatrix} E_x \\ E_y \\ E_z \end{bmatrix}, \tag{3.117}$$

where the relative permittivity is a diagonal tensor and in homogeneous, isotropic media $\varepsilon_{11} = \varepsilon_{22} = \varepsilon_{33} = \varepsilon_r \varepsilon_0$.

Electric field intensity within a real solid is extremely complicated if one examines the fields at a subatomic level. For example, in materials with an incomplete *f* shell, it is not uncommon to experimentally measure electric field intensity at nuclei in megavolts per meter even with the application of no external electric field intensity except those imposed by the neighboring atoms in a crystal lattice.

The study of electric field intensity distributions *inside* atoms or molecules is beyond the scope of this course. We will instead normally measure electric field intensity in a macroscopic sense, averaging electric field intensity distributions over many hundreds of atoms to yield a vector "mush average" of the polarization field $\vec{P}$, and will calculate the polarization due to the *local* field at an atom by using a Lorentz cavity model later in this chapter.

3.12 DIELECTRIC BOUNDARY CONDITIONS

In section 3.3 we considered the boundary conditions between a conductor and free space. In the general electromagnetic problem involving two or more materials, we need to find the relations of field intensity quantities at an interface between two different media, one or both of which is a dielectric. In this part, we derive the boundary conditions at the interface from the integral form of the field relations.

Constructing a small enclosed line integral contour C at the interface between medium 1 and medium 2 as shown in Figure 3.32, taking the limit as $\Delta t \to 0$, and considering $\oint_C \vec{E} \cdot d\vec{l} = 0$,

$$\oint_C \vec{E} \cdot d\vec{l} = 0 = \vec{E}_1 \cdot \Delta\vec{w} + \vec{E}_2 \cdot (-\Delta\vec{w}).$$

Because $\Delta\vec{w} = \hat{a}_t \Delta w$, where $\hat{a}_t$ is the unit vector in tangential direction at the interface,

$$E_{1t} = E_{2t} \quad \text{or} \quad D_{1t}/\varepsilon_1 = D_{2t}/\varepsilon_2, \tag{3.118}$$

which indicates that the tangential component of the electric field intensity, $\vec{E}$, is continuous across the interface between two media, but the electric flux density, $\vec{D}$, is not.

By constructing a Gaussian surface as an infinitesimally small pillbox at the interface, as seen in Figure 3.32, and using Gauss's theorem associated with dielectric materials, $\oint_S \vec{D} \cdot d\vec{s} = Q_{inside}$, then we have

$$\oint_S \vec{D} \cdot d\vec{s} = \left(\vec{D}_1 \cdot \hat{a}_{n2} + \vec{D}_2 \cdot \hat{a}_{n1}\right)\Delta S = \hat{a}_{n2} \cdot \left(\vec{D}_1 - \vec{D}_2\right)\Delta S = \rho_s \Delta S,$$

where the relation between two unit vectors at the interface, $\hat{a}_{n2} = -\hat{a}_{n1}$, is used. The above equation can be further simplified as

$$\hat{a}_{n2} \cdot \left(\vec{D}_1 - \vec{D}_2\right) = \rho_s \quad \text{or} \quad D_{1n} - D_{2n} = \rho_s, \tag{3.119}$$

which further indicates that the normal component of displacement field is discontinuous across an interface where a surface charge exists, and the amount of discontinuity is equal to the surface charge density.

If medium 2 is a conductor, Equation 3.119 becomes

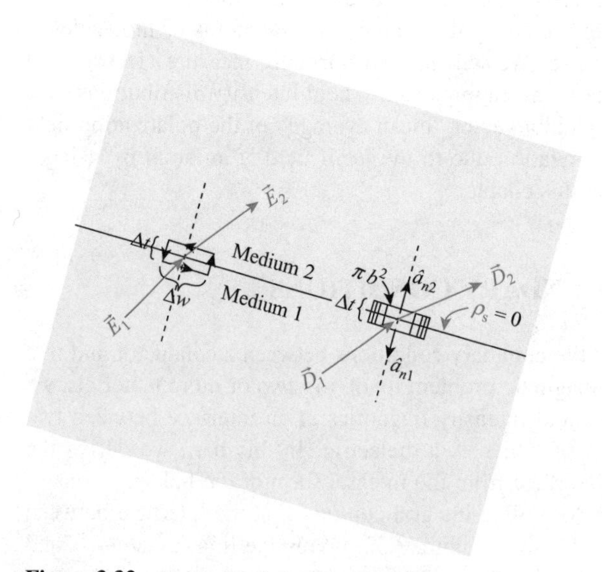

Figure 3.32 An interface between two media.

$$D_{1n} = \varepsilon_1 E_{1n} = \rho_s, \tag{3.120}$$

and, when two dielectrics are in contact with *no free charges* at their interface,

$$D_{1n} = D_{2n} \quad \text{or} \quad \varepsilon_1 E_{1n} = \varepsilon_2 E_{2n}. \tag{3.121}$$

As opposed to Equation 3.85, it is seen that, when the interface between two materials is charge free, the normal component of the electric flux density, D_n, is continuous.

3.13 ELECTROSTATIC ENERGY

Electrostatic Energy in Discrete Charge Systems

When a number of point charges $Q_1, Q_2, \ldots, Q_n$ are held fixed in space in close proximity to each other, energy is stored in the resulting field because the charges tend to move with respect to each other as a consequence of Coulomb forces.

To compute this energy, let us consider three point charges Q_1, Q_2, and Q_3 that initially are located at infinity and find the energy required to move these charges from infinity to their final positions. For simplicity, we assume that the region initially is otherwise devoid of additional charges. Bringing the first charge Q_1 from infinity to a point P_1 requires no energy because we assume that the region initially has ***no*** charge and, hence, that no force is exerted on Q_1. When bringing the second charge Q_2 from infinity to a point P_2 in the vicinity of Q_2, an energy W_{21} is required, where

$$W_{21} = Q_2 V_{21} = Q_2 \left(Q_1 / 4\pi\varepsilon_0 R_{21} \right), \tag{3.122}$$

where R_{21} is the distance between the final fixed positions of Q_1 and Q_2, and V_{21} is the electric potential at P_2 due to Q_1. Now, with the positions of Q_1 and Q_2 fixed, the energy required to move Q_3 from infinity to a point P_3 is

$$W_{31} + W_{32} = Q_3 V_{31} + Q_3 V_{32} = Q_3 \left(Q_1 / 4\pi\varepsilon_0 R_{31} \right) + Q_3 \left(Q_2 / 4\pi\varepsilon_0 R_{32} \right), \tag{3.123}$$

where $W_{ij} = Q_i [Q_j / 4\pi\varepsilon_0 R_{ij}] = Q_i V_{ij}$.

Thus, the total energy to assemble all three charges in their final position is

$$W_e = W_{21} + W_{31} + W_{32} = Q_2 V_{21} + Q_3 V_{31} + Q_3 V_{32} \tag{3.124}$$

Note that $W_{ij} = W_{ji}$, which implies the energy required to move Q_i to P_i in the presence Q_j, is the same as the energy required to move Q_j to P_j in the presence of Q_i. Therefore, Equation 3.124 can also be written as

$$W_e = W_{12} + W_{13} + W_{23} = Q_1 V_{12} + Q_1 V_{13} + Q_2 V_{23} \tag{3.125}$$

Adding Equation 3.124 to 3.125 yields

$$2W_e = Q_1 \left(V_{12} + V_{13} \right) + Q_2 \left(V_{21} + V_{23} \right) + Q_3 \left(V_{31} + V_{32} \right) \tag{3.126}$$

Note that $V_2 = V_{21} + V_{23}$ is the absolute electric potential at P_2 in the absence of Q_2 at the point as a result of the absolute electric potentials of Q_1 and Q_3, respectively. We can alternately write

$$V_i = \sum_{\substack{j=1 \\ i \neq j}}^{3} V_{ij} \tag{3.127}$$

$$W_e = (1/2)\sum_{i=1}^{3} Q_i V_i. \tag{3.128}$$

By extending this result to the assembly of an N point charge system, we have

$$W_e = (1/2)\sum_{k=1}^{N} Q_k V_k, \tag{3.129}$$

where V_k, the electric potential at Q_k, is caused by all the other charges:

$$V_k = (1/4\pi\varepsilon_0)\sum_{\substack{j=1 \\ (j \neq k)}}^{N} Q_j / R_{jk} \tag{3.130}$$

Electrostatic Energy as a Result of a Continuous Charge Distribution

For a continuous charge distribution of density ρ, the formulation for W_e in Equation 3.129 can be slightly modified. Without going through a separate proof, we replace Q_k by $\rho dv'$ and the summation by an integration to obtain

$$W_e = (1/2)\int_{V'} \rho V dv'. \tag{3.131}$$

Using Gauss's law in the point form, $\vec{\nabla}\cdot\vec{D} = \rho_v$, we can further obtain

$$W_e = (1/2)\int_{V'} \left(\vec{\nabla}\cdot\vec{D}\right) V dv'. \tag{3.132}$$

To simplify this integral, we use the identity

$$\vec{\nabla}\cdot\left(V\vec{D}\right) = V\left(\vec{\nabla}\cdot\vec{D}\right) + \vec{D}\cdot\vec{\nabla}V. \tag{3.133}$$

Substituting the vector identity, we derive

$$\begin{aligned} W_e &= (1/2)\int_{V'} \vec{\nabla}\cdot\left(V\vec{D}\right)dv' - (1/2)\int_{V'} \vec{D}\cdot\vec{\nabla}V dv' \\ &= (1/2)\oint_{S'} V\vec{D}\cdot\hat{a}_n ds' + (1/2)\int_{V'} \vec{D}\cdot\vec{E} dv' \end{aligned}, \tag{3.134}$$

where the divergence theorem has been used for the first volume integral 3.134. If we allow the volume V' in Equation 3.134 to include all space, the surface S in Equation 3.134 goes to infinity. As the distance between the source and observation point increases, the potential V decays as $1/R$, $\vec{D}$ decays as $1/R^2$, and the surface area increases as R^2. Thus, the integral in the first term in Equation 3.134 goes to zero as S tends infinity. The remaining integral yields

$$W_e = (1/2)\int_{V'} \vec{D}\cdot\vec{E}dv = (1/2)\int_{V'} \varepsilon|\vec{E}|^2 dv = (1/2)\int_{V'} |\vec{D}|^2/\varepsilon dv. \qquad (3.135)$$

By defining an *electrostatic energy density*,

$$w_e = (1/2)\vec{D}\cdot\vec{E} = (1/2)\varepsilon|\vec{E}|^2 = (1/2)|\vec{D}|^2/\varepsilon, \qquad (3.136)$$

where

$$W_e = \int_{V'} w_e dv. \qquad (3.137)$$

EXAMPLE

3.7 Static fields in a coaxial cable: A coaxial cable consists of a long, cylindrical conductor of radius r_a, an air space between r_a and r_b, a dielectric insulator between r_b and r_c, an air space between r_c and r_d, and a grounded conducting sheath between r_d and r_e, as shown in Figure 3.33.

As a function of the radius, r, plot the magnitude of the following:

a. electric field intensity, $\vec{E}$

b. electric flux density, $\vec{D}$

c. polarization field, $\vec{P}$

d. electric potential, V

e. Compute the capacitance per unit length, C_l, of a coaxial cable.

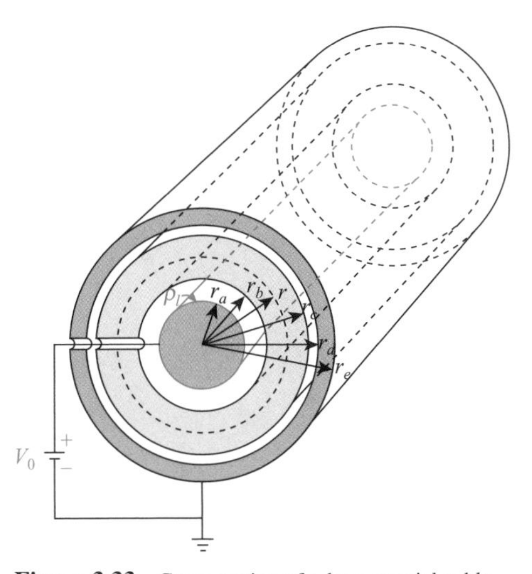

Figure 3.33 Cross section of a long coaxial cable.

SOLUTION

a. The electric field intensity inside a conductor is zero so

$$|\vec{E}| = 0 \quad \text{for} \quad 0 \le r \le r_a \quad \text{and} \quad |\vec{E}| = 0 \text{ for } r_d \le r \le r_e$$

The permittivity inside and outside of the dielectric sheath is ε_0, so, from Equation 3.14,

$$|\vec{E}| = \frac{\rho_l}{2\pi\varepsilon_0}\frac{1}{r} \quad \text{for} \quad r_a \le r \le r_b \quad \text{and for} \quad r_c \le r \le r_d.$$

The permittivity interior to the dielectric sheath is $\varepsilon_0\ \varepsilon_r$, so

$$|\vec{E}_r| = \frac{1}{\varepsilon_r}\frac{\rho_l}{2\pi\varepsilon_0}\frac{1}{r} \quad \text{for} \quad r_b \le r \le r_c.$$

The electric potential at $r = r_e$ is zero, so

$$|\vec{E}_r| = 0 \quad \text{for} \quad r_e \le r \le \infty$$

These results are shown plotted in Figure 3.34.

b. For all regions, $\vec{D} = \varepsilon_r\varepsilon_0\vec{E}$. Using the results of part a the values of electric flux density are thus shown plotted in Figure 3.35.

c. For all regions, $\vec{P} = \vec{D} - \varepsilon_0\vec{E}$ or $\vec{P} = (\varepsilon - \varepsilon_0)\vec{E}$, so $\vec{P} = (\varepsilon_r - 1)\varepsilon_0\vec{E}$. Using the results of part a and part b the values of electric polarization are thus shown plotted in Figure 3.36.

d. $V(r_e) = V(r_d) = 0$ so

$$V(r_c) = \int_{rd}^{r_c} -E(r)dr = \int_{rd}^{r_c} -\frac{\rho_l}{2\pi\varepsilon_0}\frac{dr}{r} = \frac{\rho_l}{2\pi\varepsilon_0}\ln\frac{r_d}{r_c}$$

and we can find $V(r)$ for $r_c \le r \le r_d$ by replacing r_c in this equation by r.

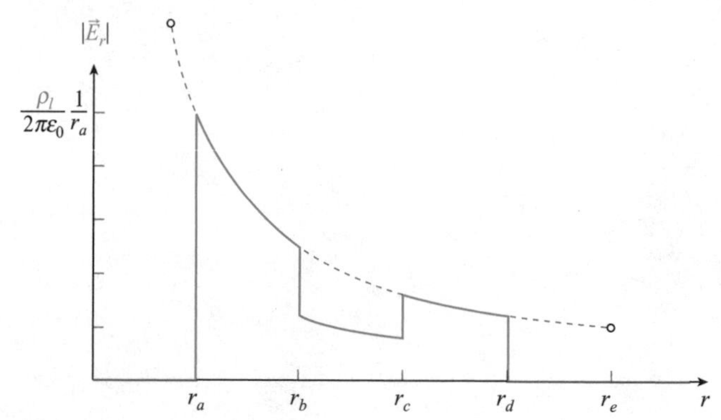

Figure 3.34 Magnitude of the electric field intensity $|\vec{E}_r|$ as a function of radius, r.

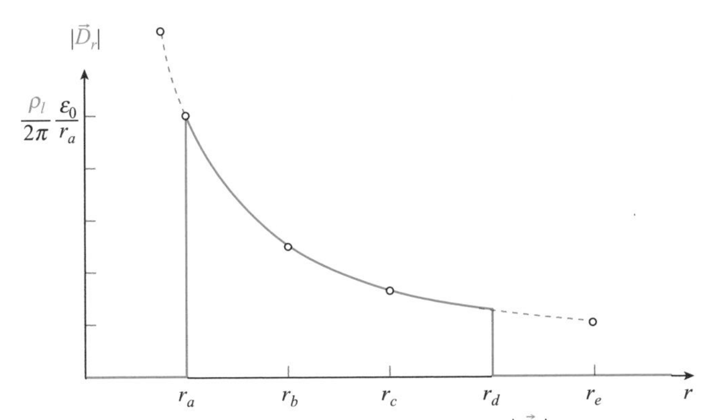

Figure 3.35 Magnitude of the electric flux density $|\vec{D}_r|$ as a function of radius, r.

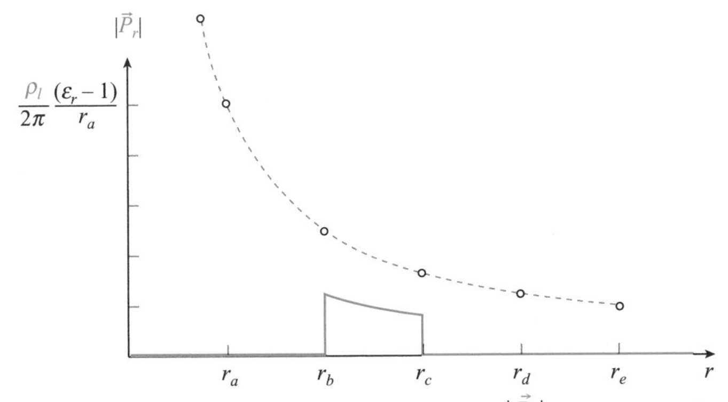

Figure 3.36 Magnitude of the polarization vector, $|\vec{P}_r|$, as a function of radius, r.

$$V(r_b) = V(r_c) - \frac{\rho_l}{2\pi\varepsilon_0\varepsilon_r}\int_{r_c}^{r_b}\frac{dr}{r} = \frac{\rho_l}{2\pi\varepsilon_0}\left[\ln\frac{r_d}{r_c} + \frac{1}{\varepsilon_r}\ln\frac{r_c}{r_b}\right]$$

$$V(r_a) = V(r_b) - \frac{\rho_l}{2\pi\varepsilon_0}\int_{r_b}^{r_a}\frac{dr}{r} = \frac{\rho_l}{2\pi\varepsilon_0}\left[\ln\frac{r_d}{r_c} + \frac{1}{\varepsilon_r}\ln\frac{r_c}{r_b} + \ln\frac{r_b}{r_a}\right]$$

These values are shown plotted in Figure 3.37.

e. Note that, since $V(r_a) = V_0$ because the battery holds the electric potential at V_0, we can compute the value of ρ_l from

$$V_0 = \frac{\rho_l}{2\pi\varepsilon_0}\left[\ln\frac{r_d}{r_c} + \frac{1}{\varepsilon_r}\ln\frac{r_c}{r_b} + \ln\frac{r_b}{r_a}\right]$$

and because $C = Q/V$ then $C_l = \frac{Q/l}{V_0} = \frac{\rho_l}{V_0} = 2\pi\varepsilon_0 \frac{1}{\left[\ln\frac{r_d}{r_c} + \frac{1}{\varepsilon_r}\ln\frac{r_c}{r_b} + \ln\frac{r_b}{r_a}\right]}$

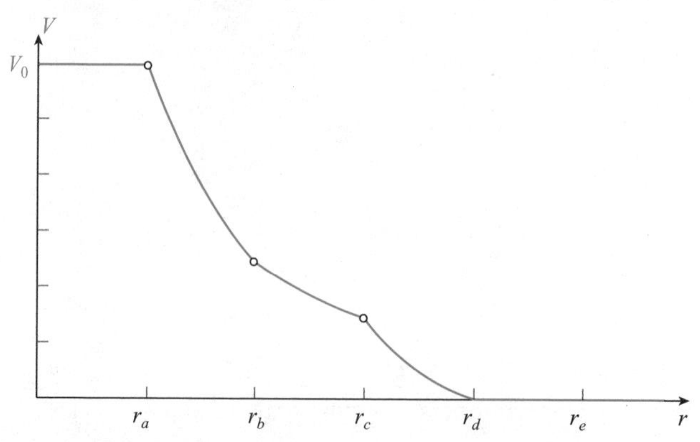

Figure 3.37 Scalar electric potential, V, as a function of radius, r.

3.14 ELECTROSTATIC FIELD IN A DIELECTRIC

In Equation 3.107, we found that the electric potential due to a polarized dielectric may be calculated from the contributions of surface and volume charge distributions having, respectively, densities $\rho_{ps} = \vec{P}\cdot\hat{a}_n$ and $\rho_V = -\vec{\nabla}\cdot\vec{P}$ and we inferred that a polarized dielectric may be replaced by an equivalent polarization surface charge density ρ_{ps} and an equivalent polarization volume charge density ρ_v for field calculations. Taking the negative gradient of Equation 3.107 and using Equation 3.100, $\vec{\nabla}(|\vec{R}-\vec{R}'|^{-1}) = -\hat{a}_{R-R'}/|\vec{R}-\vec{R}'|^2$,

$$\vec{E} = \frac{1}{4\pi\varepsilon_0}\oint_{S'}\frac{\rho_{ps}\hat{a}_{R-R'}}{|\vec{R}-\vec{R}'|^2}ds' + \frac{1}{4\pi\varepsilon_0}\int_{V'}\frac{\rho_V\hat{a}_{R-R'}}{|\vec{R}-\vec{R}'|^2}dv'. \tag{3.138}$$

The surface charge density term agrees with the physical picture of charge, as shown in Figure 3.38.

In Figure 3.38, an imaginary box has been drawn (dotted lines) with a net negative surface charge density, $\rho_s = -\vec{P}\cdot\hat{a}_n$, on the top and a net positive surface charge density, $\rho_s = \vec{P}\cdot\hat{a}_n$, on the bottom. The charge density inside the imaginary box is net zero, so (if we do not look too closely, i.e., atomically) we can argue that the average electric field intensity macroscopically inside the box can be calculated by using $\rho_V \approx 0$. Thus, the additional electric field intensity $E_1 = P/\varepsilon_0$ macroscopically in the dielectric due to these two charge density sheets in the $-\hat{a}_x$-direction and thus reduces the external electric field intensity, $E_0 = (Q/A)/\varepsilon_0$, to the macroscopic electric field:

$$\vec{E}_{Macroscopic} = \frac{(Q/A)}{\varepsilon_0}\hat{a}_x - \frac{P}{\varepsilon_0}\hat{a}_x. \tag{3.139}$$

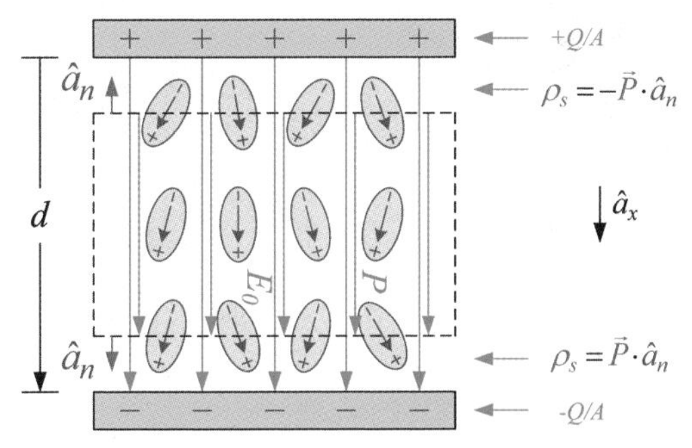

Figure 3.38 Equivalent surface charge density as a result of partial alignment of electric dipole moments by external electric field intensity.

Local Field

Lorentz found the *local* electric field intensity at an atom inside the dielectric by calculating the electric field intensity inside a spherical cavity in a uniformly polarized medium, as shown in Figure 3.39.

As shown in Figure 3.39, the surface charge density on a spherical cavity will vary as $P\cos\theta$, so the component of the electric field due to this charge density in the $\hat{a}_x$-direction will be $k_e P\cos^2\theta/r^2$. Integrating over all rings of area $(2\pi r\sin\theta)d\theta$,

$$E_P = \frac{1}{4\pi\varepsilon_0}\int_0^{\pi}\frac{(2\pi r\sin\theta)P\cos^2\theta(rd\theta)}{r^2}a_x = \frac{P}{3\varepsilon_0}.$$

NOTE The local electric field intensity at the center of a cavity does not depend on the radius of the cavity, so it holds for atoms or for empty, finite spherical cavities in a dielectric. The total electric field intensity at a *local* point P inside a dielectric sheet will thus be

$$\vec{E}_P = \left(\vec{E}_{Macroscopic} + \vec{P}/3\varepsilon_0\right). \quad (3.140)$$

Polarizability

The polarizability of an atom, α_i, is defined in terms of the electric dipole, $\vec{p}_i$, formed by the movement of electrons in the atom of type i (relative to its nucleus) by a *local* electric field intensity located at the atom at point, P, like that given in Equation 3.140:

$$\vec{p}_i = \alpha_i\vec{E}_P \quad (3.141)$$

Polarizability is thus an atomic property related to the tendency of atomic electrons to move in response to a local field. Surprisingly, electrons in a closed shell of electrons can also be displaced relative to the nucleus so that even the noble gases can have polarizability as will be shown by measurements.[i] The polarization, $\vec{P}$, of

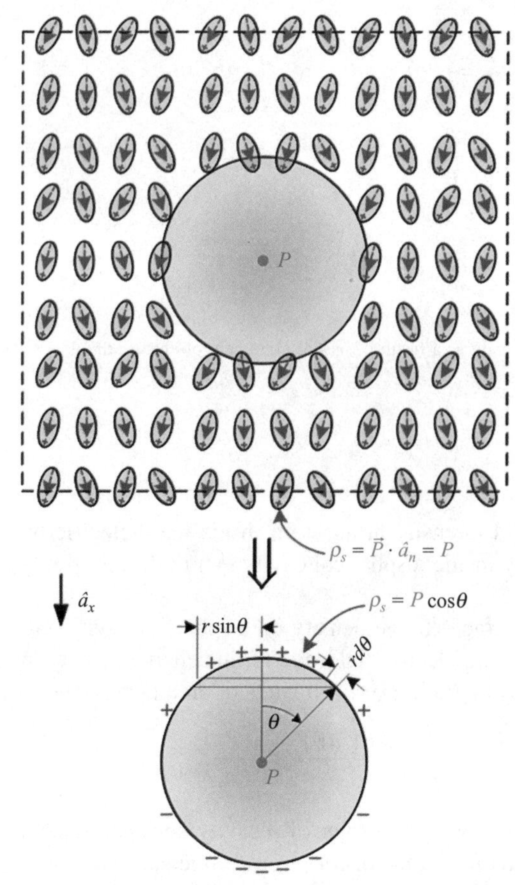

Figure 3.39 Electric field intensity at a point P inside a spherical cavity due to the surface charge induced by the alignment of electric dipoles in the dielectric.

N_i atoms per unit volume in a crystal is expressed at the product of the polarizabilities of atoms of type i times their atom concentration as

$$\vec{P} = \sum_i N_i \vec{p}_i. \tag{3.142}$$

If the local field is given by Equation 3.140, then

$$\vec{P} = \sum_i N_i \vec{p}_i = \left(\sum_i N_i \alpha_i\right)\left(\vec{E}_{Macroscopic} + \vec{P}/3\varepsilon_0\right) \tag{3.143}$$

and, if we use Equation 3.115 for the macroscopic field, $\vec{P} = \varepsilon_0 \chi_e \vec{E}_{Macroscopic}$, then

$$\chi_e = \frac{\vec{P}}{\varepsilon_0 \vec{E}_{Macroscopic}} = \frac{\sum_i N_i \alpha_i/\varepsilon_0}{1 - \sum_i N_i \alpha_i/3\varepsilon_0} \tag{3.144}$$

and because $\varepsilon_r - 1 = \chi_e$

$$\frac{1}{3\varepsilon_0}\sum_i N_i\alpha_i = \frac{\varepsilon_r - 1}{\varepsilon_r + 2}, \tag{3.145}$$

which is known as the Clausius-Mossotti relation,[ii] which relates the relative dielectric constant to the electronic polarizability, but only for crystal structures for which Equation 3.140 holds. If we view α_i as an invariable molecular constant for a single pure *i*th type molecule, the quotient $(\varepsilon_r - 1)/(\varepsilon_r + 2)$ is proportional to the material density, ρ in (gm/cm^3) as

$$\frac{\varepsilon_r - 1}{\varepsilon_r + 2} = \frac{N_i}{3\varepsilon_0}\alpha_i = \frac{N_A}{3\varepsilon_0}\frac{\rho}{M}\alpha_i, \tag{3.146}$$

where N_A is Avogadro's number (6.02217×10^{23} molecules per mole) and M is the molecular weight (gm/mole).

PROBLEM

3.5 If there are liquid water electric dipoles inside the spherical cavity, calculate the additional local field created by those dipoles and thereby find $\vec{E}_P$ for water absorbed in a fire retardant (FR-4) cavity with $\varepsilon_r \sim 4$. Let $\varepsilon_r \sim 80$ for distilled water at 20°C for this problem and consider how your answer would change if the temperature rose to 100°C. See Figures 5.13 and 5.14 from Huray, *The Foundations of Signal Integrity.*[iii] How would the answer change if the water molecules were in a vapor state?

GENERAL CHAPTER PROBLEMS

3.6 If a uniform electric field intensity, $\vec{E} = 300\hat{a}_y - 400\hat{a}_z$ (V/m), exists in air above the x–y plane, as shown in Figure 3.40 and the area below the x–y plane is water at 20°C, determine the electric field intensity, the electric flux density, and the polarization below the surface of the water.

3.7 Repeat Problem 3.6 if the temperature of the water is increased to 100°C.

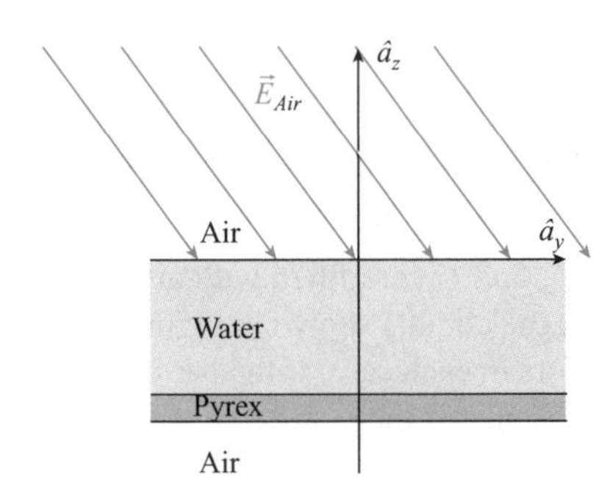

Figure 3.40 Electric field intensity above the surface of water contained by a flat pyrex dish used in Problems 3.6, 3.7, and 3.8.

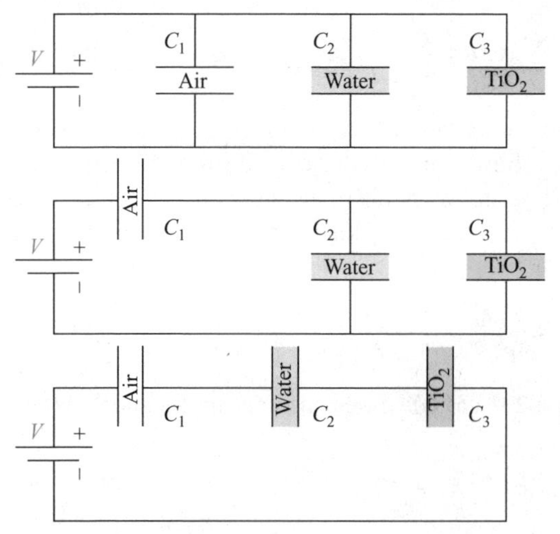

Figure 3.41 Three capacitor circuits for Problem 3.9.

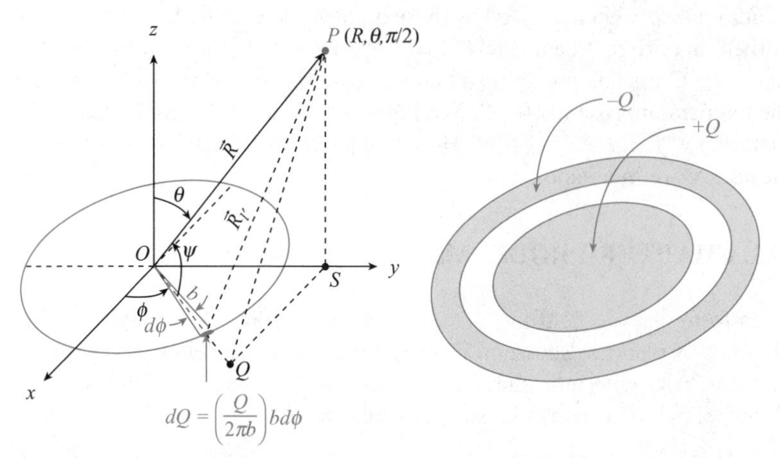

Figure 3.42 (Left) Geometry used to calculate the electric field intensity at a point P in space due to a charged ring of radius b. (Right) Thin concentric conducting ring surrounding a conducting disk as it is used in a capacitive switch.

3.8 Find the electric field intensity in the flat Pyrex ($\varepsilon_r \sim 4$) dish below the water in Problem 3.6 between z = −5 cm and z = −6 cm and in the air below z = −6 cm, and sketch the angles on the drawing in Figure 3.40.

3.9 Three otherwise identical capacitors, C_1, C_2, and C_3, are filled with air, water at 20°C, and titanium dioxide ($\varepsilon_r \sim 100$) for the circuits shown in Figure 3.41. Find the relative charges across the capacitors in each case. If the permittivity of TiO_2 is constant, how does the answer change if the water temperature is raised to 100°C? How could you use these answers to make a thermometer?

Which configuration would be the most sensitive to the temperature change? Would the thermometer work if C_2 had a fixed separation that you put across an earlobe?

3.10 Two terminals of a thin, contactless (nonmechanical) capacitor used in an elevator switch are shown on the right in Figure 3.42 as a thin conducting disk surrounded by a thin concentric annulus. If a charge $+Q$ is placed on the disk and a charge $-Q$ is placed on the annulus, use Gauss's law to show how the charge distributes itself on the conductors. Use the geometry in the left panel of Figure 3.42 to find an integral that describes the electric field intensity at a point P in space (for $R \gg b$) (***HINT*** Look Ahead to Example 6.2 in Chapter 6.) Sketch the electric field intensity in space above the plane of the ring and annulus, and explain what happens to the charge on the two conductors if a biologic sample (e.g., a finger) is placed above the ring. *Assume that a battery of constant voltage supplies the charge.* Show a circuit that would "sense" the presence of a finger using such a switch. Discuss how the size and spacing of the two conductors would be chosen to make the switch most sensitive to fingers.

ENDNOTES

i. Paul G. Huray, *The Foundations of Signal Integrity* (Hoboken, NJ: John Wiley & Sons, 2009), Chapter 12.
ii. O. F. Mosotti, Mem. Di Mathem. *e di fisica in Modena*, 24, II, 49 1850; R. Clausius, "Die mechanische warmetheorie," Vol. II, p. 62, Vieweg, 1879.
iii. Huray, *The Foundations of Signal Integrity*, Chapter 12.

Chapter 4

Solution of Electrostatic Problems

LEARNING OBJECTIVES

- Derive Poisson's and Laplace's equations and develop their solutions through separation of variables techniques for simple boundary value problems
- Understand the uniqueness theorem and its conditions
- Use the method of images for solving electric potential problems involving rectangular, cylindrical, and spherical boundary geometries
- Know how to incorporate boundary values into the analytic Green's function technique

INTRODUCTION

Electrostatic problems deal with the effects of electric charges at rest. Their solutions usually call for the determination of electric potential, electric field intensity, and electric charge distribution.

Because this chapter is an extension of Chapter 3, we mainly concentrate on Poisson's and Laplace's equations, uniqueness of electrostatic solutions, and the method of images.

4.1 POISSON'S AND LAPLACE'S EQUATIONS

The fundamental governing equations for electrostatics, from the previous chapter, are

$$\vec{\nabla}\cdot\vec{D}=\rho \tag{4.1}$$

$$\vec{\nabla}\times\vec{E}=0 \tag{4.2}$$

The electric potential

$$\vec{E}=-\vec{\nabla}V \tag{4.3}$$

In a linear and isotropic medium,

$$\vec{D} = \varepsilon\vec{E} = \varepsilon_0\varepsilon_r\vec{E} \tag{4.4}$$

Substitution of Equations 4.4 and 4.3 into Equation 4.1 leads to

$$\vec{\nabla}\cdot\varepsilon\vec{E} = \vec{\nabla}\cdot\varepsilon\left(-\vec{\nabla}V\right) = \rho_V$$

or

$$\nabla^2 V = -\rho_V/\varepsilon, \tag{4.5}$$

where the operator ∇^2 is often called *del squared*, or the Laplacian operator as studied in Chapter 2, standing for divergence of the gradient of ($\vec{\nabla}\cdot\vec{\nabla}$). Equation 4.5 is known as **Poisson's equation.**

In Cartesian coordinates,

$$\vec{\nabla} = \frac{\partial}{\partial x}\hat{a}_x + \frac{\partial}{\partial y}\hat{a}_y + \frac{\partial}{\partial z}\hat{a}_z \quad \text{so} \quad \nabla^2 V = \vec{\nabla}\cdot\vec{\nabla}V$$

or

$$\vec{\nabla}^2 V = \left(\frac{\partial}{\partial x}\hat{a}_x + \frac{\partial}{\partial y}\hat{a}_y + \frac{\partial}{\partial z}\hat{a}_z\right)\cdot\left(\frac{\partial}{\partial x}\hat{a}_x + \frac{\partial}{\partial y}\hat{a}_y + \frac{\partial}{\partial z}\hat{a}_z\right)V = -\frac{\rho_V}{\varepsilon}$$

or

$$\frac{\partial^2 V}{\partial x^2} + \frac{\partial^2 V}{\partial y^2} + \frac{\partial^2 V}{\partial z^2} = -\frac{\rho_V}{\varepsilon} \tag{4.6}$$

$$\nabla^2 \equiv \frac{\partial^2}{\partial x^2} + \frac{\partial^2}{\partial y^2} + \frac{\partial^2}{\partial z^2}.$$

Similarly, in cylindrical coordinates,

$$\nabla^2 V = \frac{1}{\rho}\frac{\partial}{\partial\rho}\left(\rho\frac{\partial V}{\partial\rho}\right) + \frac{1}{\rho^2}\frac{\partial^2 V}{\partial\varphi^2} + \frac{\partial^2 V}{\partial z^2} = -\frac{\rho_V}{\varepsilon} \tag{4.7}$$

in spherical coordinates,

$$\nabla^2 V = \frac{1}{R^2}\frac{\partial}{\partial R}\left(R^2\frac{\partial V}{\partial R}\right) + \frac{1}{R^2\sin\theta}\frac{\partial}{\partial\theta}\left(\sin\theta\frac{\partial V}{\partial\theta}\right) + \frac{1}{R^2\sin^2\theta}\frac{\partial^2 V}{\partial\varphi^2} = -\frac{\rho_V}{\varepsilon} \tag{4.8}$$

At points in a simple medium where there is no free charge density, then

$$\nabla^2 V = 0, \tag{4.9}$$

which is known as **Laplace's equation.**

4.2 SOLUTIONS TO POISSON'S AND LAPLACE'S EQUATIONS

$\nabla^2 V = \frac{\partial^2 V}{\partial x^2} + \frac{\partial^2 V}{\partial y^2} + \frac{\partial^2 V}{\partial z^2} = -\frac{\rho_V}{\varepsilon}$ in Cartesian coordinates is a second-order, first-degree (linear) inhomogeneous partial differential equation (PDE). In the special case of $\rho_V = 0$, the equation is homogeneous. The solution to the PDE is a linear combination of

1. the solution to the inhomogeneous PDE (with $-\rho_V/\varepsilon \neq 0$) and
2. the solution to the homogeneous PDE ($\nabla^2 V = 0$).

The solution to the inhomogeneous PDE is called the *particular solution* and, as we shall show in section 4.4, it is unique.

The solution to the homogeneous PDE is called the *general solution* and is often found by using the separation of variables technique as follows.

Separation of Variables Technique

Assume that the answer $V_{\text{homogenous}}(x, y, z)$ may be written as the product of three functions $X(x)Y(y)Z(z)$ and put this product into the homogeneous PDE to obtain

$$Y(y)Z(z)\partial^2 X(x)/\partial x^2 + X(x)Z(z)\partial^2 Y(y)/\partial y^2 + X(x)Y(y)\partial^2 Z(z)/\partial z^2 = 0 \quad (4.10)$$

And dividing through by $X(x)Y(y)Z(z)$, we obtain

$$\frac{\partial^2 X/\partial x^2}{X} + \frac{\partial^2 Y/\partial y^2}{Y} + \frac{\partial^2 Z/\partial z^2}{Z} = 0. \quad (4.11)$$

In this equation, we can see that all of the x dependence is in the first term, all of the y dependence is in the second term, and all of the z dependence is in the third term. The only way an x dependent term can cancel terms that depend upon y and z is that this term be at most a constant. For convenience (seen later), let us call this constant $-\alpha^2$ and recognize that α might be a complex number. Likewise, we will call the second term $-\beta^2$ and the third term $-\gamma^2$, so that Equation 4.11 becomes

$$\begin{aligned} \partial^2 X/\partial x^2 &= -\alpha^2 X \\ \partial^2 Y/\partial y^2 &= -\beta^2 Y \\ \partial^2 Z/\partial z^2 &= -\gamma^2 Z \end{aligned} \quad (4.12)$$

with the requirement that $-\alpha^2 - \beta^2 - \gamma^2 = 0$.

The technique has separated the PDE variables x, y, z into three ordinary differential equations, as shown in Equation 4.12, whose solutions (in Cartesian coordinates) are

$$X(x) = A\cos\alpha x + B\sin\alpha x = \begin{Bmatrix} \cos\alpha x \\ \sin\alpha x \end{Bmatrix}$$
$$Y(y) = C\cos\beta y + D\sin\beta y = \begin{Bmatrix} \cos\beta y \\ \sin\beta y \end{Bmatrix} \tag{4.13}$$
$$Z(z) = E\cos\gamma z + F\sin\gamma z = \begin{Bmatrix} \cos\gamma z \\ \sin\gamma z \end{Bmatrix}.$$

Subject to the restriction that

$$\alpha^2 + \beta^2 + \gamma^2 = 0. \tag{4.14}$$

In summary, we have found a solution to the homogeneous PDE of the form

$$V_{\text{homogeneous}}(x,y,z) = X(x)Y(y)Z(z) = \begin{Bmatrix} \cos\alpha x \\ \sin\alpha x \end{Bmatrix} \begin{Bmatrix} \cos\beta y \\ \sin\beta y \end{Bmatrix} \begin{Bmatrix} \cos\gamma z \\ \sin\gamma z \end{Bmatrix}. \tag{4.15}$$

Thus, the most general solution to Poisson's equation is

$$V(x,y,z) = \begin{Bmatrix} \cos\alpha x \\ \sin\alpha x \end{Bmatrix} \begin{Bmatrix} \cos\beta y \\ \sin\beta y \end{Bmatrix} \begin{Bmatrix} \cos\gamma z \\ \sin\gamma z \end{Bmatrix} + V_{\text{particular}}(x,y,z) \tag{4.16}$$

with

$$\alpha^2 + \beta^2 + \gamma^2 = 0.$$

In cylindrical coordinates,[1] the answer can be written as

$$V(x,y,z) = \begin{Bmatrix} J_m(\alpha\rho) \\ N_m(\alpha\rho) \end{Bmatrix} \begin{Bmatrix} \cos m\varphi \\ \sin m\varphi \end{Bmatrix} \begin{Bmatrix} e^{-\alpha z} \\ e^{\alpha z} \end{Bmatrix} + V_{\text{particular}}(\rho,\varphi,z). \tag{4.17}$$

In spherical coordinates,[2] the answer can be written as

$$V(x,y,z) = \begin{Bmatrix} R^l \\ R^{-l-1} \end{Bmatrix} \begin{Bmatrix} P_l^m(\cos\theta) \\ Q_l^m(\cos\theta) \end{Bmatrix} \begin{Bmatrix} e^{jm\phi} \\ e^{-jm\varphi} \end{Bmatrix} + V_{\text{particular}}(R,\theta,\phi). \tag{4.18}$$

The choice of coordinate systems usually depends on the statement of boundary conditions (BC) (e.g., the electric potential = V_0 on a cylindrical surface is easiest to express in cylindrical coordinates). Note that the solution to the homogeneous equation may be written as a linear combination of the functions (e.g., in spherical coordinates, radial terms in R with l = integer are multipole potentials, as was discussed in Chapter 3).

[1] $J_m(\alpha\rho)$ and $N_m(\alpha\rho)$ are the Bessel function and the Neumann function.

[2] $P_l^m(\cos\theta)$ and $Q_l^m(\cos\theta)$ are the associated Legendre polynomials of the first and second kind.

Application Example: Potential in a Charge-Free Box with BC

Consider above solution to the case of a rectangular box with a given electric potential specified on its sides, as shown in Figure 4.1.

BC

The electric potential on surface S is: $V(x, y, c) = V_0(x, y)$ on the surface at $z = c$ and zero everywhere else. From Equation 4.16, we can write the solution as

$$V(x,y,z) = \begin{Bmatrix} \cos \alpha x \\ \sin \alpha x \end{Bmatrix} \begin{Bmatrix} \cos \beta y \\ \sin \beta y \end{Bmatrix} \begin{Bmatrix} \cos \gamma z \\ \sin \gamma z \end{Bmatrix} + V_{\text{particular}}(x,y,z)$$

with

$$\alpha^2 + \beta^2 + \gamma^2 = 0.$$

In the hypothetical case of $V = 0$ on all faces, we could choose $V_{\text{particular}}(x, y, z) = 0$, but, if the electric potential on the six faces had been 14 V, we would have chosen $V_{\text{particular}}(x, y, z) = 14$ volts because V = constant satisfies $\nabla^2 V = 0$ and the BC. However, we have a more difficult problem to solve because we are given that $V(x, y) = V_0(x, y)$ on the surface at $z = c$. Thus, we will need to solve the homogeneous problem first and then find $V_{\text{particular}}(x, y, z)$. Note that the BC have made the problem inhomogeneous in our example.

For the homogeneous problem with homogeneous BC to be solved, $V(0, y, z) = 0$ on the face at $x = 0$, the coefficient of the cosine term in $X(x)$ can be chosen as zero. In order to make the electric potential satisfy the BC, $V(a, y, z) = 0$ on the face at $x = a$, we can see the constant α can be chosen to be an integer multiple, n, of π/a.

We can make a similar argument that the cosine term in $Y(y)$ can be chosen as zero to satisfy the homogeneous BC, $V(x, 0, z) = 0$ on the face at $y = 0$ and that the constant β can be chosen to be an integer multiple, m, of π/b to satisfy the BC, $V(x, b, z) = 0$.

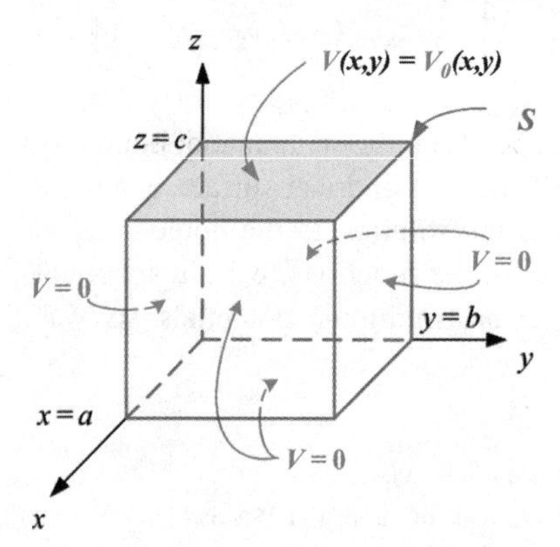

Figure 4.1 Rectangular box with Dirichlet boundary conditions specified on its surface S.

NOTE The lowest integer n permitted in $X(x)$ and m permitted in $Y(y)$, respectively, is 1 because the values $n = 0$ or $m = 0$ would give a null solution.

Finally, we can satisfy the condition in Equation 4.14 by choosing

$$\gamma = \sqrt{-\frac{n^2\pi^2}{a^2} - \frac{m^2\pi^2}{b^2}}$$

so the most general solution can be written as

$$V(x,y,z) = \sum_{n,m=1}^{\infty} A_{nm} \sin n\frac{\pi}{a}x \sin m\frac{\pi}{b}y \sin\sqrt{-\frac{n^2\pi^2}{a^2} - \frac{m^2\pi^2}{b^2}}\, z, \tag{4.19}$$

where we have taken the sum over positive integer values of n and m because there is no reason to select one value over another. We can also use $\sin j\gamma z$ with $\sinh \gamma z$ to write Equation 4.19 as

$$V(x,y,z) = \sum_{n,m=1}^{\infty} A_{nm} \sin n\frac{\pi}{a}x \sin m\frac{\pi}{b}y \sinh\sqrt{\frac{n^2\pi^2}{a^2} + \frac{m^2\pi^2}{b^2}}\, z. \tag{4.20}$$

Finally, we can use the BC at $z = c$ to find the particular solution by setting

$$V(x,y,c) = \sum_{n,m=1}^{\infty} A_{nm} \sin n\frac{\pi}{a}x \sin m\frac{\pi}{b}y \sinh\sqrt{\frac{n^2\pi^2}{a^2} + \frac{m^2\pi^2}{b^2}}\, c = V_0(x,y). \tag{4.21}$$

In Equation 4.21, we can multiply both sides by $\sin n'\frac{\pi}{a}x \sin m'\frac{\pi}{b}y$ and integrate over the ranges $0 \le x \le a$ and $0 \le y \le b$, respectively, to evaluate the coefficients A_{nm} as

$$A_{n'm'} = \frac{4}{ab\sinh\sqrt{\frac{n'^2\pi^2}{a^2} + \frac{m'^2\pi^2}{b^2}}\, c} \int_0^a \left[\int_0^b V_0(x,y) \sin n'\frac{\pi}{a}x \sin m'\frac{\pi}{b}y\, dy\right] dx. \tag{4.22}$$

The coefficients $A_{n'm'}$ found in Equation 4.22 are constants. We can put those constants into Equation 4.21 to find the most general solution to the boundary value problem.

Very Special Case

Suppose $V_0(x, y) = V_0$ = constant. In that case,

$$A_{n'm'} = \frac{4}{ab\sinh\sqrt{\frac{n'^2\pi^2}{a^2} + \frac{m'^2\pi^2}{b^2}}\, c} V_0 \left[-\frac{a}{n'\pi}\cos n'\frac{\pi}{a}x\right]_0^a \left[-\frac{b}{m'\pi}\cos m'\frac{\pi}{b}\right]_0^b \tag{4.23}$$

$$A_{n'm'} = \frac{4V_0}{n'm'\sinh\sqrt{\frac{n'^2\pi^2}{a^2} + \frac{m'^2\pi^2}{b^2}}\, c}[1 - \cos n'\pi][1 - \cos m'\pi]. \tag{4.24}$$

We can see that nonzero values of $A_{n'm'}$ only occur for n' and m' = odd, so

$$V(x,y,c)=\sum_{n,m=1,odd}^{\infty}\frac{16V_0\sin n\frac{\pi}{a}x\sin m\frac{\pi}{b}y}{nm\sinh\sqrt{\frac{n^2\pi^2}{a^2}+\frac{m^2\pi^2}{b^2}}c}\sinh\sqrt{\frac{n^2\pi^2}{a^2}+\frac{m^2\pi^2}{b^2}}z \tag{4.25}$$

$$V(x,y,c)=16V_0\sum_{n,m=1,odd}^{\infty}\left(\frac{\sin n\frac{\pi}{a}x}{n}\right)\left(\frac{\sin m\frac{\pi}{b}y}{m}\right)\frac{\sinh\sqrt{\frac{n^2\pi^2}{a^2}+\frac{m^2\pi^2}{b^2}}z}{\sinh\sqrt{\frac{n^2\pi^2}{a^2}+\frac{m^2\pi^2}{b^2}}c}. \tag{4.26}$$

NOTE If the rectangular box has different specified electric potentials on all six sides, we can use the above technique and add the answers for each of the six sides because each of the answers will satisfy Laplace's equation and the specified BC.

NOTE This technique works well for charge-free volumes (like *abc*) with specified BC on the surface, S. We could also use the technique with Equations 4.17 and 4.18 in the event that cylindrical or spherical BC are specified.

If we want to solve Poisson's equation for the above problem with an interior charge specified in the box, we will need to find $V_{\text{particular}}$ for that case. One powerful technique for solving such problems with inhomogeneous BC was found by George Green in 1824. The Green's function technique is described in the following section.

4.3 GREEN'S FUNCTIONS

Before the advent of computers, an analytic technique for solving inhomogeneous PDEs with inhomogeneous BC was described by George Green. The solutions are written as integrals in which the integrands are known or can be chosen conveniently. For that reason, the Green's function solution lends itself well to computational solutions of Poisson's equation. The technique also lends itself to solutions involving images (shown in section 4.5) and can even be used in time-dependent problems to include the effects of responses (electromagnetic fields) that travel at the speed of light in a material. The latter effects are found when we work with time-dependent problems in section 7.13.

Green's Theorems

Let us consider a vector field $\vec{A}=\phi\vec{\nabla}\psi$, where ϕ and ψ are arbitrary scalar fields. We have shown in Chapter 2 that

$$\vec{\nabla}\cdot(\phi\vec{\nabla}\psi)=\phi\vec{\nabla}\cdot\vec{\nabla}\psi+\vec{\nabla}\phi\cdot\vec{\nabla}\psi=\phi\nabla^2\psi+\vec{\nabla}\phi\cdot\vec{\nabla}\psi \tag{4.27}$$

and that $\varphi\vec{\nabla}\psi\cdot\hat{n}=\varphi\partial\psi/\partial n$, where $\partial/\partial n$ means the normal derivative at the surface S (directed outward from the volume V). Substituting these into the divergence theorem,

$$\iiint_V \vec{\nabla}\cdot(\varphi\vec{\nabla}\psi)\,dv = \iiint_V (\varphi\nabla^2\psi + \vec{\nabla}\varphi\cdot\vec{\nabla}\psi)dv = \oiint_S (\varphi\vec{\nabla}\psi)\cdot d\vec{s} \text{ or} \tag{4.28}$$

$$\iiint_V (\varphi\nabla^2\psi + \vec{\nabla}\varphi\cdot\vec{\nabla}\psi)\,dv = \oiint_S \varphi\frac{\partial\psi}{\partial n}ds \tag{4.29}$$

Equation 4.29 is known as **Green's first identity**.

Now, let us write Equation 4.29 again with ϕ and ψ exchanged as in Equation 4.30

$$\iiint_V (\psi\nabla^2\varphi + \vec{\nabla}\varphi\cdot\vec{\nabla}\psi)\,dv = \oiint_S \psi\frac{\partial\phi}{\partial n}ds \tag{4.30}$$

and subtract Equation 4.30 from Equation 4.29 to obtain

$$\iiint_V (\varphi\nabla^2\psi - \psi\nabla^2\varphi)\,dv = \oiint_S \left[\varphi\frac{\partial\psi}{\partial n} - \psi\frac{\partial\varphi}{\partial n}\right]ds. \tag{4.31}$$

Equation 4.31 is known as **Green's second identity** or as **Green's theorem**. This theorem applies to **any** two scalar fields ϕ and ψ. Green's theorem can be used to solve Poisson's equation by choosing $\psi = 1/4\pi R$ and $\varphi = V(R)$ (the scalar electric potential). Here, Equation 4.31 becomes

$$\iiint_V \left(V\nabla^2\left(\frac{1}{4\pi R}\right) - \left(\frac{1}{4\pi R}\right)\nabla^2 V\right)dv = \oiint_S \left[V\frac{\partial}{\partial n}\left(\frac{1}{4\pi R}\right) - \left(\frac{1}{4\pi R}\right)\frac{\partial V}{\partial n}\right]ds, \tag{4.32}$$

where we can substitute $\nabla^2 V = -\rho_v/\varepsilon$ in the volume integral.

Properties of the Dirac Delta Function

In spherical coordinates, we can use the Laplacian operator to find that

$$\nabla^2\left(\frac{1}{4\pi R}\right) = \frac{1}{R^2}\frac{\partial}{\partial R}\left[R^2\frac{\partial}{\partial R}\left(\frac{1}{4\pi R}\right)\right] = 0 \quad \text{for} \quad R \neq 0. \tag{4.33}$$

However, $\nabla^2(1/4\pi R)$ at $R = 0$ requires a limiting process for its evaluation. We can use the divergence theorem to see that as $R \to 0$,

$$\iiint_V \nabla^2\left(\frac{1}{4\pi R}\right)dv = \iiint_V \vec{\nabla}\cdot\vec{\nabla}\left(\frac{1}{4\pi R}\right)dv = \oiint_S \hat{n}\cdot\vec{\nabla}\left(\frac{1}{4\pi R}\right)ds. \tag{4.34}$$

And, by using the definition of $\vec{\nabla}(1/4\pi R) = \frac{\partial}{\partial R}(1/4\pi R)\hat{a}_R$ in spherical coordinates,

$$\iiint_V \nabla^2\left(\frac{1}{4\pi R}\right)dv = \lim_{R\to 0}\frac{1}{4\pi}\int_0^{2\pi}\int_0^{\pi}\left(\frac{-1}{R^2}\right)R^2\sin\theta\,d\theta\,d\phi = -1. \tag{4.35}$$

We can thus see that $\nabla^2(-1/4\pi R)$ has the property that it is zero for all finite values of R, it goes to infinity at $R = 0$, and its integral is 1. We call such a function the Dirac delta function and plot it as shown in Figure 4.2.

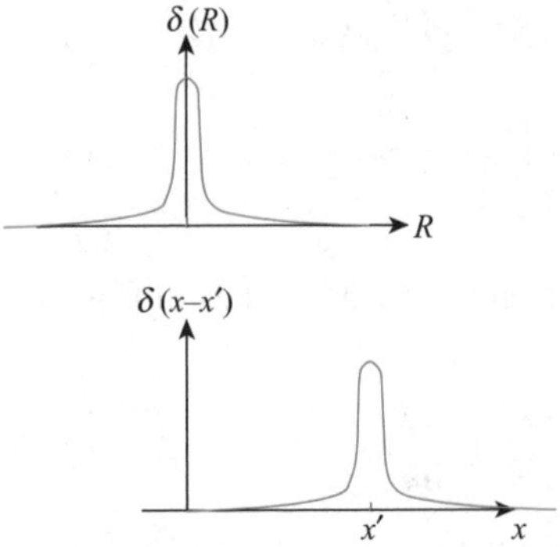

$\delta(R)$

R

$\delta(x-x')$

x' x

Figure 4.2 Properties of the Dirac delta function.

As shown in Figure 4.2, the delta function has the properties that

$$\begin{aligned} &\delta(x-x') = 0 \quad \text{for} \quad x \neq x' \\ &\delta(x-x') \to \infty \quad \text{as} \quad x \to x' \\ &\int_{-\infty}^{\infty} \delta(x-x')\,dx = 1 \quad \text{and} \\ &\int_{-\infty}^{\infty} f(x)\delta(x-x')\,dx = f(x'). \end{aligned} \tag{4.36}$$

Thus, if we use $R = |\vec{x} - \vec{x}'|$ in Equation 4.32,

$$\iiint_V \left(V[-\delta(\vec{x}-\vec{x}')] + \left(\frac{1}{4\pi|\vec{x}-\vec{x}'|}\right)\frac{\rho_V}{\varepsilon} \right) dv' = \frac{1}{4\pi} \oiint_S \left[V \frac{\partial}{\partial n'}\left(\frac{1}{R}\right) - \left(\frac{1}{R}\right)\frac{\partial V}{\partial n'} \right] ds' \tag{4.37}$$

or

$$V(\vec{x}) = \iiint_V \frac{\rho_V(\vec{x}')}{4\pi\varepsilon|\vec{x}-\vec{x}'|}\,dv' - \frac{1}{4\pi}\oiint_S \left[V \frac{\partial}{\partial n'}\left(\frac{1}{R}\right) - \left(\frac{1}{R}\right)\frac{\partial V}{\partial n'} \right] ds'. \tag{4.38}$$

This answer is very powerful because it gives the scalar electric potential at a point $\vec{x}$ as a result of a charge distribution at $\vec{x}'$ in the volume V *and* it includes the scalar electric potential and its normal derivative over the surface S that surrounds V.

However, giving the scalar electric potential and its normal derivative on a surface S overspecifies the BC. While there may be some combination of scalar electric potential or its derivative on parts of the boundary S, both are never given for the same part of the boundary.

Green got around this problem by saying that Equation 4.38 may be written as

$$V(\vec{x}) = \iiint_V G(\vec{x}-\vec{x}')\frac{\rho_V(\vec{x}')}{\varepsilon}\,dv' - \oiint_S \left[V \frac{\partial}{\partial n'}(G(\vec{x}-\vec{x}')) - G(\vec{x}-\vec{x}')\frac{\partial V}{\partial n'} \right] ds', \tag{4.39}$$

where $G(\vec{x} - \vec{x}')$ satisfies the differential equation

$$\nabla^2 G(\vec{x}-\vec{x}') = -\delta(\vec{x}-\vec{x}') \tag{4.40}$$

subject to the BC of his choice. Thus, if V is specified on the surface S, Green chose $G(\vec{x} - \vec{x}') = 0$ on the surface S, so that the last integral in Equation 4.39 vanishes. But, if $\partial V/\partial n$ is given on the surface S, Green chose $\partial G(\vec{x} - \vec{x}')/\partial n = 0$ on the surface S, so that the next to the last integral in Equation 4.39 vanishes. Many texts solve for the Green's function that satisfies Equation 4.40 and one of the two BC on a surface S (rectangular box, cylinder, or sphere), but it is beyond the scope of this book.

What Is the Point of Using Green's Functions?

We derived the mathematical Green's second identity in Equation 4.31 and chose to apply it to electrical engineering problems by choosing $\psi = 1/4\pi R$ and $\varphi = V(R)$, so that we got Equation 4.32. We then found that $\nabla^2(1/4\pi R) = -\sigma(\vec{x} - \vec{x}')$ and chose Gauss's law in the form $\nabla^2 V = -\rho_V/\varepsilon$ to deduce that

$$V(\vec{x}') = \iiint_V \frac{\rho_V(\vec{x})}{4\pi\varepsilon R}\,dv - \oiint_S \left[V_{on\ S}(R)\frac{\partial}{\partial n}\left(\frac{1}{4\pi R}\right) - \left(\frac{1}{4\pi R}\right)\frac{\partial V_{on\ S}(R)}{\partial n} \right] ds.$$

Because we normally take sources to be primed quantities and answers to be at unprimed vector locations, we exchanged primed quantities for unprimed quantities in the variable of integration to get Equation 4.38. We then let $G(\vec{x} - \vec{x}') = (1/4\pi|\vec{x} - \vec{x}'|)$ so that Equation 4.39 resulted. *This is the big answer* because we can write $(1/4\pi|\vec{x} - \vec{x}'|)$ in Cartesian, cylindrical, or spherical coordinates to solve the integrals depending on the boundaries specified in the problem.

Furthermore, if we are given Dirichlet BC, ($V_{on\ S}$), we can choose $G(\vec{x} - \vec{x}') = 0$ on the surface to remove the last surface integral in Equation 4.39. If we are given Neumann BC, ($\partial V_{on\ S}/\partial n$), we can choose $\partial G(\vec{x} - \vec{x}')/\partial n = 0$ to remove the other surface integral.

Equation 4.39 thus gives us the answer, $V(\vec{x})$, and also satisfies the BC. This is a powerful statement because we have now solved all problems for which we know charge density, $\rho_V(\vec{x}')$, in some volume of space and either the Dirichlet or Neumann BC. All we need is the solution to $\nabla^2 G(\vec{x} - \vec{x}') = -\sigma(\vec{x} - \vec{x}')$ with *homogeneous* BC, $G(\vec{x} - \vec{x}') = 0$ or $\partial G(\vec{x} - \vec{x}')/\partial n = 0$. Once we have solved these two problems (once for Dirichlet and once for Neumann BC), we have found an integral form for the scalar electric potential at any point in space for all bounded problems in which $\rho_V(\vec{x}')$ is known.

PROBLEM

4.1 Show that, if we are given mixed BC, $\alpha V_{on\ S} + \beta(\partial V_{on\ S}/\partial n)$ instead of Dirichlet or Neumann BC, the answer can be found in closed form by setting $\alpha G(|\vec{x} - \vec{x}'|) + \beta(\partial G(|\vec{x} - \vec{x}'|/\partial n) = 0$.

We will use these answers in solving future boundary value problems and will even be able to find a similar solution for problems that involve time in section 7.13 for which the speed of light must be taken into account. Such answers are often called *causal solutions* because an answer cannot exist at a vector location in space till the speed of light has permitted the scalar electric potential to arrive at that point from a primed vector location, traveling at the speed of light in that medium.

4.4 UNIQUENESS OF THE ELECTROSTATIC SOLUTION

Uniqueness Theorem

A particular solution of Poisson's equation—Laplace's equation is a special case—that satisfies the given BC is a unique solution.

Proof

Let us suppose a volume τ that is bounded outside by a surface S_0 and a number of charges interior to τ with surfaces $S_1, S_2, \ldots S_n$ that enclose each of the charges at specified scalar electric potentials, as shown in Figure 4.3. Assuming that, contrary to the uniqueness theorem, there are two solutions, V_1 and V_2, to Poisson's equation in τ' where τ' is the volume τ minus the volumes interior to the surfaces $S_l, S_2, \ldots S_n$ (i.e., there is no charge in the volume τ'),

$$\nabla^2 V_1 = -\rho_V/\varepsilon \tag{4.41a}$$

$$\nabla^2 V_2 = -\rho_V/\varepsilon. \tag{4.41b}$$

Let us also assume that both V_1 and V_2 satisfy the same BC on surfaces $S_1, S_2, \ldots, S_n$, and S_0. Now, if we define a new *difference* scalar electric potential,

$$V_d = V_1 - V_2. \tag{4.42}$$

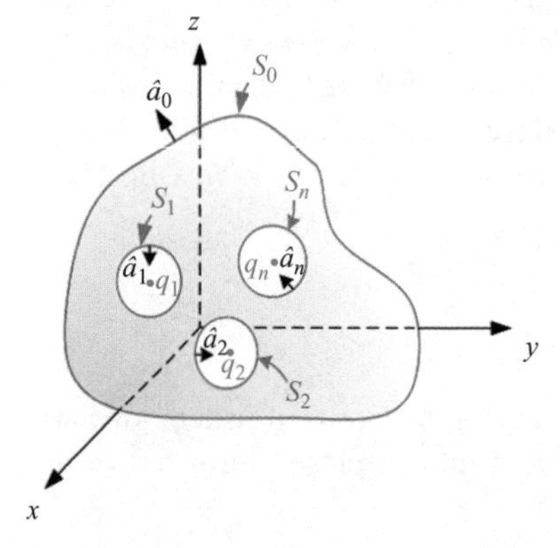

Figure 4.3 Surface S_0 enclosing volume τ that includes n-conducting charges, $q_1, q_2, \ldots q_n$. The surfaces $S_1, S_2, \ldots S_n$ are interior to S_0 and enclose each of those charges.

From Equations 4.41a and 4.41b,

$$\nabla^2 V_d = 0. \tag{4.43}$$

Recalling the vector identity,

$$\vec{\nabla}\cdot\left(f\vec{A}\right) = f\vec{\nabla}\cdot\vec{A} + \vec{A}\cdot\vec{\nabla}f.$$

Then,

$$\begin{aligned}\vec{\nabla}\cdot\left(V_d\vec{\nabla}V_d\right) &= V_d\vec{\nabla}\cdot\vec{\nabla}V_d + \vec{\nabla}V_d\cdot\vec{\nabla}V_d \\ &= V_d\nabla^2 V_d + \left|\vec{\nabla}V_d\right|^2\end{aligned}. \tag{4.44}$$

Thus,

$$\int_{\tau'}\vec{\nabla}\cdot\left(V_d\vec{\nabla}V_d\right)dv = \int_{\tau'}\left(V_d\nabla^2 V_d + \left|\vec{\nabla}V_d\right|^2\right)dv$$

where using

$$\int_{\tau'}\vec{\nabla}\cdot\vec{A} = \oint_{S'}\vec{A}\cdot d\vec{S},$$

we have

$$\oint_{S'}\left(V_d\vec{\nabla}V_d\right)\cdot\hat{a}_n ds = \int_{\tau'}\left|\vec{\nabla}V_d\right|^2 dv, \tag{4.45}$$

where $\hat{a}_n$ denotes the unit normal outward from τ'. Surface S' consists of S_0 as well as $S_1, S_2, \ldots, S_n$.

Over the conducting boundaries ($S_1, S_2, \ldots, S_n$)

$$V_d = V_1 - V_2 = 0. \tag{4.46}$$

By choosing S_0 to be very large sphere with radius R,

$$\oint_{S_0}\left(V_d\vec{\nabla}V_d\right)\cdot\hat{a}_n ds \propto \frac{1}{R}\cdot\frac{1}{R^2}\cdot R^2 \to 0. \tag{4.47}$$

Therefore,

$$\int_{\tau'}\left|\vec{\nabla}V_d\right|^2 dv = 0 \tag{4.48}$$

or

$$\left|\vec{\nabla}V_d\right| = 0 \tag{4.49}$$

considering Equation 4.46,

$$V_d = 0 \quad \text{or} \quad V_1 = V_2. \tag{4.50}$$

CONCLUSION The solution to Poisson's equation in any volume with normal BC is *unique*; that is, if we can find *one solution* that satisfies Poisson's equation and the BC, we have found the *only solution*.

4.5 METHOD OF IMAGES

There is a class of electromagnetic problems with BC on a volume τ that appear to be difficult to satisfy if the governing Laplace's equation is to be solved directly, but the condition on the boundary surfaces in these problems can be set up to produce the same conditions on the surfaces by the addition of an appropriate *image* charge, so that the potential distribution can then be determined in a straightforward manner. By finding *a solution* that satisfies Laplace's equation in the volume τ and the BC for the image problem, we have found *the only solution* (according to the uniqueness theorem).

Consider the following problem, as shown in Figure 4.4.

The formal procedure for solving the problem is to solve Laplace's equation in Cartesian coordinates:

$$\nabla^2 V = \frac{\partial^2 V}{\partial x^2} + \frac{\partial^2 V}{\partial y^2} + \frac{\partial^2 V}{\partial z^2} = 0, \tag{4.51}$$

which should be valid for $y > 0$ except at the position of the point charge.

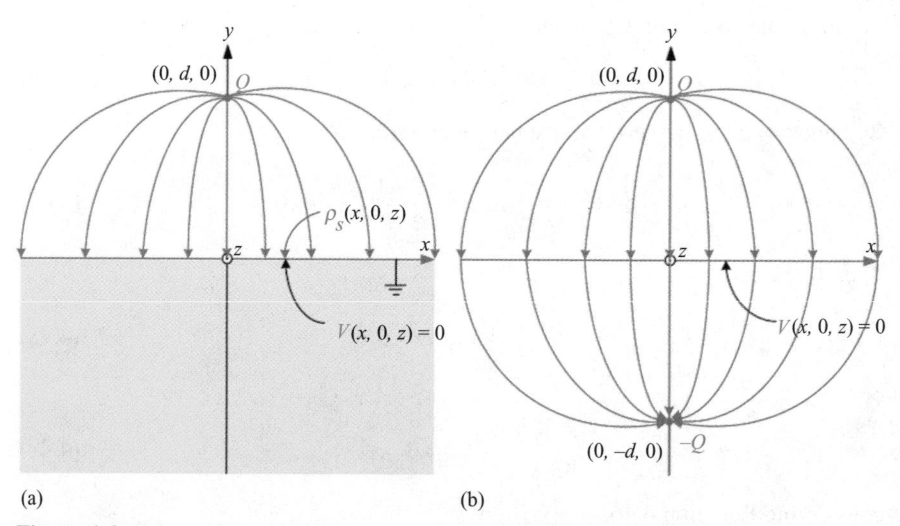

Figure 4.4 Point charge and ground plane conductor. (a) Physical arrangement; (b) equivalent image charge arrangement.

The solution $V(x, y, z)$ should satisfy the following conditions:

1. $V(x, 0, z) = 0$.
2. Because $V \to Q/4\pi\varepsilon_0 R$ as $R \to 0$ at points very far from Q; that is,

$$(x \to \pm\infty, y \to \infty, z \to \pm\infty),$$

   the scalar electric potential approaches zero on an infinite hemisphere above the x,z plane.
3. The electric potential function is even with respect to x and z coordinates; that is,

$$V(x,y,z) = V(-x,y,z) = V(x,y,-z) \quad \text{(Symmetry)}$$

It appears difficult to construct a solution for V that will satisfy all these conditions. From another point of view, we may reason that the presence of a positive charge Q at $y = d$ would induce negative charges on the surface of the conducting plane, resulting in a surface charge density $\rho_S(x, 0, z)$. Hence,

$$V(x,y,z) = (Q/4\pi\varepsilon_0)\left[x^2 + (y-d)^2 + z^2\right]^{-1/2} + (1/4\pi\varepsilon_0)\int_S [\rho_S(x',0,z')/R_1]ds', \quad (4.52)$$

where R_1 is the distance from the differential area ds' to the field point under consideration (x, y, z), and S is the surface of the entire conducting plane at $y = 0$.

However, for the scalar electric potential, $V(x, y, z)$ to be calculated, the surface charge density $\rho_S(x, 0, z)$ must be known in Equation 4.52. Without the method of images (below), this would be difficult to find.

Point Charge near a Conducting Plane

In the above problem, if we remove the conductor and replace it by an image point charge→Q at $y = -d$, then the electric potential at a point $P(x, y, z)$ in the $y \geq 0$ region is

$$V(x,y,z) = (Q/4\pi\varepsilon_0)(1/R_+ - 1/R_-) \quad (4.53)$$

where

$$R_+ = \sqrt{x^2 + (y-d)^2 + z^2}, \quad \text{and} \quad R_- = \sqrt{x^2 + (y+d)^2 + z^2} \quad (4.54)$$

and

$$\nabla^2 V = 0.$$

It is easy to prove that Equation 4.53 satisfies Laplace's equation and all four conditions listed above. We have found a solution in the region above the $y = 0$ plane that satisfies Laplace's equation and the BC on the infinite hemisphere above the x, y plane, so we have found the only solution.

EXAMPLE

4.1 Find the surface charge density on a conducting plane near a charge Q.

We can compute the electric field intensity in the region $y \geq 0$ from Equation 4.53 for either the charge Q and its image or for the charge Q and the grounded conducting plate at $y = 0$ because they are the same:

$$\vec{E}(x,0,z) = -\vec{\nabla}V\Big|_{y=0} = -\partial V/\partial y\big|_{y=0}\,\hat{a}_y \tag{4.55}$$

or

$$\vec{E}(x,0,z) = -(Q/4\pi\varepsilon_0)\left\{\frac{-(y-d)}{\left[x^2+(y-d)^2+z^2\right]^{3/2}} + \frac{(y+d)}{\left[x^2+(y+d)^2+z^2\right]^{3/2}}\right\}\Bigg|_{y=0}\hat{a}_y,$$

so if we evaluate the electric field intensity at $y = 0$,

$$\vec{E}(x,0,z) = -(2dQ/4\pi\varepsilon_0)\left(x^2+d^2+z^2\right)^{-3/2}\hat{a}_y. \tag{4.56}$$

This electric field intensity has no component in the x or the z direction as expected from Equation 3.84. However, from Equation 3.85,

$$\rho_S = \varepsilon_0 E_n = (-2Qd/4\pi)\left(x^2+d^2+z^2\right)^{-3/2}. \tag{4.57}$$

We see that this function is a maximum for $x = 0$, $z = 0$, where $\rho_S = -2Q/4\pi d^2$, and it falls off with $\rho = \sqrt{x^2+z^2}$ as

$$\rho_S = \left(-2Q/4\pi d^2\right)\left(1+\rho^2/d^2\right)^{-3/2}, \tag{4.58}$$

as shown in Figure 4.5.

CHECK Let us integrate ρ_S over the surface at the $y = 0$ plane to determine the total charge induced on the conducting surface. So the charge induced is

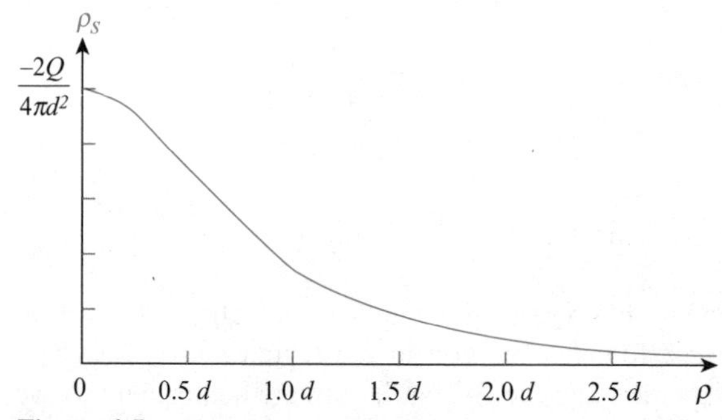

Figure 4.5 Charge density distribution induced on a conducting plane near a charge Q.

$$\int_0^{2\pi}\int_0^{\infty}\left(-2Q/4\pi d^2\right)\rho\left(1+\rho^2/d^2\right)^{-1} d\rho d\phi=\left(-Q/d^2\right)\int_0^{\infty}\left(1+\rho^2/d^2\right)^{-3/2} 2\rho d\rho \tag{4.59}$$

or

$$\text{Charge Induced}=\left(-Q/d^2\right)\left[-d^2(1+\rho^2/d^2)^{-1/2}\right]\Big|_0^{\infty}=-Q. \tag{4.60}$$

The total charge induced on the conducting plane at $y = 0$ due to the charge Q at $y = d$ is $-Q$.

PROBLEM

4.2 Find the force on a charge, Q, if it is located a distance, d, from an infinite, conducting grounded plane.

EXAMPLE

4.2 Point charge near two conducting planes: As shown in Figure 4.6, a positive point charge Q is located at distances d_1 and d_2, respectively, from two-grounded, perpendicular, conducting half-planes. Determine the force on Q caused by the surface charges induced on the planes.

In this case, we can argue that three images charges would be required to cause the scalar electric potential on the surface at $y = 0$ and the surface at

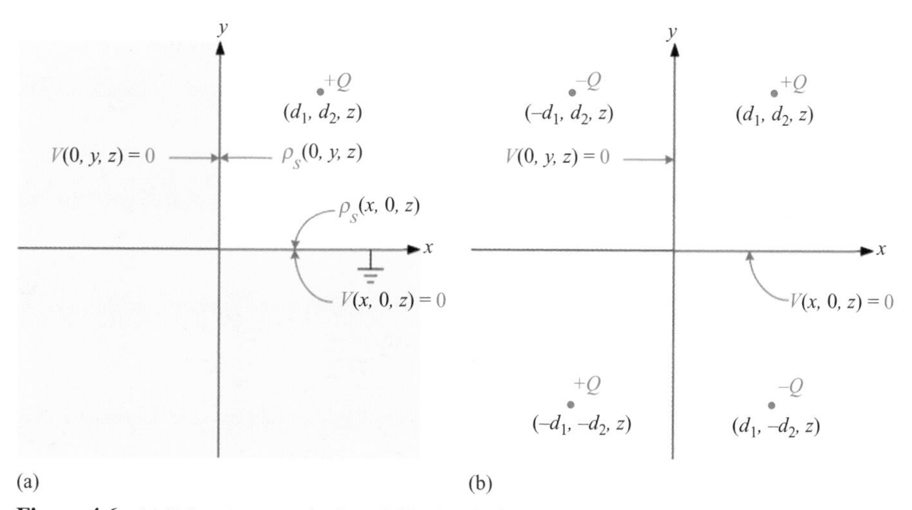

Figure 4.6 (a) Point charge, Q, in the neighborhood of two perpendicular conducting planes; (b) equivalent image arrangement of four charges that give the same scalar electric potential in the first quadrant.

$x = 0$ to be zero. This can be deduced by considering the electric potential of a test charge, q_t, as it approaches the configuration of four charges from infinity along the x- or y-axis.

Given three image charges, we thus have for the net force on Q:

$$\vec{F} = \vec{F}_1 + \vec{F}_2 + \vec{F}_3 \quad \text{where} \tag{4.61}$$

$$\vec{F}_1 = -\hat{a}_y\left[Q^2/4\pi\varepsilon_0(2d_2)^2\right] \tag{4.62a}$$

$$\vec{F}_2 = -\hat{a}_x\left[Q^2/4\pi\varepsilon_0(2d_1)^2\right] \tag{4.62b}$$

$$\vec{F}_3 = \left(Q^2/4\pi\varepsilon_0\right)\left[(2d_1)^2 + (2d_2)^2\right]^{-3/2}(\hat{a}_x 2d_1 + \hat{a}_y 2d_2). \tag{4.62c}$$

Therefore,

$$\vec{F} = \frac{Q^2}{16\pi\varepsilon_0}\left(\hat{a}_x\left\{\frac{d_1}{\left[(d_1)^2 + (d_2)^2\right]^{3/2}} - \frac{1}{d_1^2}\right\} + \hat{a}_y\left\{\frac{d_2}{\left[(d_1)^2 + (d_2)^2\right]^{3/2}} - \frac{1}{d_2^2}\right\}\right). \tag{4.63}$$

PROBLEM

4.3 Find the scalar electric potential at a point (x, y, z) in the first quadrant, and calculate the induced surface charge density on the two plane surfaces at $x = 0$ and $y = 0$. Explain what happens at the point $(0, 0, z)$.

EXAMPLE

4.3 Line charge near a conducting plane: Suppose a line charge, ρ_l, parallel to the z-axis is located above a conducting, grounded plane, as shown in Figure 4.7.

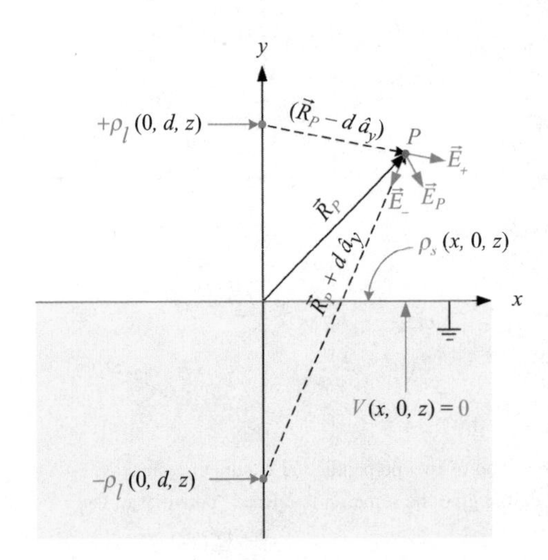

Figure 4.7 Orthogonal view of an infinitely long line of charge at height, $y = d$, above the grounded, conducting plane at $y = 0$ and its image line charge (of charge density, $-\rho_l$) located at $y = -d$ below the x, z plane.

According to the uniqueness theorem, the scalar electric potential in the charge-free region above the x, z plane is the same for the conducting line charge in the neighborhood of a grounded conducting plane at $y = 0$ and for the line charge and its image because the potential at $y = 0$ and at $R = \infty$ is the same for both problems, and the electric potential satisfies Laplace's equation in both problems. We will thus choose to calculate the potential here using the method of images to find the electric field intensity at the point P located at arbitrary point x, y above the $y = 0$ plane.

Using the results of Example 3.5 for a single line of charge density, we can see from Figure 4.7 that the electric field intensity at point $P(x, y)$ for the line of charge and its image is given by

$$\vec{E}_P = \frac{\rho_l}{2\pi\varepsilon_0}\frac{(\vec{R} - d\hat{a}_y)}{|\vec{R} - d\hat{a}_y|^2} - \frac{\rho_l}{2\pi\varepsilon_0}\frac{(\vec{R} + d\hat{a}_y)}{|\vec{R} + d\hat{a}_y|^2}. \tag{4.64}$$

Note that this configuration of two line charges produces zero scalar electric potential on the $y = 0$ plane and because charges are static, $\vec{\nabla} \times \vec{E} = 0$, so the electric field intensity is a conservative field. Thus, the path integral between any point P_0 on the $y = 0$ plane to the point $P(x, y)$

$$V(x, y) = -\int_{P_0}^{P(x,y)} \vec{E} \cdot d\vec{l} \tag{4.65}$$

is independent of the path taken. Furthermore, it does not matter whether the point P_0 is on the x, z plane because this plane is an equipotential. If we choose the point P_0 to be the point directly below the point $P(x, y)$ on the $y = 0$ plane and choose the path of integration to be $d\vec{l} = dy'\hat{a}_y$, as shown in Figure 4.8, we will evaluate the scalar electric potential at $P(x, y)$ relative to the $y = 0$ plane (which is at zero potential).

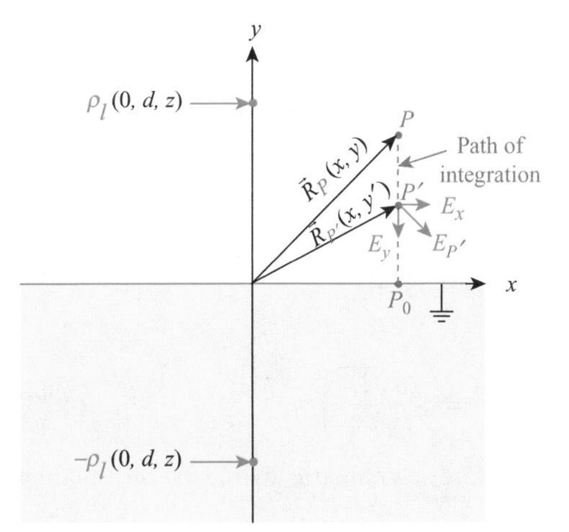

Figure 4.8 Path of integration for Equation 4.65 between the point P_0 and the point $P(x, y)$.

For the path of integration in Figure 4.8, we need the y-component of the electric field intensity at every point along the path because $\vec{E}\cdot dy'\hat{a}_y = E_y dy'$. We can find the components of $\vec{E}_R$ at each point $\vec{R} = x\hat{a}_x + y'\hat{a}_y$ along the path of integration from Equation 4.64 as

$$\vec{E}_R(x,y') = \frac{-\rho_l}{2\pi\varepsilon_0}\frac{(x\hat{a}_x + y'\hat{a}_y - d\hat{a}_y)}{\left|x\hat{a}_x + y'\hat{a}_y - d\hat{a}_y\right|^2} - \frac{\rho_l}{2\pi\varepsilon_0}\frac{(x\hat{a}_x + y'\hat{a}_y + d\hat{a}_y)}{\left|x\hat{a}_x + y'\hat{a}_y + d\hat{a}_y\right|^2} \quad (4.66)$$

so that

$$E_{R,y}(x,y') = \frac{\rho_l}{2\pi\varepsilon_0}\frac{(y'-d)}{x^2+(y'-d)^2} - \frac{\rho_l}{2\pi\varepsilon_0}\frac{(y'+d)}{x^2+(y'+d)^2} \quad (4.67)$$

and

$$V(x,y) = \frac{\rho_l}{2\pi\varepsilon_0}\int_0^y\left[\frac{(y'-d)}{x^2+(y'-d)^2} - \frac{(y'+d)}{x^2+(y'+d)^2}\right]dy'. \quad (4.68)$$

We can separate the two terms into two integrals and make a substitution of variables in the first using $y'' = y' - d$ so that $dy'' = dy'$ and in the second using $y'' = y' + d$ so that $dy'' = dy'$ in which case Equation 4.68 becomes

$$V(x,y) = \frac{-\rho_l}{2\pi\varepsilon_0}\left\{\int_{-d}^{y-d}\frac{y''}{x^2+y''^2}dy'' - \int_{d}^{y+d}\frac{y''}{x^2+y''^2}dy''\right\} \quad (4.69)$$

and because

$$\int\frac{y''}{x^2+y''^2}dy'' = \frac{1}{2}\ln\left|x^2+y''^2\right|$$

$$V(x,y) = \frac{\rho_l}{4\pi\varepsilon_0}\ln\frac{(y+d)^2+x^2}{(y-d)^2+x^2}. \quad (4.70)$$

4.4 Charged cylinder near a conducting plane: We can use Equation 4.70 to compute the electric potential on the surface of a cylinder by recognizing that the equipotential lines for the parallel wires are found when $V(x, y) = V_1$, where V_1 is a constant. The values of x and y that satisfy this condition are

$$D_1 = e^{4\pi\varepsilon_0 V_1/\rho_l} = \frac{(y+d)^2+x^2}{(y-d)^2+x^2} \quad (4.71)$$

or

$$\left(y - d\frac{D_1+1}{D_1-1}\right)^2 + x^2 = \left(\frac{2d\sqrt{D_1}}{D_1-1}\right)^2 \quad (4.72)$$

We conclude that equipotentials are asymmetric cylinders surrounding the line of charge, as shown in Figure 4.9.

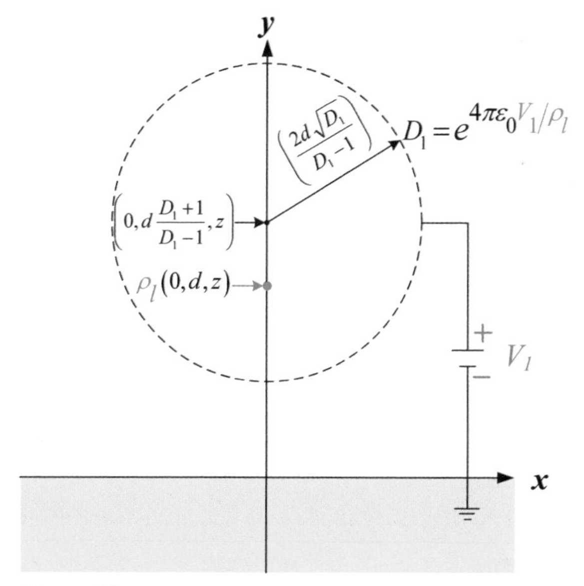

Figure 4.9 Equipotential surfaces in the neighborhood of a line of charge near a grounded conducting plane.

Points on the equipotential surface are those on the surface of a cylinder whose electric potential is determined by an external voltage source.

PROBLEM

4.4 Given a cylinder of radius a at scalar electric potential V_1, with its center located at a distance b from a grounded conducting plane, show that the parameter D_1, the equivalent charge per unit length of a charged wire, and the capacitance per unit length are

$$D_1 = \left(2b^2/a^2 - 1\right) + 2b/a\sqrt{\left(b^2/a^2 - 1\right)} \tag{4.73a}$$

$$\rho_l = 4\pi\varepsilon_0 V/\ln D_1 \tag{4.73b}$$

$$C_l = \rho_l/V_1 = 4\pi\varepsilon_0/\ln D_1\,. \tag{4.73c}$$

EXAMPLE

4.5 Charges near spherical conductors: Consider the "spherical capacitor" shown in Figure 4.10. By using Gauss's law for the Gaussian surface of radius r,

$$\iiint_V \vec{\nabla}\cdot\vec{D}\,d^3x = \oiint_S \vec{D}\cdot d\vec{s} = Q \tag{4.64}$$

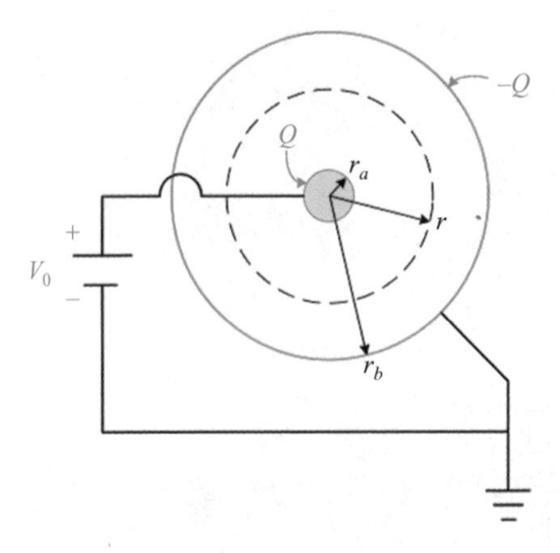

Figure 4.10 Cross section of a charged spherical conductor inside of (and concentric to) another grounded spherical conductor.

and, recognizing that $|\vec{D}|$ is of constant magnitude on S and always normal to the differential surface ds,

$$|\vec{D}_R| 4\pi R^2 = Q$$

so

$$\vec{D} = (Q/4\pi R^2)\hat{a}_R \tag{4.74}$$

and

$$V_{a,b} = \int_{r_b}^{r_a} -\vec{E}\cdot d\vec{l} = -(Q/4\pi\varepsilon_0)\int_{r_b}^{r_a} R^{-2} dR = (Q/4\pi\varepsilon_0)(1/r_a - 1/r_b)$$

The capacitance can be thus be written as

$$C_{a,b} = Q/V_{a,b} = 4\pi\varepsilon_0 [r_a r_b/(r_b - r_a)] \tag{4.75}$$

and in the limit as $r_b \to \infty$,

$$C_a = 4\pi\varepsilon_0 r_a, \tag{4.76}$$

which we can call the capacitance of an "isolated sphere."

Nonconcentric Charges near a Grounded Conducting Sphere

The more general case is for a charge Q that is not concentric with an outer sphere. Here, we can use the method of images to find the electric field intensity. Figure 4.11 shows a charge Q located a distance z from the concentric center of a spherical conductor of radius R_a.

Because of spherical symmetry, we have chosen the z-axis to lie along the direction between the sphere center and the charge Q. Now, let us locate an image charge $-Q'$ on the z-axis at point z' so that it produces a zero scalar electric potential on the spherical surface at R_a. We will thus want

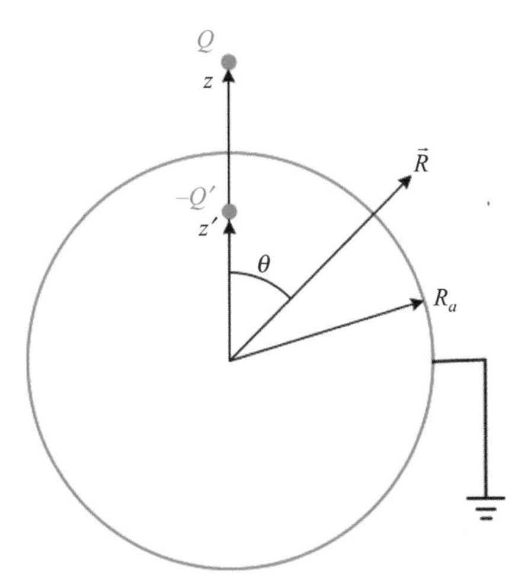

Figure 4.11 Potential at point $\vec{R}$ due to a charge Q outside a grounded conducting sphere.

$$V(R,\theta)=\frac{Q}{4\pi\varepsilon_0}\frac{1}{\left|\vec{R}-z\hat{a}_z\right|}-\frac{Q'}{4\pi\varepsilon_0}\frac{1}{\left|\vec{R}-z'\hat{a}_z\right|} \tag{4.77}$$

to produce zero potential when $\vec{R}$ is located anywhere on the sphere at R_a:

$$V(R_a,\theta)=\frac{Q}{4\pi\varepsilon_0}\frac{1}{\left|\vec{R}_a-z\hat{a}_z\right|}-\frac{Q'}{4\pi\varepsilon_0}\frac{1}{\left|\vec{R}_a-z'\hat{a}_z\right|} \tag{4.78}$$

To make this happen, we can factor R_a out of the first term in Eqaution 4.78 and z' out of the second term to get

$$V(R_a,\theta)=\frac{Q}{4\pi\varepsilon_0 R_a}\frac{1}{\left|\hat{a}_R-(z/R_a)\hat{a}_z\right|}-\frac{Q'}{4\pi\varepsilon_0 z'}\frac{1}{\left|(\vec{R}_a/z')-\hat{a}_z\right|}. \tag{4.79}$$

We can make the two terms cancel by choosing

$$\frac{Q}{R_a}=\frac{Q'}{z'} \quad \text{and} \quad \frac{z}{R_a}=\frac{R_a}{z'}. \tag{4.80}$$

Thus, the magnitude and position of the image charge should be

$$Q'=Q\frac{R_a}{z} \quad \text{and} \quad z'=\frac{R_a^2}{z} \tag{4.81}$$

so that the scalar electric potential at a general point R is

$$V(R,\theta)=\frac{Q}{4\pi\varepsilon_0}\frac{1}{\left|\vec{R}-z\hat{a}_z\right|}-\frac{Q}{4\pi\varepsilon_0}\frac{R_a/z}{\left|\vec{R}-(R_a^2/z)\hat{a}_z\right|}. \tag{4.82}$$

PROBLEMS

4.5 Show that the surface charge density on the conducting sphere is given by

$$\rho_S = -\frac{Q}{4\pi R_a^2}\frac{\left(1-R_a^2/z^2\right)}{\left(1+R_a^2/z^2-2R_a/z\cos\theta\right)} \tag{4.83}$$

and plot the charge density as a function of θ.

4.6 Show that the force exerted on the charge Q by the induced charge density on the sphere is

$$\left|\vec{F}\right| = \frac{Q^2}{4\pi\varepsilon_0 R_a^2}\left(\frac{R_a}{z}\right)^3\left(1-\frac{R_a^2}{z^2}\right)^{-2}. \tag{4.84}$$

4.7 Find the scalar electric potential *inside* a grounded conducting sphere if a charge Q is located at z *inside* the sphere.

Nonconcentric Charges near an Insulated Conducting Sphere

We can also use the method of images to calculate the scalar electric potential due to an insulated (as opposed to grounded) conducting sphere. Suppose we want to find the potential of a conducting sphere with total charge Q'' in the presence of a point charge Q. We can think of the process to construct the situation by first calculating the field due to a charge Q in the neighborhood of a grounded conducting sphere as above. This process will produce a net charge $-QR_a/z$ on the grounded conducting sphere. Then, we can disconnect the ground wire from the sphere. The scalar electric potential at the conducting surface will still be zero. Now, if we add charge $Q'' + QR_a/z$ to the conducting sphere (distributed uniformly over its surface), the net charge on the sphere will be Q'', and the scalar electric potential at a point $\boldsymbol{R}$ in space will be

$$V(R,\theta) = \frac{Q}{4\pi\varepsilon_0}\frac{1}{\left|\vec{R}-z\hat{a}_z\right|} - \frac{Q}{4\pi\varepsilon_0}\frac{R_a/z}{\left|\vec{R}-\left(R_a^2/z\right)\hat{a}_z\right|} + \frac{(Q''+QR_a/z)}{4\pi\varepsilon_0\left|\vec{R}\right|} \tag{4.85}$$

And the force on the charge Q will be

$$\left|\vec{F}\right| = \frac{Q^2}{4\pi\varepsilon_0 R_a^2}\left(\frac{R_a}{z}\right)^3\left(1-\frac{R_a^2}{z^2}\right)^{-2} + \frac{(Q''+QR_a/z)}{4\pi\varepsilon_0\left|\vec{R}\right|^2}. \tag{4.86}$$

Nonconcentric Charges near a Conducting Sphere at Potential V_0

We can also use the method of images to calculate the scalar electric potential due to a conducting sphere held at electric potential V_0 in the presence of a point charge Q. The technique is the same as the one above, except the total charge on the sphere will be $Q'' = 4\pi\varepsilon_0 V_0 R_a$ from Equation 4.76. Thus, the scalar electric potential and forces due to this configuration become

$$V(R,\theta) = \frac{Q}{4\pi\varepsilon_0}\frac{1}{\left|\vec{R} - z\hat{a}_z\right|} - \frac{Q}{4\pi\varepsilon_0}\frac{R_a/z}{\left|\vec{R} - \left(R_a^2/z\right)\hat{a}_z\right|} + \frac{\left(4\pi\varepsilon_0 V_0 R_a + Q\,R_a/z\right)}{4\pi\varepsilon_0\left|\vec{R}\right|} \tag{4.87}$$

$$\left|\vec{F}\right| = \frac{Q^2}{4\pi\varepsilon_0 R_a^2}\left(\frac{R_a}{z}\right)^3\left(1 - \frac{R_a^2}{z^2}\right)^{-2} + \frac{\left(4\pi\varepsilon_0 V_0 R_a + Q\,R_a/z\right)}{4\pi\varepsilon_0\left|\vec{R}\right|^2}. \tag{4.88}$$

EXAMPLE

4.6 Dipoles in the neighborhood of conducting planes: We can also consider a configuration of charges in the form of vertical or horizontal electric dipoles as shown in Figure 4.12.

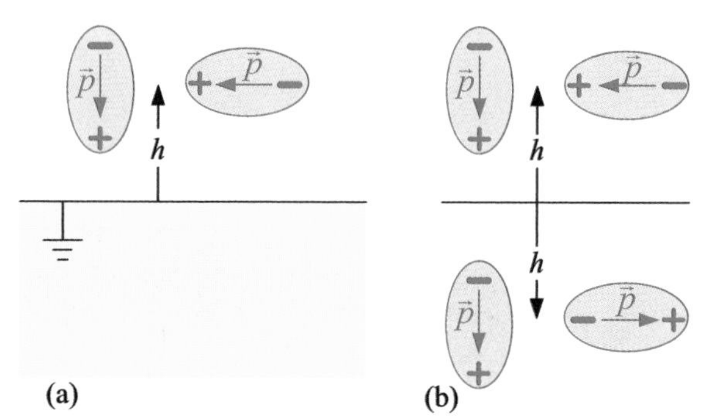

Figure 4.12 (a) Vertical and horizontal electric dipoles above a grounded conducting plane and (b) the equivalent electric dipoles with their respective electric dipole images.

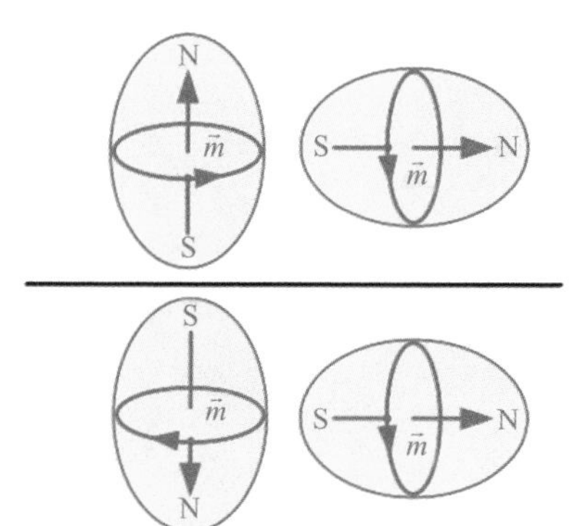

Figure 4.13 Equivalent vertical and horizontal magnetic dipole moment pairs that yield the same magnetic field intensity in space as a magnetic dipole and the induced surface current density on an infinite conducting plane.

PROBLEM

4.8 Justify the image electric dipoles shown in Figure 4.12, and find the image electric dipole for an electric dipole above a grounded conducting plane if it makes angle θ with respect to the horizontal.

With a similar analysis, we could deduce the equivalent image linear vertical or horizontal quadrupole charges that would produce the same scalar electric potential above a grounded conducting plane.

Does it thus make sense that we could produce a similar analysis for a multipole configuration of charges above a grounded conducting plane? If so, and if the general distribution of charges can be written as a linear combination of monopole, dipole, quadrupole, ..., charges, is it possible to give the solution for the general configuration of charges in the neighborhood of a grounded conducting plane?

LOOK AHEAD We will perform a similar analysis for the case of magnetic dipoles below a grounded conducting plane with the result shown in Figure 4.13. This figure is given here for completeness.

Chapter 5

Steady Electric Currents

LEARNING OBJECTIVES

- Relate electric field quantities to currents in circuits
- Understand Ohm's current law, Kirchhoff's current law, and Joule's law from the point of view of fields
- Have a basic appreciation of superconductors and the role they are likely to play in future technologies
- Understand the basic theory of a free electron gas and the development of band gaps in materials
- Derive the boundary conditions for current density in a conductor

INTRODUCTION

Previously, we dealt with electrostatic problems, where field problems are associated with electric charges at rest. We now consider charges in motion that constitute current flow. There two types of electric current caused by the motion of free charges:

a. Conduction currents in conductors and semiconductors that are caused by the drift motion of conduction electrons, holes, and ions

b. Convection currents that result from the motion of electrons and/or ions in a vacuum, gas, or insulating medium

In this chapter, we will deal mainly with currents in conductors or semiconductors.

5.1 CURRENT DENSITY AND OHM'S LAW

Consider the steady motion of N charge carriers per unit volume, each of charge q (negative for electrons), and average drift velocity, $\vec{u}$, across an element of surface, $\Delta\vec{s}$, in a long, uniform wire as shown Figure 5.1.

Maxwell's Equations, by Paul G. Huray

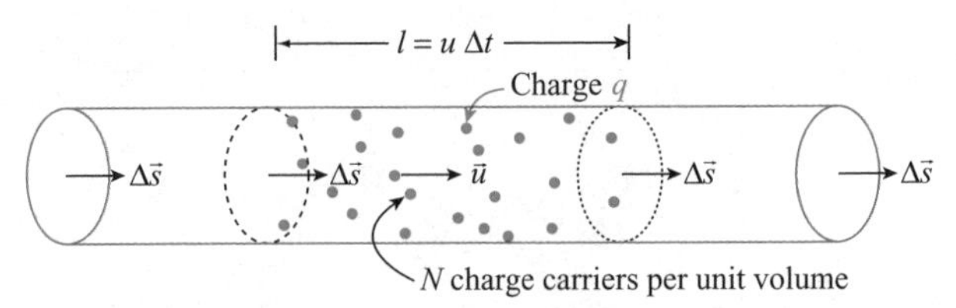

Figure 5.1 Charge carriers within a length, l, of a uniform wire of cross-sectional area, $\Delta\vec{s}$.

We can see that in time Δt, an amount of charge

$$\Delta Q = \rho_V \vec{u}\Delta t \cdot \Delta\vec{s} = qN\vec{u} \cdot \Delta\vec{s}\Delta t\,(\mathrm{C}) \tag{5.1}$$

will flow across the cross-sectional area $\Delta\vec{s}$. Because current is the time rate of change of charge flowing across $\Delta\vec{s}$, we have

$$\Delta I \equiv \frac{\Delta Q}{\Delta t} = \frac{qN\vec{u} \cdot \Delta\vec{s}\Delta t}{\Delta t} = Nq\vec{u} \cdot \Delta\vec{s}\,(\mathrm{A}), \tag{5.2}$$

where $\Delta\vec{s} = \hat{a}_n \Delta s$ is a vector quantity.

By defining a vector point function, $\vec{J}$, to be the current density, in amperes per square meter (A/m^2),

$$\vec{J} = Nq\vec{u}\,\left(\mathrm{A/m^2}\right) \tag{5.3}$$

then,

$$\Delta I = Nq\vec{u} \cdot \Delta\vec{s} = \vec{J} \cdot \Delta\vec{s}\,(\mathrm{A}). \tag{5.4}$$

The total current I following through an arbitrary surface S is then the integral of the current through S:

$$I = \int_S \vec{J} \cdot d\vec{s}\,(\mathrm{A}). \tag{5.5}$$

Noting that the product Nq is free charge per unit volume, we can rewrite

$$\vec{J} = Nq\vec{u} = \rho_V \vec{u}\,\left(\mathrm{A/m^2}\right), \tag{5.6}$$

which is the relation between the conduction current density and the average velocity of the charge carriers.

Conduction Current

In the case of conduction currents, there may be more than one kind of charge carrier (electrons, holes, and ions) drifting, with different average velocities, in which case

$$\vec{J} = \sum_i N_i q_i \vec{u}_i\,\left(\mathrm{A/m^2}\right). \tag{5.7}$$

n-*Type Materials*

If the conduction current is predominantly a result of the *movement of negative charges*, such as in most conductors, the conducting material is said to be *n*-type. In the case of negatively charged electrons, the average velocity of each of conduction charges, $-e$, is directly proportional to (and opposite from) the electric field intensity applied to that charge;[1]

$$\vec{u} = -\mu_e \vec{E} \ (\mathrm{m/s}), \tag{5.8}$$

where μ_e is the electron mobility measured in (m^2/Vs). Therefore,

$$\vec{J} = Nq\vec{u} = -\rho_V \mu_e \vec{E} \ (\mathrm{A/m^2}), \tag{5.9}$$

where ρ_V is the charge density of the electrons.

For conductors, for example copper with $\mu_e = 3.2 \times 10^{-3}$ ($\mathrm{m^2/Vs}$), the outermost electrons of the constituent atoms are the primary charge carriers because they are less massive than ions or holes and are not necessarily associated with their parent atom but are relatively "free" to move to their adjacent neighbors. We often say these conduction electrons form a "free electron gas" spread uniformly within a large conductor volume of atoms so that the net electrical charge density remains neutral, while the electrons are able to move. As we will see when we study band theory in the next section, this mental picture is an oversimplification of the details but in the end produces the same phenomenological results.

We can also rewrite Equation 5.9 as

$$\vec{J} = -\rho_V \mu_e \vec{E} = -\sigma_e \vec{E} \ (\mathrm{A/m^2}), \tag{5.10}$$

where the constant $\sigma_e = \rho_V \mu_e$ is a macroscopic parameter of the conducting medium called the *electrical conductivity*.

p-*Type Materials*

If the conduction current is predominantly a result of the *movement of positive charges*, such as in some semiconductors or insulators, the conducting material is said to be *p*-type. In the case of positively charged *holes* (e.g., atoms from which an electron has been removed), the average velocity of each of conduction charges, $+e$, is directly proportional to the electric field intensity applied to that charge:

$$\vec{u} = \mu_h \vec{E} \ (\mathrm{m/s}), \tag{5.11}$$

where μ_h is the hole mobility measured in $\mathrm{m^2/Vs}$.

[1] Note that this behavior is in contradiction to Newton's second law, $\vec{F} = m\vec{a}$ or $\vec{a} = q\vec{E}/m = (q/m)\vec{E}$ because the charge carriers are accelerating, bumping into obstacles (which causes their deceleration), reaccelerating, and so forth. Paul Drude proposed this model in 1900 to explain ohmic conduction in a homogeneous, isotropic, linear, local time invariant material. Under a static electric field, Equation 5.8 states the charge carriers will have an average *velocity* that takes into account these multiple scatterings. We will study these scattering effects in detail later in this chapter.

Therefore,

$$\vec{J} = Nq\vec{u} = \rho_{h,v}\mu_h\vec{E}\ (\mathrm{A/m^2}), \tag{5.12}$$

where $\rho_{h,v}$ is the volume charge density of the holes.

Net Conduction Current

For semiconductors, both electrons and holes can cause conduction. For example, in silicon with $\mu_e = 0.12\ \mathrm{m^2/Vs}$ and $\mu_h = 0.03\ \mathrm{m^2/Vs}$, both the outermost electrons of the constituent atoms and the holes are relatively "free" to move to their adjacent neighbors. In accordance with Equations 5.7, 5.10, and 5.12,

$$\vec{J} = -N_e|e|\vec{u}_e + N_h|e|\vec{u}_h = -\rho_e\mu_e\vec{E} + \rho_{h,v}\mu_h\vec{E} = \sigma\vec{E}\ (\mathrm{A/m^2}). \tag{5.13}$$

For pure silicon at room temperature, $\sigma = 0.0016$ A/Vm or siemens per meter (S/m). By comparison, for copper, $\sigma = 5.8 \times 10^7$ S/m and, for fused quartz, $\sigma = 1.0 \times 10^{-17}$ S/m. The conductivity of materials is clearly one of the most variable quantities in nature; its contributions coming from the density of competing carriers, their respective motilities, from impurities in the material, its electric potential relative to an adjacent material, and from its temperature. The value of σ taken from a table of parameters for various materials is thus likely to be quite different from that for a particular specimen.

Equation 5.13 is a constitutive relation of a conducting medium. Isotropic materials for which the linear relation 5.13 holds are called ohmic media. The conductivity, σ, is large and relatively constant for practical conducting materials, but, in semiconductors and insulators, it may depend highly upon external effects such as the temperature of the material, the amount of water absorbed, and the frequency of the applied electric field intensity (considered constant in this chapter) in which case it may even have an imaginary component.

Drift Velocity

As noted above, the force on a "free" electron is caused by the applied electric field intensity and by the collisions with other electrons, impurities, ions, or the thermal vibrations of the atomic lattice. For an electron in a typical conductor, the average time between collisions is very short (e.g., 10^{-14} s for copper). Unless the frequency of the external electric field intensity is very high (e.g., 100 terahertz for copper), collisions of the conduction electrons will be a dominant factor in their movement. For frequencies much less than 10^{14} Hz, an electron suffers many scattering events per driving cycle and ohmic loss is apparent.

Electromigration

The scattering of electrons from positive atomic ions in a lattice can cause a differential collisional pressure on ion cores at high current densities (typically above $10^4\ \mathrm{A\ cm^{-2}}$). If there is a differential pressure on two different types of cores (e.g.,

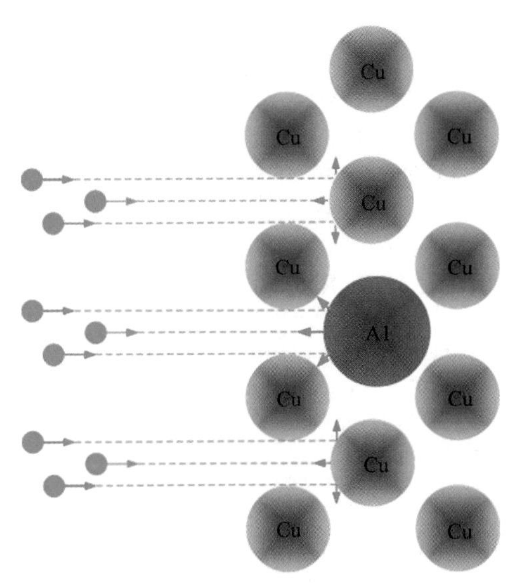

Figure 5.2 Preferential back scattering of an electron current from an Al ion core in a Cu ion lattice.

Al in Cu) because of a difference in their scattering cross section, the differential pressure can displace one type of metal ion relative to the other. This displacement generally takes place slowly but, over months or years, can cause the transport of mass (positive ions) in the same direction as the electron current. This mass transport is called *electromigration* and is indicated by the two-dimensional sketch shown in Figure 5.2. The effect is usually ignored, but, if small features in a device (such as a man-made barrier of a concentration of atoms in a transistor) involves a region of high concentration of one type, its relative proportion may be physically spread over time. Electromigration can result in the failure of an electronic system over a long period of time, so it is often advisable to estimate the time required for such failure compared with its useful replacement time or to reduce the incident current density.

If the differential cross section of an Al ion core (in a vertical set of Cu ions) is larger than that of a Cu ion core as shown in this figure, more electrons will be back scattered from the Al, and, thus, the momentum transfer, $\Delta\vec{p}_{Al}$, will be larger than that for Cu ions, $\Delta\vec{p}_{Cu}$. If this momentum transfer occurs in a time, Δt, then a differential force will be experienced by the Al ion, $\Delta\vec{F} = (\Delta\vec{p}_{Al} - \Delta\vec{p}_{Cu})/\Delta t$. The collisions will be a statistical average over time, so, most of the time, the differential force will be small compared with the force needed to overcome the potential barrier for a displacement of the Al ion one atomic lattice displacement to the right. On occasion, however, there may be a statistically large number of collisions on the Al ion that will be sufficient for the Al ion to be displaced, and the Al will migrate to the right.

The Al ion is shown in this figure as an impurity in a Cu lattice but it might also be part of an Al superlattice that forms some man-made geometric structure. In that case, the migration of one of the Al ions will disrupt the planned structure and potentially disrupt the intent of the structure (i.e., a barrier caused by a line of Al ions). For example, aluminum-based alloys (Al-Cu, Al-Si, ...) are widely used in interconnects for integrated circuits. One of the problems associated with Al-based alloys is their poor resistance to electromigration in terms of their time-to-failure of a device.

5.2 RELATION TO CIRCUIT PARAMETERS

Ohm's law from circuit theory states that the voltage V_{12} across a resistance R, in which a current I follows from point 1 to point 2, is equal to RI as shown in Figure 5.3:

In basic circuit courses, we have learned that

$$\Delta V_{12} = R_{12} I \tag{5.14}$$

gives the voltage (potential) drop across a finite length wire of length $(l_2–l_1)$, but Equation 5.14 is not a point relation. Although there is little resemblance between Equations 5.13 and 5.14, the former is generally referred to as the point form of Ohm's law and holds at all points in space.

We can use the point form of Ohm's law to derive the voltage–current relationship of a wire of homogeneous material of conductivity, σ, length, $(l_2–l_1)$, and uniform cross section, ΔS, from Figure 5.3. The potential difference (voltage) between the two ends is

$$\begin{aligned} \Delta V_{12} &= \int_1^2 dV = -\int_1^2 \vec{E}\cdot d\vec{l} \\ \Delta V_{12} &= -\int_1^2 (\vec{J}/\sigma)\cdot d\vec{l} = -(J/\sigma)(l_2 - l_1) \\ \Delta V_{12} &= I\,(l_1 - l_2)/\sigma\Delta S \end{aligned} \tag{5.15}$$

and, if we define

$$R_{12} \equiv (l_2 - l_1)/\sigma\Delta S \tag{5.16}$$

We can see that the point Equation 5.13 yields the macroscopic Ohm's Law, Equation 5.14. Hence, we have the formula for the resistance of a straight piece of

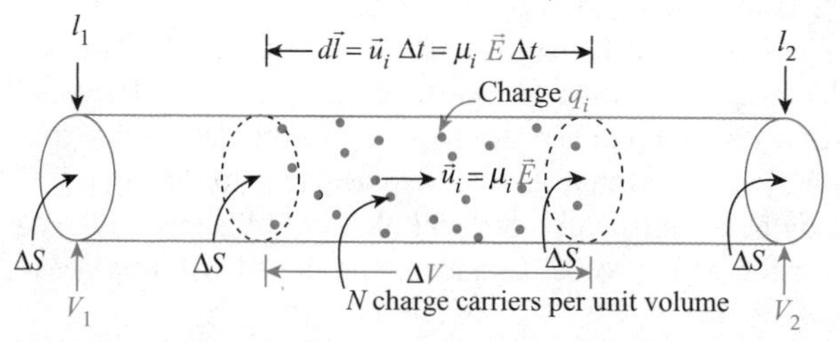

Figure 5.3 Voltage $\Delta V_{12} = V_2 - V_1$, across a long, homogeneous wire of resistance R, in which a uniform current I flows from point 1 to point 2.

homogeneous material of uniform cross section for steady current (DC). Electrical resistivity, ρ, is defined as

$$\rho \equiv 1/\sigma \tag{5.17}$$

and electrical conductance, G, is defined as

$$G \equiv 1/R \text{ (Mhos)}. \tag{5.18}$$

Because the electrical conductivity, σ, depends upon temperature, the electrical resistivity also depends upon the temperature of the material. A convenient linear relationship near room temperature is often written as

$$\rho = \rho_0[1+\alpha(T-T_0)]. \tag{5.19}$$

Here, T_0 is a selected reference temperature, ρ_0 is the resistivity at that temperature, and α is the temperature coefficient of resistivity. Table 5.1 gives the electrical resistivity, ρ_0, of some common materials at room temperature and gives their temperature coefficient of resistivity, α, at that same temperature.

Figure 5.4 shows measured values of resistivity for copper as a function of temperature. It is seen from this figure that the temperature dependence is not quite linear but that at room temperature the resistivity is $\rho_0 = 1.69 \times 10^{-8}\ \Omega$m, and the slope is $\alpha = 4.3 \times 10^{-3}\ \text{K}^{-1}$. If we examine the resistivity near the absolute zero of temperature, we will find that the resistivity does not go to zero, as implied by Figure 5.4, but approaches a small limiting amount, called the residual resistivity, determined in part by the trace impurities in the sample. Most normal conductors exhibit similar behavior.

Table 5.1 Electrical Resistivity, ρ_0, and the Temperature Coefficient of Resistivity, α, of Some Common Materials at Room Temperature

Material	ρ_0 (Ωm) at 293 K	α (K^{-1}) at 293 K
Silver	1.62×10^{-8}	4.1×10^{-3}
Copper	1.69×10^{-8}	4.3×10^{-3}
Aluminum	2.75×10^{-8}	4.4×10^{-3}
Iron	9.68×10^{-8}	6.5×10^{-3}
Platinum	10.6×10^{-8}	3.9×10^{-3}
Silicon (*p*-type)[a]	2.8×10^{-3}	—
Silicon (*n*-type)[b]	8.7×10^{-4}	—
Silicon (pure)	2.5×10^{3}	-70×10^{-3}
Glass	$10^{10} \times 10^{14}$	—
Fused quartz	2×10^{17}	—

[a] Pure silicon doped with aluminum impurities to a charge carrier density of 10^{23} m^{-3}.

[b] Pure silicon doped with phosphorus impurities to a charge carrier density of 10^{23} m^{-3}.

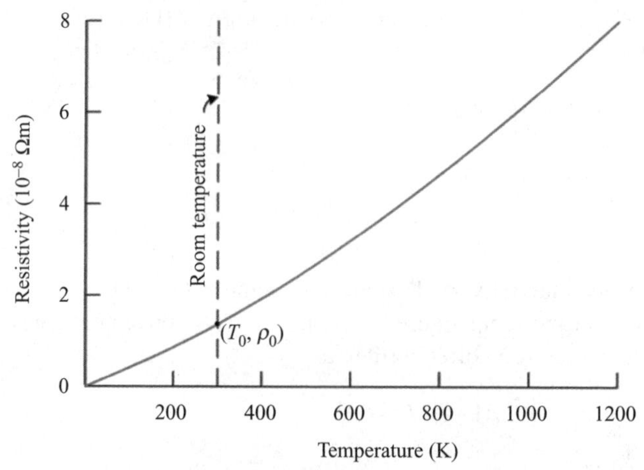

Figure 5.4 Resistivity of copper as a function of temperature.

5.3 SUPERCONDUCTIVITY

In 1898, James Dewar succeeded in liquefying hydrogen at a temperature of 20 K, and, in 1908, Heike Kamerlingh Onnes succeeded in liquefying helium at a temperature of 4.2 K. Onnes used his (then unique) low-temperature capability to measure the electrical resistance of a variety of metals at low temperatures. Onnes began his studies with platinum and gold but felt there were trace impurities left in those materials. In purifying his samples, he reasoned that, because mercury is liquid at room temperature, he could use a standard distillation process to obtain the purest sample possible. In 1911, to his surprise, an assistant, Gilles Holst, immersed a thin capillary tube of mercury into a helium bath and measured the resistance curve shown in Figure 5.5.

Onnes had surmised that, without impurities, the resistivity of a very pure metal like mercury would approach zero smoothly with decreasing temperature. He repeated his measurements many times before he became convinced that the resistance of mercury dropped suddenly at 4.2 K to an unmeasurable value (the lower limit of his measurement equipment was 10^{-6} Ω). This critical temperature is today called T_c and it denotes the transition to the superconducting state for a material. In 1913, Onnes won the Nobel Prize in physics for his research. That year, he also discovered that there was a "threshold value" of the current density, J_c, that can be carried by a superconducting sample before it reverts to its normal (metallic) state and that J_c increased as the temperature of the superconductor was lowered below T_c. In 1914, Onnes reported that an applied magnetic field intensity, H_c, can also destroy the superconducting state and that the critical field was also temperature dependent. Based on a fit of the empirical data, it was possible to quantitate this temperature dependence as

$$H_c(T) \approx H_{c0}\left[1-(T/T_c)^2\right]. \tag{5.21}$$

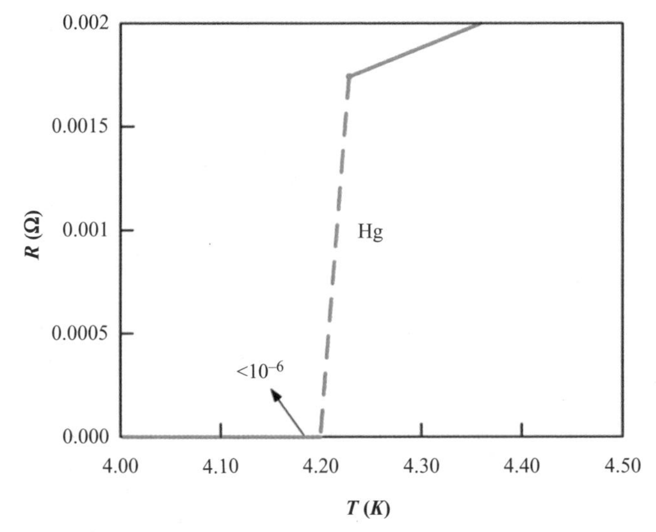

Figure 5.5 Resistance versus temperature curve for Hg that Kamerlingh Onnes reported to announce the discovery of superconductivity in 1911.

The best measurements that have been carried out to date conclude that the DC resistance of superconductors is dead zero (not just a very low value). We can also conclude that it is **not correct** to define a superconductor as a medium that satisfies the nondispersive constitutive relation $\vec{J} = \sigma_0 \vec{E}$ with $\sigma_0 \to \infty$ because electric field intensity cannot exist in a perfect conductor.

Two-Fluid Model

In 1934, Cornelius Gorter and H. B. G. Casimir found that many of the properties of superconductors, including the Meissner effect (which states that magnetic flux density inside a superconductor is zero), can be explained by a constitutive relation for the current density in a material in the superconducting state:

$$\vec{J} = \vec{J}_n + \vec{J}_S, \tag{5.22}$$

where $\vec{J}_n$ and $\vec{J}_S$ are the current density of normal electrons and superconducting electrons, respectively. In this model, the total number of conduction electrons in a material, n_{tot}, can be written as

$$n_{tot} = n(T) + 2n^*(T) \quad \text{for all } T, \tag{5.23}$$

where n is the number of unpaired (normal electrons) and n^* is the number of Cooper pairs (superconducting electrons). The microscopic theory proposed by John Bardeen, Leon Cooper, and Robert Schrieffer in 1957 (known as the BCS theory) showed how electrons could become paired when they interact with the lattice of parent atoms and one another. These electron pairs are called *Bosons* because they

1 H																	2 He
3 Li	4 Be 26 K											5 B	6 C	7 N	8 O	9 F	10 Ne
11 Na	12 Mg											13 Al 1.14K	14 Si	15 P	16 S	17 Cl	18 Ar
19 k	20 Ca	21 Sc	22 Ti 0.39K	23 V 5.38K	24 Cr	25 Mn	26 Fe	27 Co	28 Ni	29 Cu	30 Zn 0.88K	31 Ga 1.09K	32 Ge	33 As	34 Se	35 Br	36 Kr
37 Rb	38 Sr	39 Y	40 Zr 0.55K	41 Nb 9.50K	42 Mo 0.92K	43 Tc 7.77K	44 Ru 0.51K	45 Rh 0.3m K	46 Pd	47 Ag	48 Cd 0.56K	49 In 3.40K	50 Sn 3.72K	51 Sb	52 Te	53 I	54 Xe
55 Cs	56 Ba		72 Hf 0.12K	73 Ta 4.48K	74 W 12mK	75 Re 1.4K	76 Os 0.66K	77 Ir 0.14K	78 Pt	79 Au	80 Hg 4.15 K	81 T1 2.39 K	82 Pb 7.19 K	83 Bi	84 Po	85 At	86 Rn
87 Fr	88 Ra																
			57 La 6.00K	58 Ce	59 Pr	60 Nd	61 Pm	62 Sm	63 Eu	64 Gd	65 Tb	66 Dy	67 Ho	68 Er	69 Tm	70 Yb	71 Lu 0.1K
			89 Ac	90 Th 1.37K	91 Pa 1.4K	92 U	93 Np	94 Pu	95 Am	96 Cm	97 Bk	98 Cf	99 Es	100 Fm	101 Md	102 No	103 Lr

Figure 5.6 Periodic table of the superconducting elements.

have integer spin and are not subject to the Pauli exclusion law for half-integer spin particles that no two electrons can have all their quantum numbers identical. The superelectron pairs have the charge and mass of two electrons and can carry absolutely lossless supercurrents, whereas the normal electrons produce lossy currents like a normal conductor. Many elemental metals are superconductors, as shown in Figure 5.6.

At least one of the 5 f elements, Am, is also a superconductor, but studies of others are difficult because of their limited quantities available for measurement. Many alloys or compounds of pure elements are also superconductors, including some between elements that are not superconductors as pure materials (e.g., MgB_2).

The theory of superconductivity seemed to be reasonably complete until 1987 when Paul Chu and Maw-Kuen Wu demonstrated that $YBa_2Cu_3O_7$ was a superconducting material with a critical temperature of *95 K*. Note that none of these elements by themselves are superconductors (as is seen in Figure 5.6). The importance of this discovery is enormous because it concluded that superconductivity had a high-enough transition temperature to be cooled by inexpensive liquid nitrogen (which boils at 77 K). There are other copper oxide materials that retain their superconducting properties above 120 K, but their mechanical brittleness makes them difficult to form into long wires. A compromise material, magnesium diboride with a T_C of 39 K is cheap to form into wires and can be cooled by liquid hydrogen (boiling point

20.2 K). It thus appears that commercial applications of superconductivity may soon become more widespread and common.

The BCS theory of superconductivity does not explain the properties of high-temperature superconductors, so we will not give further details about their properties. A student of electrical engineering today will likely use these materials in practical applications in his or her professional career, so critical currents, temperatures, and magnetic fields are topics worthy of regular review.

5.4 FREE ELECTRON GAS THEORY

The "free electron gas" model provides a first-order approximation to the behavior of conduction electrons in which we view electrons as waves through a probability wave function, $\psi(\vec{x}, t)$. In 1905, Albert Einstein postulated that all *electromagnetic radiation* could be viewed as a collection of *particles* known as photons whose energy, ε, and momentum, $\vec{p}$, can be written as

$$\varepsilon = \hbar\omega \quad \text{and} \quad \vec{p} = \hbar\vec{k}, \tag{5.24}$$

In 1924, Louis de Broglie took the inverse position by describing *particles* as *matter waves* in his doctoral thesis. Equations 5.24 are known as the Einstein–de Broglie relations. Here, the total energy, ε, of a quantum particle is related to its frequency of oscillation, ω, and its momentum, $\vec{p}$, to the wave vector $\vec{k}$, where the magnitude of $\vec{k}$ is related to the matter wavelength, λ, in the usual way:

$$|\vec{k}| = 2\pi/\lambda \quad \text{and} \quad \hbar = h/2\pi, \tag{5.25}$$

where $h = 6.62606876(52) \times 10^{-34}$ J s or $4.13566727(16) \times 10^{-15}$ eV s). The Einstein–de Broglie relations let us go back and forth between the wavelike and the particle-like behavior of matter.

With these ideas in mind, let us find an equation of motion for the simplest quantum system: a "free" electron. We will begin by writing the classical expression for conservation of energy as

$$K.E. + P.E. = \varepsilon_{total} \tag{5.26a}$$

$$(1/2)mv^2 + qV(\vec{x}) = \varepsilon_{total} \tag{5.26b}$$

$$(1/2m)m\vec{v} \cdot m\vec{v} + qV(\vec{x}) = \varepsilon_{total} \tag{5.26c}$$

$$\vec{p} \cdot \vec{p}/2m - eV(\vec{x}) = \varepsilon_{total}, \tag{5.27}$$

where $V(\vec{x})$ is the electric potential. Now, if we substitute Equation 5.24 for $\vec{p}$ into this equation for a "free" electron (i.e., $V[\vec{x}] = 0$), then

$$\hbar^2|\vec{k}|^2/2m = (\hbar^2/2m)(k_x^2 + k_y^2 + k_z^2) = \varepsilon_{total}. \tag{5.28}$$

This equation says that the energy of a free electron is related to the magnitude of its wave vector $\vec{k}$ and is called the *dispersion relation* for matter waves.

In section 7.7, on E&M waves (i.e., nonstatic or time-dependent variations of electric and magnetic waves), we will see that the electric field intensity and magnetic field intensity in *charge-free* regions of space obey time-dependent vector wave equations:

$$\nabla^2 \vec{E} = \frac{1}{c^2}\frac{\partial^2 \vec{E}}{\partial t^2} \quad \text{or} \quad \nabla^2 \vec{H} = \frac{1}{c^2}\frac{\partial^2 \vec{H}}{\partial t^2}, \tag{5.29}$$

where the quantity c is called the speed of the wave in the medium. In Cartesian coordinates, each of these equations represents three scalar equations for each component of the electric field intensity or magnetic field intensity. If we say that the quantity $\psi(\vec{x}, t)$ can represent any of the Cartesian components of the electric or magnetic field intensity, we can write the general form of the wave equation as

$$\nabla^2 \psi(\vec{x}, t) = \frac{1}{c^2}\frac{\partial^2 \psi(\vec{x}, t)}{\partial t^2}. \tag{5.30}$$

This equation may be interpreted as saying that each of the Cartesian components of the electric field intensity and magnetic field intensity behave as waves because each satisfy the wave equation. As we shall see, the solution to the wave equation can be assumed to consist of a spatial quantity, $\hat{\psi}(\vec{x})$, times a time quantity, $e^{j\omega t}$, where ω is called the angular frequency of the wave measured in rad/s.

This is a method of solving second-order partial differential equations known as the separation of variables technique, described previously in section 4.2. If we make this assumption, we can take the second derivative with respect to time on the right-hand side of the equation to see that

$$\nabla^2 \hat{\psi}(\vec{x}) e^{j\omega t} = -(\omega^2/c^2)\hat{\psi}(\vec{x}) e^{j\omega t} \tag{5.31}$$

and canceling out the time-dependent terms,

$$\nabla^2 \hat{\psi}(\vec{x}) + (\omega^2/c^2)\hat{\psi}(\vec{x}) = 0. \tag{5.32}$$

The latter equation is called the vector Helmholtz equation and describes the *spatial* variation of either electric field intensity or magnetic field intensity. Electrical engineers call the quantity, $\psi(\vec{x})$, the *phasor* descriptor for the wave. We often set the ratio $\omega^2/c^2 = k^2$ because it is just a constant property of the wave (called the wave number) determined by the frequency and the speed of the wave in a medium. For a wave, we can also show that $k = 2\pi/\lambda$, where λ is called the wavelength. We can also interpret $k^2 = |\vec{k}|^2 = \vec{k} \cdot \vec{k}$, where the vector, $\vec{k}$, is called the wave vector because it describes the direction of propagation of the wave as $\hat{\psi}_{\vec{k}}(\vec{x}) = \psi_0 e^{j\vec{k}\cdot\vec{x}}$ in *charge-free* space. We can use the de Broglie formulation above, $\varepsilon = \hbar\omega$ and $\vec{p} = \hbar\vec{k}$, to reinterpret electric and magnetic field intensity waves as having both energy and momentum as if they were particles.

Suppose we adopt the duality interpretation of matter by saying that electrons may be considered as waves instead of particles. If that is the case, electrons should also satisfy a wave equation, and the uniform plane wave that describes them should be of the form

$$\psi_{\vec{k},\, electrons}(\vec{x}, t) = \psi_0 e^{i(\vec{k}\cdot\vec{x} - \omega t)} \tag{5.33}$$

that satisfies the *dispersion relation.*[2] From this point on, we will be talking about the wave description of electrons, so we will drop the subscript electrons and the carrot on the quantity ψ with the understanding that it describes the phasor property of electrons.

Schrödinger noticed the similarity between these equations and saw that the interpretation

$$-\left(\hbar^2/2m\right)\nabla^2\psi_{\vec{k}} = \left(\hbar^2/2m\right)\left|\vec{k}\right|^2\psi_{\vec{k}} \tag{5.34}$$

in Equation 5.28 led to an expression that described the evolution of free wave-particles in space and time that looked like the Helmholtz equation. Schrödinger wrote this equation as

$$-\frac{\hbar^2}{2m}\nabla^2\psi_{\vec{k}} = -\frac{\hbar^2}{2m}\left(\frac{\partial^2}{\partial x^2}+\frac{\partial^2}{\partial y^2}+\frac{\partial^2}{\partial z^2}\right)\psi_{\vec{k}}(\vec{x}) = \frac{\hbar^2 k^2}{2m}\psi_{\vec{k}}(\vec{x}) = \varepsilon_{\vec{k}}\psi_{\vec{k}}(\vec{x}). \tag{5.35}$$

In this equation, we have interpreted the total energy of the "free" electrons as

$$\varepsilon_{total} = \varepsilon_{\vec{k}} = \hbar^2 k^2/2m = \hbar^2\vec{k}\cdot\vec{k}/2m. \tag{5.36}$$

Schrödinger's equation is a second-order, first-degree, homogeneous PDE. If it is to meet the periodic boundary conditions, the answer must be periodic in x, y, z with period L, so

$$\psi_{\vec{k}}(x+L, y, z) = \psi_{\vec{k}}(x, y, z) \tag{5.37}$$

and similarly for y and z. By using a normalization factor over the volume $V = L^3$ so there will be one electron found in each unit cell of volume V, Equation 5.37 yields the same answer as Equation 5.34 if

$$\psi_{\vec{k}}(\vec{x}, t) = \psi_0 e^{i\left(\vec{k}\cdot\vec{x}-\omega t\right)} = e^{i\left(\vec{k}\cdot\vec{x}-\omega t\right)}/\sqrt{V} \tag{5.38}$$

and the values of k are restricted to

$$k_x = 0;\ k_x = \pm 2\pi/L;\ k_x = \pm 4\pi/L; \ldots\ k_x = n_x\, 2\pi/L \quad \text{where } n_x = \text{integer} \tag{5.39}$$

and similarly for k_y and k_z; that is, any component of $\vec{k}$ is of the form $2n\pi/L$, where n is a positive or negative integer. The components of $\vec{k}$ are the quantum numbers for the problem, along with the quantum number m_s for the spin direction.

Pauli Exclusion Principle

No two fermions (half-integer spin particles, e.g., electrons) can have the same quantum numbers. By Equation 5.32, this also means that no two electrons can have

[2] Note that we are using the physics notation that $i = \sqrt{-1}$ rather than the electrical engineering convention of j so the equations will look like those in physics journals. Physicists conventionally use $e^{-i\omega t}$ for time dependence, while electrical engineers use $e^{j\omega t}$.

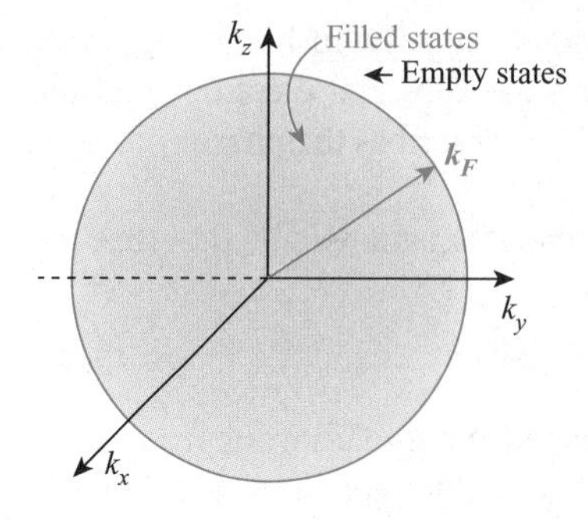

Figure 5.7 At temperature $T = 0$, N-free electrons in the lowest occupied energy states fill up the available energy states from the zero of energy to the highest possible energy (the Fermi energy).

the same wave function (they cannot exist at the same place at the same time with the same spin).

We have now found the wave function for the dispersion equation (equation of state) of free electrons in a box of volume $V = L^3$. There are some very important constraints on the wave function that we must apply, one of which is

$$\varepsilon_{\bar{k}} = \frac{\hbar^2 k^2}{2m} = \frac{\hbar^2}{2m}\left(k_x^2 + k_y^2 + k_z^2\right) = \frac{\hbar^2}{2m}\left(n_x^2 + n_y^2 + n_z^2\right)\frac{4\pi^2}{L^2}. \tag{5.40}$$

Equation 5.40 says that there are only discrete energy states that the "free" electrons can occupy (i.e., the electron energy states are quantized). There are only two electrons in each of these energy states (one with spin-up and one with spin-down). Furthermore, if we want to put N electrons into the box of volume $V = L^3$, we can add them two at a time, with the first two going into the lowest energy state ($n_x = 0$, $n_y = 0$, $n_z = 0$). The next six will go into the next highest energy states (one of the $n_i = 1$ and the others $= 0$), and so forth, until we have put all N electrons into the available lowest energy states. When the N electrons have all been added to the box, they will fill up all of the available (discrete) values of wave vectors, as shown in Figure 5.7. Although there are only specific points in k-space that can be occupied, they are so close together that they appear to be a continuum of states.

The highest filled energy states are at energy $\varepsilon_F = \hbar^2 k_F^2/2m$. In the sphere of volume $4\pi k_F^3/3$, the total allowed number of states is $N/2$. Thus,

$$\left(4\pi k_F^3/3\right)/\left(2\pi/L\right)^3 = Vk_F^3/6\pi^2 = N/2. \tag{5.41}$$

We can solve Equation 5.41 to get

$$k_F = \left(3\pi^2 N/V\right)^{1/3}, \tag{5.42}$$

which depends only on the particle concentration (and not the mass). Thus, the Fermi energy is

$$\varepsilon_F = \left(\hbar^2/2m\right)\left(3\pi^2 N/V\right)^{2/3} \tag{5.43}$$

Table 5.2 Fermi Surface Parameters for Electrons in Monovalent Metals

Material	Electron concentration, N/V ($\times 10^{28}$ m^{-3})	Fermi wavevector, k_F ($\times 10^{10}$ m^{-1})	Fermi velocity, v_F ($\times 10^6$ m/s)	Fermi energy, ε_F (eV)
Li	4.6	1.1	1.3	4.7
Na	2.5	0.90	1.1	3.1
K	1.34	0.73	0.85	2.1
Rb	1.08	0.68	0.79	1.8
Cs	0.86	0.63	0.73	1.5
Cu	8.50	1.35	1.56	7.0
Ag	5.76	1.19	1.38	5.5
Au	5.90	1.20	1.39	5.5

and the electron velocity, v_F, for electrons at the Fermi surface is

$$v_F = \hbar k_F / m = (\hbar/m)\left(3\pi^2 N/V\right)^{1/3}. \tag{5.44}$$

Calculated values for v_F and ε_F for a variety of monovalent metals are given in Table 5.2.

Effects of Temperature

If the free electron gas interacts with atoms that are above the absolute zero of temperature, it comes into thermal equilibrium with the atoms. Electrons that have energy below the Fermi energy are in a filled quantum state with specific values of n_x, n_y, n_z. A nearby energy state (e.g., $n'_x = n_x$, $n'_y = n_y$, $n'_z = n_z + 1$) will be filled, so the addition of thermal energy to the electron at n_x, n_y, n_z cannot make it jump to the energy state n'_x, n'_y, n'_z because the higher energy state is already filled and the Pauli exclusion principle does not permit two electrons to have the same quantum numbers. However, electrons just below the Fermi level can absorb thermal energy and move to a higher (unoccupied) state with energy above the Fermi energy without violating the Pauli exclusion principle. The Fermi–Dirac distribution

$$f(\varepsilon) = \left[e^{(\varepsilon-\mu)/k_B T} + 1\right]^{-1} \tag{5.45}$$

gives the probability that the energy level at energy ε will be populated.[i] The quantity μ is called the *chemical potential* and is itself a slight function of temperature chosen in such a way that the integral under the curve remains constant, but, at all temperatures, $f(\varepsilon)$ is equal to ½ when $\varepsilon = \mu$. k_B is the Boltzmann constant ($k_B = 1.3806503[24] \times 10^{-23}$ J/K) we reviewed in Chapter 1 that expresses energy as a temperature. When this probability function is multiplied by the Fermi distribution of states at $T = 0$, as shown in Figure 5.7, the product describes the filled and unfilled states. It can be interpreted that some states below the Fermi level have a probability of being

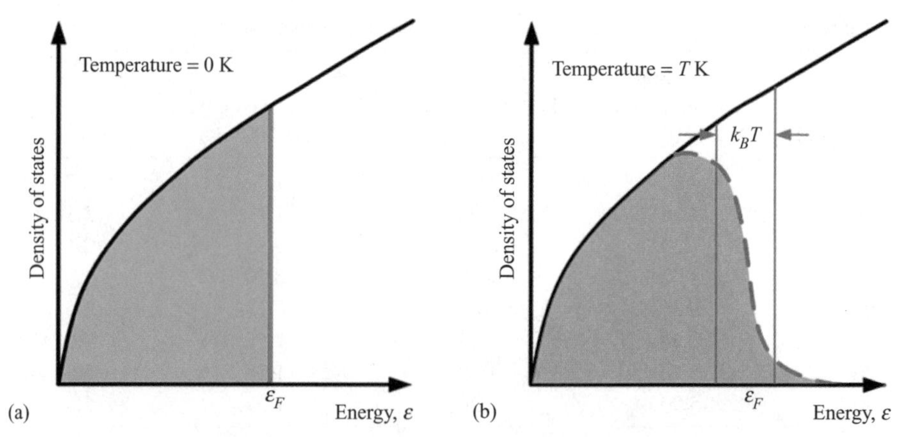

Figure 5.8 Density of available conduction electron states as a function of energy (parabolic solid line). The shaded red area under it shows the occupied states at (a) $T = 0$ and (b) for general temperature T.

unoccupied and that some states above the Fermi level have a probability of being occupied. In a sense, multiplying the Fermi distribution of Figure 5.7 by the Fermi–Dirac distribution 5.45 makes the Fermi sphere of occupied states become fuzzy. This effect can be shown quantitatively by calculating the number of permitted conduction electron states per unit energy range,

$$D(\varepsilon) = dN/d\varepsilon = \frac{d}{d\varepsilon}\left[\left(V/3\pi^2\right)\left(2m\varepsilon/\hbar^2\right)^{3/2}\right] = \left(V/2\pi^2\right)\left(2m/\hbar^2\right)^{3/2}\varepsilon^{1/2} \tag{5.46}$$

called the Density of States, as is shown in Figure 5.8.

For $T = 0$, only states below ε_F are occupied. The curve (b) to the right is the parabola multiplied by the Fermi–Dirac function, the probability that the state is full at that energy (shown as the blue dashed line). The area under the blue dashed line shows the occupied states for a finite temperature T (i.e., some of the states below ε_F are empty, and some of the states above ε_F are occupied).

The high-energy tail of the distribution, when $(\varepsilon - \mu) >> k_BT$, is called the Boltzmann or Maxwell distribution, and, when multiplied by the Fermi distribution of states at $T = 0$, as shown in Figure 5.8, it gives the probability of a filled energy state above the Fermi energy as an exponentially decreasing (but finite) population

$$f(\varepsilon) = e^{(\mu-\varepsilon)/k_BT}. \tag{5.47}$$

This result can be interpreted as showing how the proportion of filled states below the $T = 0$ Fermi energy can absorb thermal energy at temperature T to move to a higher energy state above the Fermi energy. In this sense, the Fermi sphere of occupied states at $T = 0$ becomes a fuzzy Fermi sphere of occupied states at temperature T.

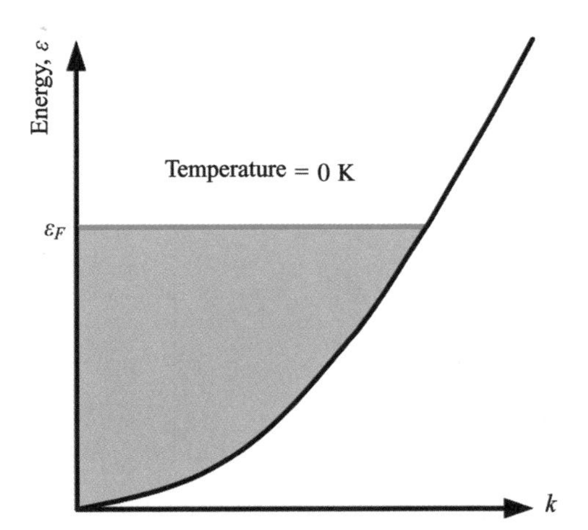

Figure 5.9 Free electron gas energy states allowed at temperature $T = 0$ K. The pink energy band below ε_F indicates that those states are populated. The white energy band above ε_F indicates that those states are vacant.

Figure 5.8 is often shown in a rotated orientation in which the filled states are shown in energy, which we showed in Equation 5.40 $\varepsilon_{\vec{k}} = \hbar k^2/2m$. This is plotted in Figure 5.9.

5.5 BAND THEORY

The free electron model gives good results for the electrical conductivity of metals and also gives good results for thermal conductivity, heat capacity, magnetic susceptibility and electrodynamics of metals. However, the model fails to give a satisfactory explanation of the electrical properties of semiconductors and insulators and cannot explain the large temperature dependence of the conductivity of semiconductors like germanium and silicon, as is shown in Figure 5.10.

For example, the probability distribution of Equation 5.41 can be approximated by a Taylor series in $1/k_BT$ (shown in Figure 5.8). As the temperature increases, the number of filled permitted electron states near the Fermi level with an adjacent empty permitted electron states grows, increasing the conductivity of the material and suggesting no qualitative difference between metals, semimetals, and insulators. Comparing the temperature dependence of Figure 5.4 for a metal with Figure 5.10 for a semiconductor, we see that there is, in fact, a very different mechanism at work between these materials.

The explanation of these fundamental differences lies in the crude approximation that was made for conduction electrons as constituting a "free electron gas." For example, in assuming a "free electron gas," we knew that we were neglecting the interaction of the conduction electrons (outer electrons) with the potential atomic core potentials in a lattice, as shown in Figure 5.11.

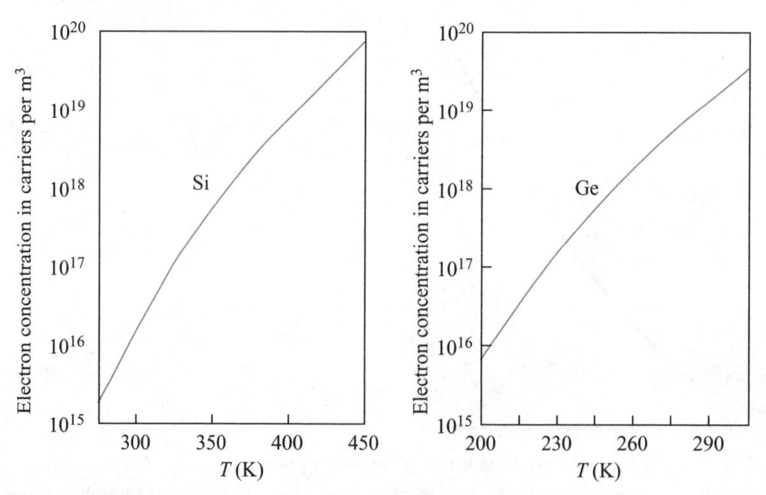

Figure 5.10 Conduction electron concentration for silicon and germanium as a function of temperature.

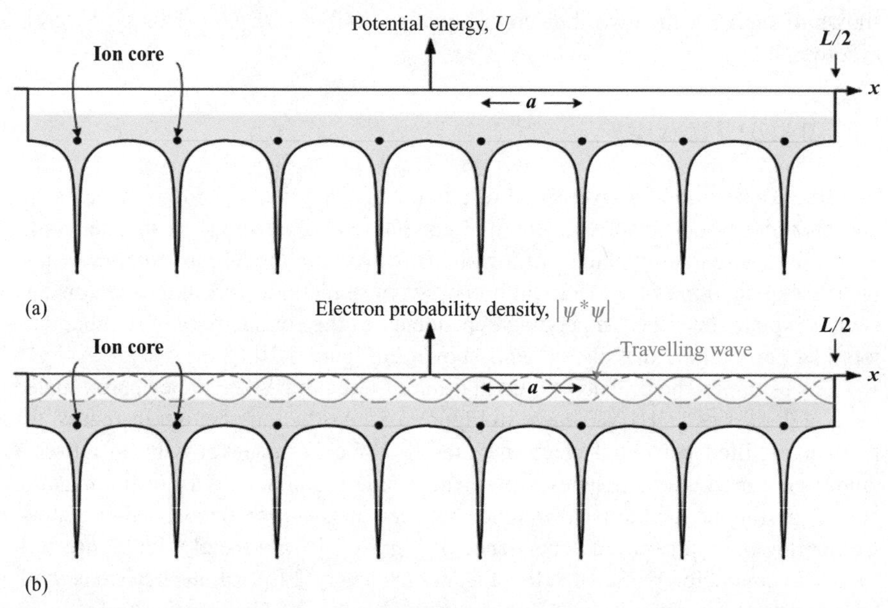

Figure 5.11 (a) Potential energy function surrounding the atomic cores in a material lattice; (b) interaction of a propagating conduction electron ($\lambda = a$) with the atomic cores.

Because the atomic cores of the atoms are effectively screened by the other conduction electrons, it was argued in the early part of the twentieth century that the effective potentials at the atomic cores were of very short range (compared with the atomic lattice dimensions) and that they were shallow enough to ignore. That approximation becomes questionable when the periodic structure of the material permits the constructive interference between waves that are scattered from many atomic cores. This is especially important when the wave vector of a conduction electron produces a wavelength, $\lambda = na$, that is an integer multiple of the lattice spacing a; that is, we can expect problems with the free electron model when

$$\left|\vec{k}\right| a = n\pi. \tag{5.48}$$

Except for a simple cubic lattice, there are several wave vector directions in k-space for which multiple integer of π condition 5.48 can occur. These directions are called the Bragg reflection directions for electrons in crystals and are the subject of a much deeper discussion than is possible in this text. We will note that, for those k values that satisfy Equation 5.48, the energy versus k curves is modified, as is shown in Figure 5.12. Here, we see that the permitted values of k are relatively continuous (except for the fact that they are very densely quantized) but that there are values of

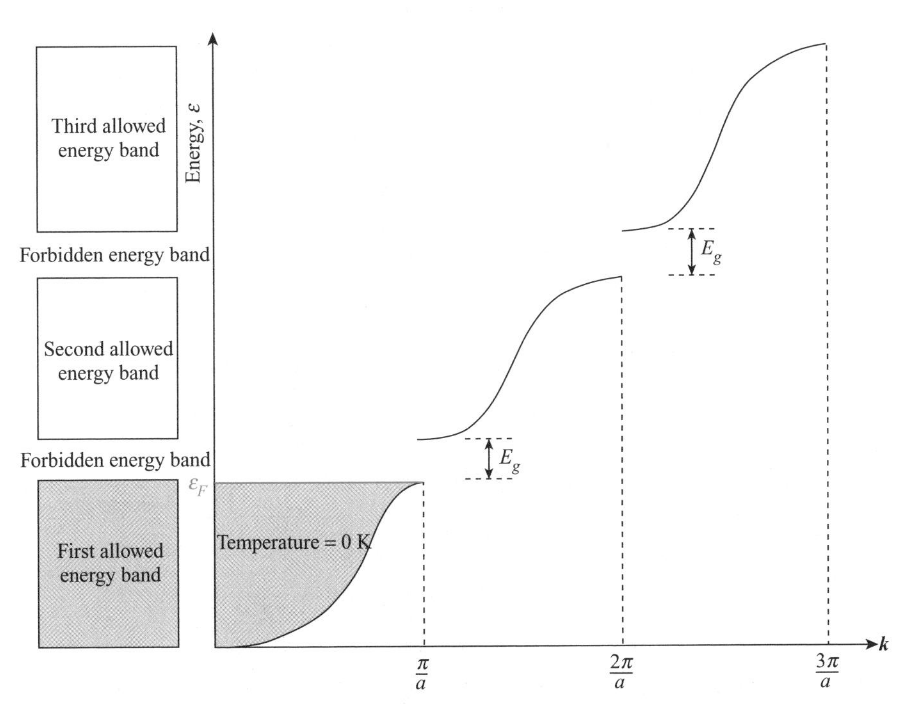

Figure 5.12 Plot of energy versus wave vector magnitude, k, for a linear lattice of square well potentials with lattice spacing a.

energy that are forbidden when condition 5.48 is met. This view is called the extended zone scheme for square well potentials.

Some texts show only the rectangular boxes in Figure 5.12 (allowed energy bands) to the left of the ε versus k solutions as they fill all or part of the lowest boxes to the level ε_F with a background shade to indicate the electron filling of the box to the Fermi energy at temperature $T = 0$. As one would expect, the details of the ε versus k solutions will change with the kind of well potential for the core ions, but the location of the forbidden energy bands continues to satisfy condition 5.48. The separation between the allowed energy bands is usually called an *energy gap*, E_g.

We know that, at temperature T, the population of allowed states near the Fermi energy will become fuzzy as some of the states below ε_F become vacant and some of the states above ε_F become populated. The width or spread of the distribution will be proportional to k_BT. However, if there are no allowed states above the Fermi energy and the temperature is insufficient to raise a populated state to the bottom of the next highest allowed energy band, there will be only a statistically low probability that any electrons will be promoted to a higher band. Thus, the density of states at the Fermi level will be very small, and the conductivity of the material will be very low. This will be the case for an insulator, as shown in the left boxes in Figure 5.13.

If the Fermi energy falls within an allowed energy band, and the temperature is T, there can be a substantial spreading (fuzziness) of the populated states near the Fermi level, and hence there can be many electrons ready to move under the influence of an external applied electric field. We say that the density of states is high in that case, and the conductivity of the material will be relatively high. This will be the case for a metal, as shown in the second set of boxes in Figure 5.13. Here,

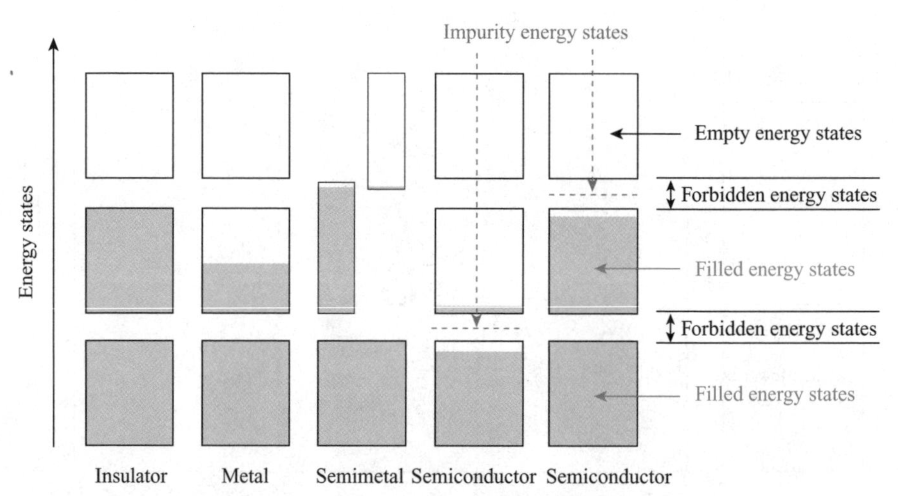

Figure 5.13 Ranges of available energy states for conduction electrons in various classes of materials. Energies within the boxes are bands that are allowed, but energies between the boxes are forbidden. The forbidden energies are said to constitute band gaps.

the allowed band is said to be half-full as would be the case for copper in which two 4 s electrons are required to fill the lowest allowed band, but only one 4 s electron is available.

For a particular crystal structure of the core ions, the details of the ε versus k solutions will also depend upon the direction of the vector $\vec{k}$. Solutions for a particular value of $\vec{k}$ might have an allowed first energy band that extends above the bottom of the second allowed energy band in a different direction. In that case, care would have to be taken when measuring the energy bands for a single crystal of sample. For a polycrystalline sample (the normal situation for conductors on printed circuit boards (PCBs), we would have to produce an angular average over all possible angles for single crystals. This average process can permit adjacent bands to overlap one another in the energy variable, as shown in the case of a semimetal, in Figure 5.13 in which the bands have been folded back in a "reduced zone scheme."

Semiconductors can be even more complicated if the material is composed of a compound material or if the pure material has had impurity atoms inserted in a pure crystal in small quantities. In this case, the temperature may be able to raise the energy of populated states just below the filled band to the next allowed band by making smaller jumps to allowed impurity levels that reside between the bands. This case is shown in the next to last set of boxes in Figure 5.13, but the small number of impurity levels between the allowed bands is not shown. Finally, some semiconductors are produced by "draining off" the highest energy electrons just below the Fermi energy at the top of an otherwise full energy band. The process to drain off the highest energy electrons can be accomplished by a different nearby material, by an applied electric potential, or both. Such a situation is shown in the set of boxes on the right-hand side of Figure 5.13.

Summary

The shaded areas of Figure 5.13 indicate that energy states that are filled with conduction electrons and the white areas are empty. States at the upper energy level (the Fermi Energy) can move to an empty adjacent state with an incremental addition of energy (such as that provided by the temperature of the material). In an insulator, a large amount of energy, E_g, is required to overcome the band gap to boost a conduction electron into a permitted state, so Equation 5.45 would provide only a small fraction of those states, and the material would be a poor electrical conductor. By comparison, a metal with a half-filled band has electrons at the Fermi level that are easily moved to the next highest empty level, and the density of those states given in Equation 5.46 is large, so the material is a very good electrical conductor. A semimetal may have two energy bands that are partially filled in different momentum directions relative to the crystal structure, so a small amount of energy is needed to scatter the electrons between the bands, but the density of available states may can also be small, resulting in a conductivity that is intermediate between metals and insulators. Semiconductors are so influenced by their impurities that they require a separate discussion.

5.6 EQUATION OF CONTINUITY

Conservation of charge is a fundamental postulate that, to our knowledge, has never been seen to be violated. Simply stated, electric charge may not be created or destroyed. Unless there is a source of charge inside a volume ΔV, indicated schematically in Figure 2.17, the total charge that enters the volume from any direction is the same as the charge that leaves.

However, this is a statement that must be evaluated over a long period of time because a net charge can build up inside the volume in a short time Δt. We have examined this principle previously as a specific example of the divergence theorem.

We concluded that, because charge, q, is conserved in any volume element of space,

$$\partial \rho_V / \partial t + \vec{\nabla} \cdot \vec{J} = 0 \quad \text{(Continuity Equation)}, \tag{2.58}$$

where we have defined $\rho_V = \lim_{\Delta x, \Delta y, \Delta z \to 0} q_{in\ \Delta x \Delta y \Delta z} / \Delta x \Delta y \Delta z$ as the volume charge density at x, y, z. In Chapter 3, we further derived the point form of Gauss's law to be

$$\vec{\nabla} \cdot \vec{E} = (Q/V)/\varepsilon_0 = \rho_V/\varepsilon_0. \tag{3.6}$$

So, if we apply Equation 5.13, $\vec{J} = \sigma \vec{E}$, to Equation 3.6, we get

$$\vec{\nabla} \cdot (\vec{J}/\sigma) = \rho_V/\varepsilon_0 \tag{5.49}$$

in which case Equation 2.58 becomes

$$\partial \rho_V / \partial t + \sigma \rho_V / \varepsilon_0 = 0 \tag{5.50}$$

Equation 5.50 is a first-order, linear, homogeneous, ordinary differential equation with solution

$$\rho_V(t) = \rho_V(0) e^{-\sigma t/\varepsilon_0}, \tag{5.51}$$

where both ρ_V(t) and ρ_V(0) can be functions of the space coordinates.

EXAMPLE

5.1 Use Equation 5.51 to find the relaxation time for free charge in a conductor like copper.

SOLUTION We can rewrite Equation 5.51 in the form $\rho_V(t) = \rho_V(0)e^{-t/\tau}$, where $\tau = \varepsilon_0/\sigma$ and use the value of conductivity from Table 5.1, $1/\sigma(293\text{ K}) = 1.69 \times 10^{-8}\ \Omega\text{m}$ and $\varepsilon_0 = 8.85 \times 10^{-12}$ F/m to find $\tau_{\text{copper}} = 1.5 \times 10^{-19}$ s. Thus, we conclude that, in the time 1.5×10^{-19} s, the conduction electrons in copper rearrange themselves to reduce externally applied electric field intensity to 1/e (36.8%) of its original value. This time is so short; we can state that conduction electrons inside a good conductor will move to reduce externally applied electric field intensity to zero for practical times; good conductors will behave like static conductors (with zero internal electric field intensity) unless times are extremely short or frequencies are very high (>10^{18} Hz). This is called the "quasistatic approximation."

PROBLEMS

5.1 Find the relaxation time for a good insulator like fused quartz at room temperature. What does this imply about static charge that is deposited on a good insulator?

5.2 Find the relaxation time for *n*-type and *p*-type doped silicon at room temperature. What does this imply about the ultimate speed of a transistor made of these materials?

5.7 MICROSCOPIC VIEW OF OHM'S LAW

Electrical resistivity in metals can arise from scattering of the conduction electrons from impurities or vacancies in the lattice or from deformations caused by grain boundaries as shown schematically for a two-dimensional lattice in Figure 5.14.

Note that the nearest neighbor atoms to the vacancies (lower left) or to the impurities (red, upper right) have been relocated to account for a local lattice strain. Of course, second nearest neighbors or further could also shift ± from their periodic locations. We could approximately account for these deviations in a periodic potential function, $V(\vec{r}_i)$, in Equation 5.32 by adding a differential potential, $V'(\vec{r})$, to the periodic potential. Thus, to first-order, Equation 5.27 could be written as

$$-(\hbar^2/2m)\nabla^2\psi_{\vec{k}} + \sum_i V(\vec{r}_i)\psi_{\vec{k}}(\vec{r}_i) + V'(\vec{r})\psi_{\vec{k}}(\vec{r}) = \varepsilon_{\vec{k}}\psi_{\vec{k}}(\vec{r}). \tag{5.52}$$

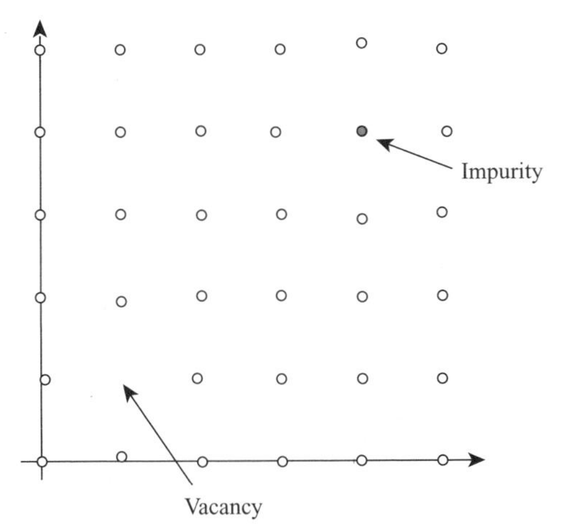

Figure 5.14 Two-dimensional view of potentials caused by a periodic metal lattice showing the deviations from simple periodicity caused by impurities or vacancies in the lattice.

An approximate solution to Equation 5.52 can be found by first solving Equation 5.52 with $V'(\vec{r}) = 0$ to find the solution to a pure metal as in section 5.6 above and then using those solutions for $\psi^0_{\vec{k}}(\vec{r})$ to measure their scattering from the vacancy or impurity (including distortions from periodicity by nearest neighbors). Such a technique would produce a first-order approximate solution called the first Born approximation:

$$\psi_{\vec{k}}(\vec{r}) \approx \psi^0_{\vec{k}}(\vec{r}) + \psi^{Scatt}_{\vec{k}}(\vec{r}). \tag{5.53}$$

One popular technique in solving such problems is then putting the first-order approximate solution back into Euqation 5.52 and recalculating a new scattered wave function in a second-order approximation called the second Born approximation. The process may be repeated until one computes a limiting answer that differs by an arbitrary amount (say, 1%) from the previous one. Such techniques are popular when solving Equation 5.52 computationally. But the students must be warned that, while the process may yield a convergent solution, they should ask, "Is the convergent solution the correct solution?"

Electrical resistivity in metals can also arise from scattering of the conduction electrons from lattice vibrations as shown schematically for a two-dimensional lattice in Figure 5.15.

The permanent alterations from the periodic potential shown schematically in a one-dimensional lattice in Figure 5.9 clearly depend on the sample temperature, and the lattice deformations will change with time as the atom structure vibrates. Deformations may be along the *x*-, *y*-, or *z*-axis or along any other axis of the crystalline structure, and each of the deformations will have a characteristic frequency of oscillation that depends on the direction of deformation because of the coupling

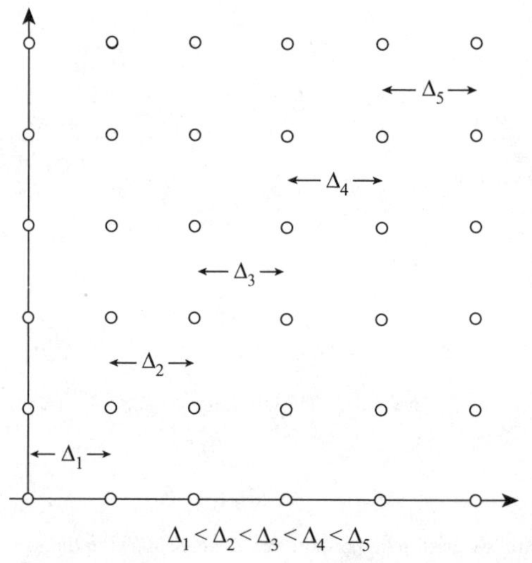

Figure 5.15 Two-dimensional view of potentials caused by a periodic metal lattice showing the deviations from simple periodicity (along the horizontal axis) caused by thermal vibrations.

interactions among the atoms. Such periodic time variations of the lattice are called *phonons* because they are similar in nature to vibrations caused by acoustic vibrations and represent distortion waves among the periodic atomic structures. Of course, at a finite temperature, we would expect all possible vibrational modes (in arbitrary directions $\vec{k}$) to be present and that the amplitude of a particular mode to depend on the energy of that mode in accordance with the Fermi–Dirac population statistics of Equation 5.26, $f(\varepsilon) = [e^{(\varepsilon-\mu)/k_BT} + 1]^{-1}$. The assemblage of all such phonon modes of oscillation is called a *phonon gas* and represents the ensemble of all of the thermal vibrations of atoms. In the symmetry of nature as postulated by Einstein and DeBroglie, we can also think of the wavelike vibrations of the atoms in terms of phonon particles that are moving with momentum and energy through the lattice. We can see that these phonon particles would also be influenced by impurities and vacancies and might themselves be scattered just like the electrons in a conduction electron gas.

The solution of Schrödinger's equation, for these periodic variations in potential is the subject of a solid state physics course and is complicated enough to defer to alternate consideration. We can, however, consider the scattering process of electrons and phonons through a classical argument that explains the wide variations in the conduction electron velocities at the Fermi level and the velocity caused by the drift in a charge current as discussed in the following sections.

Electron Scattering from Impurities, Vacancies, and Lattice Vibrations

We have found in Equation 5.8 that conduction electrons have a drift velocity $\vec{u} = -\mu_e\vec{E}$ under the application of an electric field by an external potential, and, for copper, the electron mobility is $\mu_e = 3.2 \times 10^{-3}$ (m²/Vs). For an internal field of 0.5 V/m, this would result in a drift velocity of $|\vec{u}|_{drift} = 1.6 \times 10^{-3}$ m/s.

PROBLEM

5.3 Calculate the current caused by an electric field intensity of 0.5 V/m in a #12 copper wire.

We have also seen from Table 5.2 that the velocity of a conduction electron at the Fermi level for copper is $v_f = 1.56 \times 10^6$ m/s. While there are many electrons in a free electron gas with velocities below the Fermi velocity, the difference between electrons at the Fermi velocity in copper and the electron drift velocity for a field of 0.5 V/m is nine orders of magnitude! Classically, this enormous difference is explained by the theory of scattering in which conduction electrons travel many lattice locations before they encounter an impurity, a vacancy, or a lattice deformation caused by a phonon, as indicated in Figure 5.16.

In Figure 5.16, the presence of electric field intensity causes a gradual drift of the electron in the horizontal direction. Were this electric field intensity 0.5 V/m, in this figure, the drift velocity would be very small compared with the Fermi velocity, so the difference in the green and red trajectories would be very small.

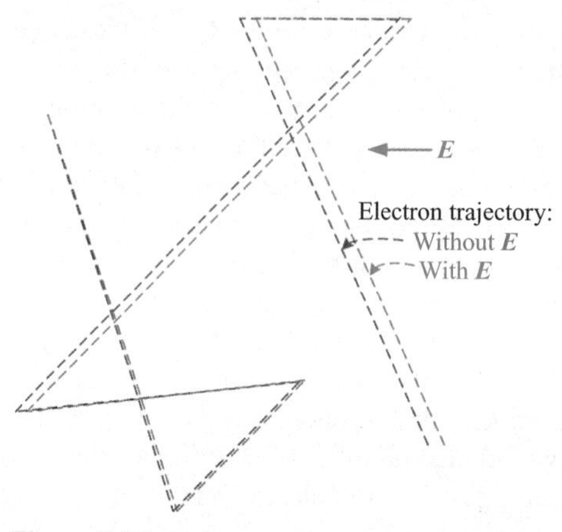

Figure 5.16 Trajectory of an electron as it moves in a conductor between a random sequence of scattering events (green line). The same electron trajectory is shown schematically as it moves between the same random scattering events in the presence of an external electric field intensity (red line).

Classical Numerical Analysis

An electron of mass, m_e, placed in an electric field intensity, $\vec{E}$, will experience an acceleration $\vec{a} = \vec{F}/m_e = -e\vec{E}/m_e$. In the average time, τ, between collisions with a scattering point, the electron will change direction randomly but will experience a drift speed of $u_e = |\vec{a}|\,\tau = (-e\vec{E}/m_e)\tau$. Because $J = eNu_e = (e^2N\tau/m_e)E = \sigma E$, we can solve for τ:

$$\tau = m_e\sigma/e^2N = m_e/e^2N\rho. \tag{5.54}$$

For copper, $1/\sigma = \rho = 1.6 \times 10^{-8}$ Ωm, $e = 1.6 \times 10^{-19}$ C, $m_e = 9.1 \times 10^{-31}$ kg/e, and $N = \dfrac{\text{density}}{\text{molecular mass}} = \dfrac{9.0\times10^3\ \text{kg/m}^3}{64\times10^{-3}\ \text{kg/mole}}6.023\times10^{23}\dfrac{e}{\text{mole}}$ (assuming there is one free e per copper atom), so we get $\tau = 2.5 \times 10^{-4}$ s as the average time between collisions. For electrons traveling at the Fermi velocity of 1.56×10^6 m/s, they will move $\lambda = v_F\tau$ or $\lambda = 3.87 \times 10^{-8}$ m = 387 Å in this time (about 150 times the distance between nearest-neighbor copper atoms).

CONCLUSION We can conclude from the above numbers that conduction electrons move very fast, collide with scattering points in a short time (but a long relative distance between scattering potentials), and change their velocity on the average in time as a result of the random scattering events (in the presence of external electric field intensity) by a very small amount of drift velocity.

5.8 POWER DISSIPATION AND JOULE'S LAW

Collisions of electrons with the conductor grain boundaries, lattice vibrations, impurities, vacancies, or interstitials result in thermal heating of the conductor. The work, Δw_i, carried out by electric field intensity, $\vec{E}$, on a conduction charge, q_i, in moving through a displacement, $\Delta\vec{l}_i$, is $\Delta w_i = q_i\vec{E}\cdot\Delta\vec{l}_i$, and the power (rate of work per unit time) is

$$P_i = \Delta w_i/\Delta t = q_i\vec{E}\cdot\Delta\vec{l}_i/\Delta t = q_i\vec{E}\cdot\vec{u}_i, \tag{5.55}$$

where $\vec{u}_i = d\vec{l}_i/dt = \lim_{\Delta t\to 0}\Delta\vec{l}_i/\Delta t$ is the drift velocity of the ith charge. The amount of power delivered into the volume $d\tau = S\Delta l$, as was shown in Figure 5.3.

The total power delivered to all the charge carriers in a volume $d\tau$ is

$$dP = \sum_i dP_i = \vec{E}\cdot\left(\sum_i N_i q_i \vec{u}_i\right)d\tau, \tag{5.56}$$

where N_i is the number of charges of type i per unit volume, which, by virtue of Equation 5.7, is

$$dP = \vec{E}\cdot\vec{J}dv \tag{5.57}$$

or

$$dP/dv = \vec{E}\cdot\vec{J}. \tag{5.58}$$

Thus, the point function $\vec{E}\cdot\vec{J}$ is a power density under steady-current conditions. For a given volume, V, the total electric power converted into heat is

$$P = \int_V \vec{E}\cdot\vec{J}dv\,(\mathrm{W}). \tag{5.59}$$

Equation 5.59 is known as Joule's law, but power is expressed in watts (not joules).

In a conductor of a constant cross section, ΔS, as shown in Figure 5.3, $dv = \Delta S dl$, with dl measured in the direction $\vec{J}$, Equation 5.59 can be written as

$$P = \underbrace{\int_L \vec{E}\cdot d\vec{l}}_{V}\ \underbrace{\int_S \vec{J}\cdot d\vec{s}}_{I} = VI \tag{5.60}$$

and because $V = RI$,

$$P = I^2R. \tag{5.61}$$

5.9 BOUNDARY CONDITION FOR CURRENT DENSITY

When current obliquely crosses an interface between media with different conductivities, the vector current density, $\vec{J}$, changes in both direction and magnitude. A set of boundary conditions can be derived for $\vec{J}$ from the above equations for current density, $\vec{J}$, in the absence of a source or sink of charges (Table 5.3): where the curl equation was obtained by combing Ohm's law $\vec{J} = \sigma\vec{E}$ with $\vec{\nabla}\times\vec{E} = 0$.

Table 5.3 Summary Current Density Equations

Differential form	Integral form
$\nabla \cdot \vec{J} = -\partial \rho_V / \partial t$ (Continuity equation)	$\oint_S \vec{J} \cdot d\vec{s} = -dQ_{inside\ S} / dt$
$\vec{\nabla} \cdot \vec{J} = 0$ (special case for steady currents)	$\oint_S \vec{J} \cdot d\vec{s} = 0$ ($\vec{J}$ is solenoidal for steady currents)
$\nabla \times (\vec{J}/\sigma) = 0$ (irrotational flow)	$\oint_C (\vec{J}/\sigma) \cdot d\vec{l} = 0$ (note $\vec{J}$ itself is not conservative)

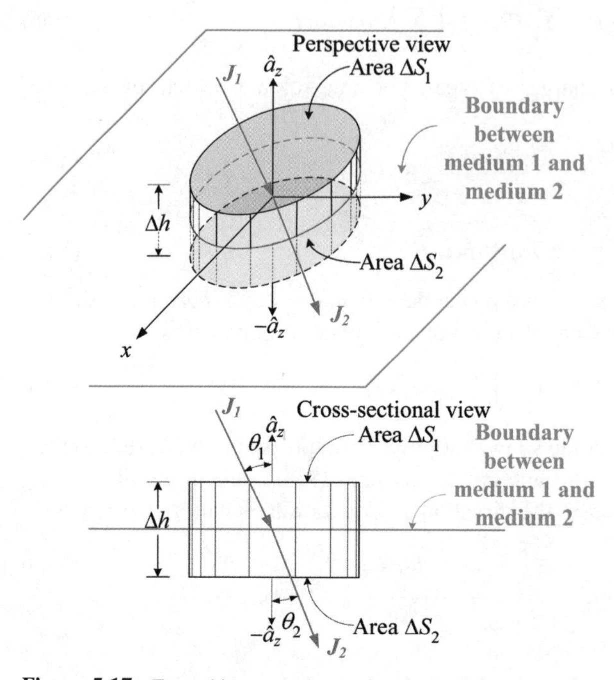

Figure 5.17 Two arbitrary conductors in a lossy dielectric medium.

We can use these equations to determine the current density change at the boundary between two media, 1 and 2, with conductivities, σ_1 and σ_2, respectively as is shown in Figure 5.18. From $\oint_S \vec{J} \cdot d\vec{s} = 0$, we can conclude

$$J_{1n} = J_{2n} \tag{5.62}$$

just as we did for Gauss's law. We can also conclude from $\oint_C (\vec{J}/\sigma) \cdot d\vec{l} = 0$ that

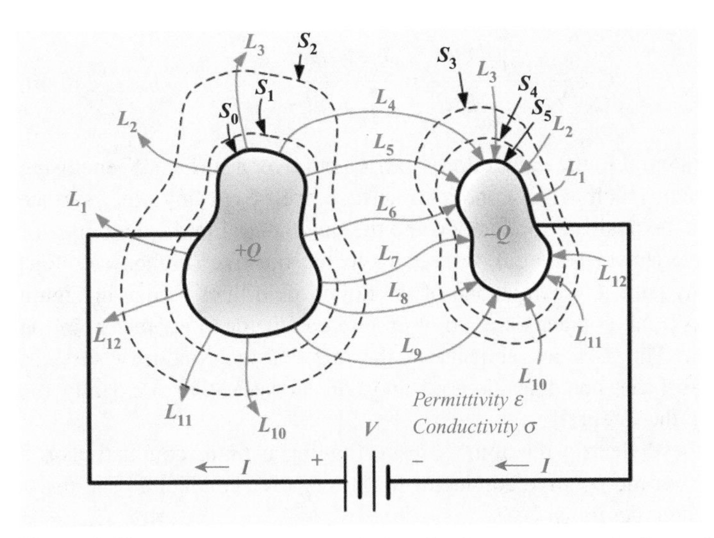

Figure 5.18 Boundary conditions at the interface between two conducting media (1 and 2).

$$J_{1t}/J_{2t} = \sigma_1/\sigma_2, \tag{5.63}$$

which states that the ratio of the tangential components of the current density at two sides of an interface is equal to the ratio of the conductivities in those two media.

Special Case

For a homogeneous conducting medium, with a constant value of σ everywhere, we can further deduce that

$$\vec{\nabla} \times \vec{J} = 0. \tag{5.64}$$

So, if we let $\vec{J} = -\vec{\nabla}\psi$, and substitute this into $\vec{\nabla} \cdot \vec{J} = 0$, we get Laplace's equation

$$\nabla^2 \psi = 0, \tag{5.65}$$

where Ψ is called the current density potential. A problem of steady-current flow in a homogeneous conducting medium can therefore be solved by determining solutions of Laplace's Equation 5.65.

5.10 RESISTANCE/CAPACITANCE CALCULATIONS

As discussed previously, the capacitance between two conductors separated by a dielectric medium can be expressed as

$$C = \frac{Q}{V} = \frac{\oint_S \vec{D} \cdot d\vec{s}}{-\int_L \vec{E} \cdot d\vec{l}} = \frac{\oint_S \varepsilon \vec{E} \cdot d\vec{s}}{-\int_L \vec{E} \cdot d\vec{l}}, \tag{5.66}$$

where the surface integral in the numerator is carried out over a surface, S, enclosing the positive conductor (such as the conductor surface itself S_0 or any other surface S_1, S_2, that encloses the positive conductor), and the line integral in the denominator is from the negative (lower-potential) conductor to the positive (higher-potential) conductor along any path, L (such as one of the dotted field lines shown in Figure 5.17, L_1, L_2, ... , L_{12}), that connects any point on one conductor to another point on the other conductor. This is a consequence of the fact that the conductor surfaces are equipotential surfaces and that electric fields are conservative (i.e., only the endpoints matter in the integral).

Likewise, if the dielectric medium is lossy (having a finite conductivity), a current will flow from the positive conductor to the negative conductor. The resistance between the conductors is

$$R = \frac{V}{I} = \frac{-\int_L \vec{E} \cdot d\vec{l}}{\oint_S \vec{J} \cdot d\vec{s}} = \frac{-\int_L \vec{E} \cdot d\vec{l}}{\oint_S \sigma \vec{E} \cdot d\vec{s}}, \tag{5.67}$$

where the line and surface integrals can be taken to be the same (L and S) as those chosen in Equation 5.66. Using Equations 5.66 and 5.67, we can thus see that

$$RC = \frac{-\int_L \vec{E} \cdot d\vec{l}}{\oint_S \sigma \vec{E} \cdot d\vec{s}} \cdot \frac{\oint \varepsilon \vec{E} \cdot d\vec{s}}{-\int_L \vec{E} \cdot d\vec{l}} = \frac{\varepsilon}{\sigma}. \tag{5.68}$$

CONCLUSION If we know ε and σ for a dielectric medium, we can compute R if we know C (or vice versa) from

$$RC = C/G = \varepsilon/\sigma. \tag{5.69}$$

ENDNOTE

i. Charles Kittel, *Introduction to Solid State Physics*, 7th ed. (Hoboken, NJ: John Wiley & Sons, 1996), 146.

Chapter 6

Static Magnetic Fields

LEARNING OBJECTIVES

- Develop the origins of the magnetic field intensity, $\vec{H}$, magnetic flux density, $\vec{B}$, magnetic vector potential, $\vec{A}$, magnetization, $\vec{M}$, and magnetic permeability, μ
- Define the Lorentz force equation and use it to develop the Biot–Savart law
- Show differential and integral forms of fundamental postulates that specify divergence and curl of magnetic flux density, $\vec{B}$
- Derive tangential and normal boundary conditions for $\vec{H}$ and $\vec{B}$
- Use magnetic monopole charges and currents to describe equivalent magnetic fields
- Develop definitions of inductance and show how to use them in examples
- Describe magnetic moments in terms of quantum numbers and populations

INTRODUCTION

We began the mathematical analysis of electric fields by examining fundamental force "laws" that were discovered experimentally by Coulomb and others. In a similar fashion, the mathematical analysis of magnetic fields follows from the experimental "laws" of magnetic forces. In both cases, we recognize our limited ability to measure the laws because we are restricted to make measurements in a macroscopic manner at the scale of humans. For example, we only infer the electric and magnetic forces that exist at a nucleus from applications of electromagnetic theory as it shifts energy levels, or we compute the electrostatic potential or magnetic fields of a neutron star by applying theory to extreme circumstances that prevent it from collapsing into a black hole; we cannot measure them directly. In addition, we idealize our experimental setups by modeling the conductors and insulators as perfectly flat, homogeneous materials when we know that this is not the case. For this reason, The Foundations of Signal Integrity[i] discusses the ideal model versus the real world, limits of experimental measurements, numerical simulation of complex boundary conditions, and finally perturbations of the theory to partially account for the microscopic variations in geometry. But Maxwell's equations are fundamentally based on two force laws, so we begin the study of magnetic fields with experimentally observed forces.

6.1 MAGNETIC FORCE

Electric Force Summary

Previously, we have learned that the following two fundamental equations form the basis for electrostatic fields:

$$\vec{\nabla}\cdot\vec{D}=\rho_V \tag{6.1}$$
$$\vec{\nabla}\times\vec{E}=0. \tag{6.2}$$

In the special case that the permittivity tensor is diagonal and all of the diagonal components are the same (i.e., ε is a scalar), the electric field intensity $\vec{E}$ and the electric flux density (or electric displacement) $\vec{D}$ are related by

$$\vec{D}=\varepsilon\vec{E}. \tag{6.3}$$

When a small test charge q is placed in an electric field, $\vec{E}$, it experiences an electric force $\vec{F}_e$, which is a function of q:

$$\vec{F}_e=q\vec{E}. \tag{6.4}$$

Magnetic Force Symmetry

When a test charge is in motion in a magnetic field characterized by a magnetic flux density, $\vec{B}$, experiments show that it experiences another force, $\vec{F}_m$, in which

- the magnitude of $\vec{F}_m$ is proportional to q; and
- the direction of $\vec{F}_m$ at any point is at right angles to both the velocity vector, $\vec{v}$, of the test charge and the normal component of the magnetic flux density, $\vec{B}$, at that point ($\vec{F}_m \propto \vec{v}\times\vec{B}$), as shown in Figure 6.1.

$\vec{F}_m$, the magnetic force, can be described by defining the magnetic flux density, $\vec{B}$, that specifies the direction and constant of proportionality in International System of Units (SI) as

$$\vec{F}_m=q\vec{v}\times\vec{B}, \tag{6.5}$$

where $\vec{v}$ (m/s) is the test particle velocity vector and $\vec{B}$ is measured in webers per square meter (Wb/m^2) or tesla (T). In Equation 6.5, choosing the constant of proportionality to be unity defines the magnitude of the magnetic flux density in SI units from measured quantities of force, charge, and velocity in SI units.

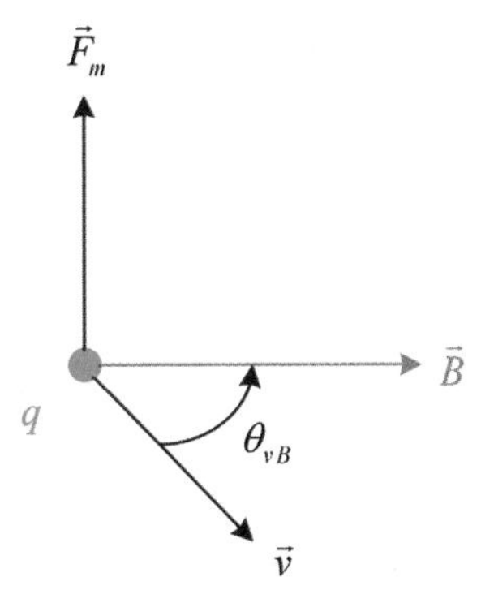

Figure 6.1 Direction of the force, $\vec{F}_m$, on a charged particle, q, with velocity, $\vec{v}$, due to a magnetic flux density, $\vec{B}$, at that point.

Combined Electric and Magnetic Forces

The *total* electromagnetic force on a charge due to electric and magnetic fields is

$$\vec{F} = q\left(\vec{E} + \vec{v} \times \vec{B}\right), \tag{6.6}$$

which is called the Lorentz force equation.

6.2 MAGNETOSTATICS IN FREE SPACE

Magneto*statics* means steady (static) magnetic fields (or as we shall see, magnetic fields produced by steady electric currents). Magneto*static* equations in free space that mirror the electro*static* Equations 6.1 and 6.2 are

$$\vec{\nabla} \cdot \vec{B} = 0 \tag{6.7}$$

$$\vec{\nabla} \times \vec{B} = \mu_0 \vec{J}, \tag{6.8}$$

where μ_0 is the magnetic permeability in the free space and

$$\mu_0 \equiv 4\pi \times 10^{-7}\ (\mathrm{H/m})\ \text{or}\ (\mathrm{Tm/A}). \tag{6.9}$$

From Equation 6.8, we can use the equation of continuity, $\vec{\nabla} \cdot \vec{J} = -\partial \rho_V / \partial t$, to deduce that, for steady charge densities,

$$\vec{\nabla} \cdot \vec{\nabla} \times \vec{B} = \vec{\nabla} \cdot \left(\mu_0 \vec{J}\right) = 0 \text{ for } \vec{\nabla} \cdot \vec{J} = 0. \tag{6.10}$$

Taking the volume integral of Equation 6.7 and applying the divergence theorem, we have

$$\iiint_V \vec{\nabla} \cdot \vec{B} dv = \oiint_S \vec{B} \cdot d\vec{s} = 0, \tag{6.11}$$

where the surface integral is carried out over the surface boundary, *S*, of the arbitrary volume, *V*. By comparison to Gauss's law for electric flux, Equation 6.11 indicates that there are no magnetic charges, q, to mirror the divergence equation for electrical fields and electric charge, q, and, thus, magnetic flux lines always close upon themselves; that is, there are no sources or sinks of magnetic flux.

Symmetry Note

If we choose to make Equations 6.7 and 6.8 completely symmetric to Equations 6.1 and 6.2, we would logically use the symmetry shown in Table 6.1.

These equations would be mathematically more satisfying and would lead to a more symmetric form of Maxwell's equations. However, the symmetry in the divergence equations implies there is a magnetic charge density per unit volume, ρ_v, that mirrors the electric charge density, ρ_v, and the symmetry in the curl equations implies that there is a magnetic charge current density, $-\vec{J}$, that mirrors the electric charge current density, $\vec{J}$. In terms of images of bare electric charges (which we called *electric monopoles*), the divergence equation thus implies that there is a class of magnetic particles called *magnetic monopoles*. Note that we have chosen the color to mirror the time derivative of magnetic charge, as in the case of electric change. To date, no one has detected the existence of isolated magnetic monopoles although there have been many searches[ii] for them. If there are no isolated magnetic monopoles, then it would be difficult to have a current density of magnetic monopoles, so we would conclude that both $\rho_v = 0$ and $\vec{J} = 0$. Some electrical engineering texts (e.g., Balanis and Harrington) prefer to use the symmetric form of the equations because they lead to a simpler set of combined partial differential equations that can be more easily solved. Then, after the solution, the fact that $\rho_v = 0$ can be imposed on the solution.

As we shall see in section 6.7, it is possible to use the concept of magnetic dipoles aligned in a magnetic field to imagine that an excess of north charges appears on one surface of a material and that an excess of south charges appears on another surface of a material. Mathematically, this is equivalent to saying that virtual magnetic north poles exist on one surface, and virtual magnetic south poles exist on the other surface. In a macroscopic sense, we do not need to insist that magnetic monopoles do not exist and, in fact, can permit their existence because they yield an equivalent electromagnetic field distribution for a permanent magnet.

If the virtual north and south poles can move, they can produce virtual magnetic monopole current densities. Furthermore, as an electromagnetic wave propagates

Table 6.1 Symmetric Forms of the Divergence and Curl Equations for Electro*static* and Magneto*static* Fields

	Electro*static* equations	Magneto*static* equations
Divergence	$\vec{\nabla}\cdot\vec{D} = \rho_v$	$\vec{\nabla}\cdot\vec{B} = \rho_v$
Curl	$\vec{\nabla}\times\vec{E} = -\vec{J}$	$\vec{\nabla}\times\vec{H} = \vec{J}$

from one end of a material with magnetic dipoles to the other, it will undergo an energy loss in aligning the dipoles as it proceeds. This loss can be measured as a kind of resistance to virtual magnetic charge flow not included in Maxwell's quaternion equations of section 1.9. This concept was developed by Leon Chu in 1971, and leads to the concept of *memristance*.

The traditional designation of north and south poles in a permanent bar magnet does not imply that an isolated positive magnetic charge exists at the north pole and a corresponding amount of isolated negative magnetic charge exists at the south pole. As shown in Figure 6.2, magnetic poles cannot be isolated.[1]

Integral Forms

The integral form of the curl in Equation 6.8 can be obtained by applying Stokes's theorem

$$\iint_S \vec{\nabla} \times \vec{B} \cdot d\vec{s} = \mu_0 \iint_S \vec{J} \cdot d\vec{s} \quad \text{or}$$

$$\oint_C \vec{B} \cdot d\vec{l} = \mu_0 I, \tag{6.12}$$

where the path C for the line integral is the contour bounding the surface, S, and I is the total electric current passing through the surface S. The direction of the path

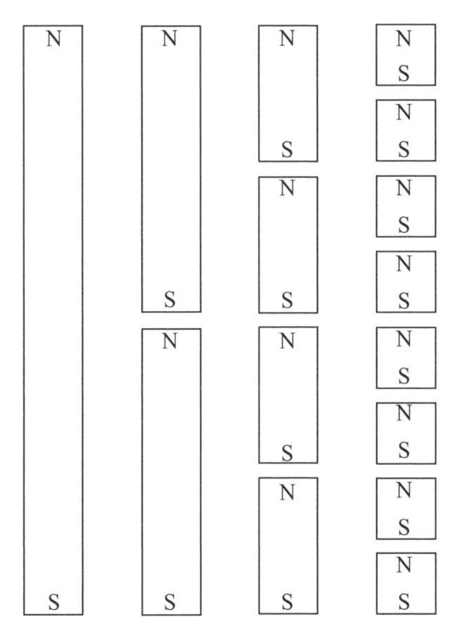

Figure 6.2 Successive division of the poles in a bar magnet.

[1] A list of searches for the magnetic monopole can be found at http://arxiv.org/abs/hep-ex/0302011. To date, accelerator searches have been conducted to 1.8 TeV, cosmic ray searches have been conducted to 10^{17} GeV, and matter searches have been conducted in meteorites, iron ore and aerosols, manganese, and seawater. Several positive results have failed to be repeated and remain unexplained.

Table 6.2 Assymetric Forms of the Divergence and Curl Equations for Magneto*static* Fields in Free Space

Differential form	Integral form
$\vec{\nabla}\cdot\vec{B}=0$	$\oiint_S \vec{B}\cdot d\vec{s}=0$
$\vec{\nabla}\times\vec{B}=\mu_0\vec{J}$	$\oint_C \vec{B}\cdot d\vec{l}=\mu_0 I$

C (counterclockwise) and the direction of electric current flow follows the right-hand rule.

Equation 6.12 is the form of Ampere's circuital law, which states that the circulation of the magnetic flux density in free space around any closed path is equal to μ_0 times the total electric current flowing through the surface bounded by the path (Table 6.2).

EXAMPLE

6.1 An infinitely long, straight, solid conductor with a circular cross section of radius a carries a steady uniform electric current, I (Figure 6.3). Determine the magnetic flux density, $\vec{B}$, inside and outside the conductor.

SOLUTION

a. Inside the conductor, $r < a$: $\vec{B} = B_\phi\hat{a}_\phi$ and $d\vec{l} = \hat{a}_\phi r d\phi$ so

$$\oint_{Path\ a} \vec{B}\cdot d\vec{l} = \int_0^{2\pi} B_\phi r d\phi = 2\pi r B_\phi.$$

With a uniform electric current density, the electric current through the area enclosed by path a is

$$I_a = \left(\pi r^2/\pi a^2\right)I = (r/a)^2 I.$$

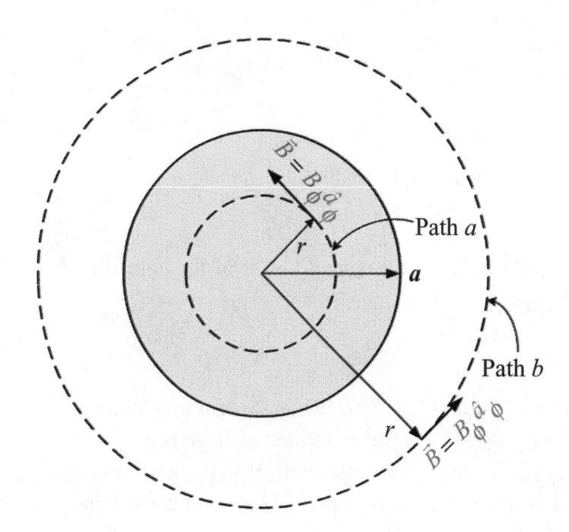

Figure 6.3 Integration paths for finding the magnetic flux density for an infinitely long, circular cylinder (shown in cross-section).

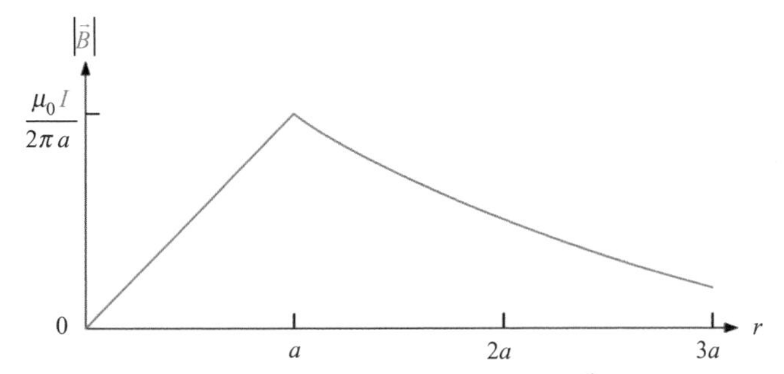

Figure 6.4 Magnitude of the magnetic flux density, $|\vec{B}|$, as a function of radius, r, for a conductor of radius, a, carrying a uniform electric current density, $J = I/(\pi a^2)$.

Thus, from Ampere's circuital law,

$$\oint_{Path\ a} \vec{B} \cdot d\vec{l} = 2\pi r B_\phi = \mu_0 (r^2/a^2) I$$

or

$$\vec{B} = \hat{a}_\varphi B_\varphi = \hat{a}_\varphi (\mu_0 r / 2\pi a^2) I.$$

b. Outside the conductor $r > a$,

$$\vec{B} = B_\varphi \hat{a}_\varphi \quad \text{and} \quad d\vec{l} = \hat{a}_\varphi r d\varphi,$$

so

$$\oint_{Path\ b} \vec{B} \cdot d\vec{l} = \int_0^{2\pi} B_\phi r d\phi = 2\pi r B_\phi.$$

The electric current through the area enclosed by path b is $I_b = I$. Thus, from Ampere's law,

$$\oint_{Path\ b} \vec{B} \cdot d\vec{l} = 2\pi r B_\phi = \mu_0 I_b, \text{ or } \vec{B} = \hat{a}_\varphi B_\varphi = \hat{a}_\varphi (\mu_0 / 2\pi r) I.$$

CHECKPOINT The solution to parts a (for $r = a$) and b is $\vec{B} = \hat{a}_\varphi(\mu_0/2\pi a)I$. The magnitude of the magnetic flux density is shown plotted in Figure 6.4.

6.3 MAGNETIC VECTOR POTENTIAL

The relation $\vec{\nabla} \cdot \vec{B} = 0$ assures that $\vec{B}$ is *solenoidal*, and we showed in Chapter 2 that *the divergence of the curl of any vector field is identically zero*. As a consequence, we can express $\vec{B}$ as the curl of another vector field called $\vec{A}$, such that

$$\vec{B} \equiv \vec{\nabla} \times \vec{A} \ (\text{Wb/m}^2) \text{ or (T)}. \tag{6.13}$$

Defined in this way, the vector field $\vec{A}$ (Wb/m) is called the magnetic vector potential. Thus, if we can find $\vec{A}$ for a given current distribution, $\vec{B}$ can be obtained from Equation 6.13. Unfortunately, there are an infinite number[2] of vector fields $\vec{A}$ that satisfy Equation 6.13, so a unique definition of this vector field requires an additional specification. Coulomb made one such specification by also defining the divergence of $\vec{A}$ to be zero as shown below.

Combining Equation 6.8 with Equation 6.13, we can see

$$\vec{\nabla}\times\vec{B}=\vec{\nabla}\times\vec{\nabla}\times\vec{A}=\mu_0\vec{J}, \tag{6.14}$$

and, by using the vector identity

$$\vec{\nabla}\times\vec{\nabla}\times\vec{A}=\vec{\nabla}\left(\vec{\nabla}\cdot\vec{A}\right)-\nabla^2\vec{A}, \tag{6.15}$$

Equation 6.14 becomes

$$\vec{\nabla}\left(\vec{\nabla}\cdot\vec{A}\right)-\nabla^2\vec{A}=\mu_0\vec{J} \tag{6.16}$$

To simplify Equation 6.16, Coulomb chose the additional constraint

$$\vec{\nabla}\cdot\vec{A}=0, \tag{6.17}$$

which is now called the *Coulomb gauge*,[3] and it reduces Equation 6.16 to

$$\nabla^2\vec{A}=-\mu_0\vec{J}. \tag{6.18}$$

This is a vector form of Poisson's equation, which is, in fact, three equations when we separate the left- and right-hand sides in Cartesian coordinates. The three equations are

$$\nabla^2 A_x=-\mu_0 J_x \tag{6.19a}$$
$$\nabla^2 A_y=-\mu_0 J_y \tag{6.19b}$$
$$\nabla^2 A_z=-\mu_0 J_z. \tag{6.19c}$$

To obtain the solution to these equations in free space, we can compare them with the electric potential solutions of Poisson's Equation 4.5 in free space:

$$\nabla^2 V=-\rho_V/\varepsilon_0, \tag{4.5}$$

which we have shown (Equation 4.37) to have the particular solution

$$V(\vec{x})=\iiint_{V'}\frac{\rho_V(\vec{x}')}{4\pi\varepsilon_0|\vec{x}-\vec{x}'|}dv'=\frac{1}{4\pi\varepsilon_0}\iiint_{V'}\frac{\rho_V}{R}dv'. \tag{4.37}$$

[2] For example, if $\vec{A}$ yields $\vec{B}=\vec{\nabla}\times\vec{A}$ then $\vec{A}+\vec{C}$ with $\vec{C}=$ constant will yield the same $\vec{B}$.

[3] The *Coulomb gauge* is used with static fields. As we will see in Chapter 7 when we deal with time-dependent fields, an alternate definition called the *Lorenz gauge* will be more convenient.

Hence, we can write the particular solutions for all of the Equations 6.19 as

$$A_x = \frac{\mu_0}{4\pi}\iiint_{V'} \frac{J_x}{R} dv' \tag{6.20a}$$

$$A_y = \frac{\mu_0}{4\pi}\iiint_{V'} \frac{J_y}{R} dv' \tag{6.20b}$$

$$A_z = \frac{\mu_0}{4\pi}\iiint_{V'} \frac{J_z}{R} dv' \tag{6.20c}$$

or

$$\vec{A} = A_x\hat{a}_x + A_y\hat{a}_y + A_z\hat{a}_z = \frac{\mu_0}{4\pi}\iiint_{V'} \frac{J_x\hat{a}_x + J_y\hat{a}_y + J_z\hat{a}_z}{R} dv'$$
$$\vec{A} = \frac{\mu_0}{4\pi}\iiint_{V'} \frac{\vec{J}}{R} dv' \tag{6.21}$$

From Equation 6.21, we can now find $\vec{B}$ via $\vec{B} = \vec{\nabla}\times\vec{A}$ and can find the magnetic flux Φ (Wb) through a given area S that is bounded by contour C via

$$\Phi = \iint_S \vec{B}\cdot d\vec{s} = \iint_S (\vec{\nabla}\times\vec{A})\cdot d\vec{s} = \oint_C \vec{A}\cdot d\vec{l}. \tag{6.22}$$

PROBLEM

6.1 Show that $\vec{\nabla}\cdot(\vec{\nabla}\times\vec{A})$, where $\vec{A}$ is given by Equation 6.21 yields 0.

6.4 THE BIOT–SAVART LAW

In applications involving current carrying wires, we are interested in magnetic fields such as that shown in Figure 6.5.

For a thin wire with cross-sectional area $\Delta\vec{s}$ and length L with $dv' = \Delta\vec{s}\cdot d\vec{l}'$, we have

$$\vec{J}dV' = \vec{J}\cdot\Delta\vec{s}d\vec{l}' = Id\vec{l}', \tag{6.23}$$

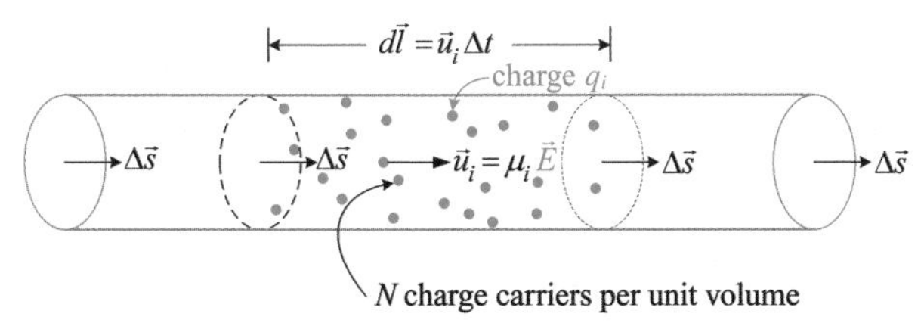

Figure 6.5 Long, thin wire carrying a uniform electric current.

so Equation 6.21 becomes

$$\vec{A} = \frac{\mu_0}{4\pi}\iiint_{V'}\frac{\vec{J}}{R}dV' = \frac{\mu_0 I}{4\pi}\int_L \frac{d\vec{l}'}{R} \tag{6.24}$$

and

$$\vec{B} = \vec{\nabla}\times\vec{A} = \vec{\nabla}\times\left[\frac{\mu_0 I}{4\pi}\int_L \frac{d\vec{l}'}{R}\right] = \frac{\mu_0 I}{4\pi}\int_L \vec{\nabla}\times\frac{d\vec{l}'}{R}, \tag{6.25}$$

where

$$R = \left[(x-x')^2 + (y-y')^2 + (z-z')^2\right]^{1/2}. \tag{6.26}$$

Here, the curl operation implies differentiation with regard to the unprimed space coordinates of the field point, and the integral operation is with respect to the primed source coordinates, so that $\vec{\nabla}$ operates only on the R function. Using

$$\vec{\nabla}\times\left(f\vec{G}\right) = f\vec{\nabla}\times\vec{G} + \left(\vec{\nabla}f\right)\times\vec{G} \tag{6.27}$$

with $f = 1/R$ and $\vec{G} = d\vec{l}$, we have

$$\vec{B} = (\mu_0 I/4\pi)\int_L \left[(1/R)\vec{\nabla}\times d\vec{l}' + \vec{\nabla}(1/R)\times d\vec{l}'\right] \tag{6.28}$$

and the first integral is zero. Thus, because

$$1/R = \left[(x-x')^2 + (y-y')^2 + (z-z')^2\right]^{-1/2} \tag{6.29}$$

$$\vec{\nabla}(1/R) = \hat{a}_x\partial(1/R)/\partial x + \hat{a}_y\partial(1/R)/\partial y + \hat{a}_z\partial(1/R)/\partial z$$

$$\vec{\nabla}(1/R) = -\frac{\hat{a}_x(x-x') + \hat{a}_y(y-y') + \hat{a}_z(z-z')}{\left[(x-x')^2 + (y-y')^2 + (z-z')^2\right]^{3/2}} = -\frac{\vec{R}}{R^3}, \tag{6.30}$$

$$\vec{\nabla}(1/R) = -\hat{a}_R/R^2$$

where $\hat{a}_R$ is the unit vector directed from the source point to the field point. Finally,

$$\vec{B} = \frac{\mu_0 I}{4\pi}\int_L \frac{d\vec{l}'\times\hat{a}_R}{R^2}. \tag{6.31}$$

The above equation is known as **Biot–Savart law**. It is the formula for determining $\vec{B}$ caused by a current I in a path L. Some texts write Equation 6.31 as

$$\vec{B} = \int_L d\vec{B} \tag{6.32}$$

in which

$$d\vec{B} = \frac{\mu_0 I}{4\pi}\left(\frac{d\vec{l}'\times\hat{a}_R}{R^2}\right) = \frac{\mu_0 I}{4\pi}\left(\frac{d\vec{l}'\times\vec{R}}{R^3}\right) \tag{6.33}$$

and then call the differential form, Equation 6.33, the Biot–Savart law.

EXAMPLE

6.2 Find the magnetic flux density, $\vec{B}$, produced by a circular loop of radius b that carries a current I, as shown in Figure 6.6.

SOLUTION We first find the magnetic vector potential $\vec{A}$ by the relation

$$\vec{A} = \frac{\mu_0 I}{4\pi} \oint_{C'} \frac{d\vec{l}'}{R_{l'}}, \tag{6.34}$$

where $R_{l'}$ is the distance between the source element $Id\vec{l}'$ and the field point P, as shown in Figure 6.6. Because of symmetry, the magnetic flux density is obviously independent of angle ϕ (the field point). For convenience we can pick $P(R, \theta, \pi/2)$ in the y–z plane:

$$Id\vec{l}' = I(-\hat{a}_x \sin\varphi' + \hat{a}_y \cos\varphi')\, bd\varphi'. \tag{6.35a}$$

For every $Id\vec{l}'$, there is another symmetrically located differential electric current element on the other side of the y-axis that will contribute an equal amount to $\vec{A}$ but will cancel the contribution of $Id\vec{l}'$ in the $\hat{a}_y$ direction, namely,

$$Id\vec{l}'' = I(-\hat{a}_x \sin\varphi' - \hat{a}_y \cos\varphi')\, bd\varphi'. \tag{6.35b}$$

Equation 6.34 can thus be written as

$$\vec{A} = -\hat{a}_x \frac{\mu_0 I}{4\pi} \int_0^{2\pi} \frac{b \sin\phi'}{R_{l'}} d\phi' \text{ or} \tag{6.36a}$$

$$\vec{A} = \hat{a}_\phi \frac{\mu_0 I}{4\pi} \int_{-\pi/2}^{\pi/2} \frac{b \sin\phi'}{R_{l'}} d\phi'. \tag{6.36b}$$

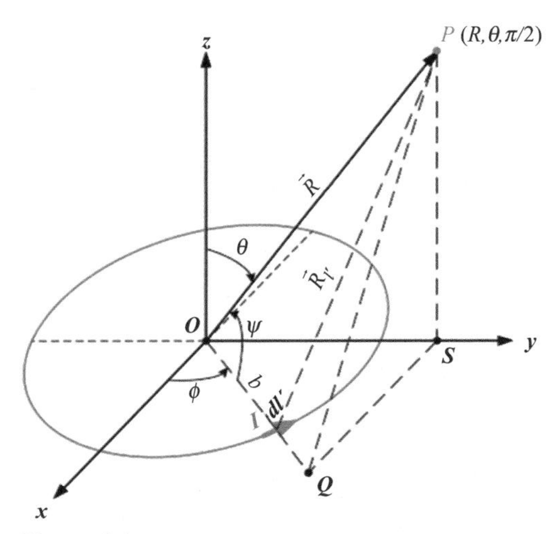

Figure 6.6 Parameters involved in the calculation of the magnetic vector potential due to a loop of wire of radius b carrying current i.

Using the law of cosines as shown in Figure 6.6, we have

$$R_{l'} = \left(R^2 + b^2 - 2bR\cos\psi\right)^{1/2}, \tag{6.37}$$

where $R\cos\psi$ is the projection of R on the radius OQ, which is the same as the projection of OS ($OS = R\sin\theta$) on OQ. Hence,

$$R_{l'} = \left[R^2 + b^2 - 2bR\sin\theta\sin\phi'\right]^{1/2} \text{ and } 1/R_{l'} = (1/R)\left[1 + b^2/R^2 - (2b/R)\sin\theta\sin\phi'\right]^{-1/2}$$

When $R \gg b$, we have

$$1/R_{l'} \approx [1 - (2b/R)\sin\theta\sin\phi']^{-1/2}/R \approx [1 + (b/R)\sin\theta\sin\phi']/R.$$

Thus,

$$\vec{A} \approx \hat{a}_\phi(\mu_0/2\pi)(Ib/R)\int_{-\pi/2}^{\pi/2}[1 + (b/R)\sin\theta\sin\phi']\sin\phi' d\phi',$$

which yields

$$\vec{A} \approx \hat{a}_\phi(\mu_0/4\pi)\left(I\pi b^2/R^2\right)\sin\theta. \tag{6.38}$$

Finally,

$$\vec{B} \approx \nabla \times \vec{A} = (\mu_0/4\pi)\left(I\pi b^2/R^3\right)(\hat{a}_R 2\cos\theta + \hat{a}_\theta \sin\theta). \tag{6.39}$$

Let us now rearrange the expression of the magnetic vector potential in Equation 6.38 as

$$\vec{A} \approx \hat{a}_\phi(\mu_0/4\pi)\left(I\pi b^2/R^2\right)\sin\theta \tag{6.40a}$$

or

$$\vec{A} \approx (\mu_0/4\pi)\vec{m} \times \hat{a}_R/R^2, \tag{6.40b}$$

where

$$\vec{m} \equiv \hat{a}_z IS = \hat{a}_z I\pi b^2 \tag{6.41}$$

is defined as the magnetic dipole moment. Correspondingly, we have

$$\vec{B}_{dipole} \approx (\mu_0/4\pi)\left(m/R^3\right)(\hat{a}_R 2\cos\theta + \hat{a}_\theta \sin\theta). \tag{6.42}$$

6.5 HISTORICAL CONCLUSIONS

The similarity of Equation 6.42 to that of the electric field intensity in Equation 3.67 is

$$\vec{E}_{dipole} = (1/4\pi\varepsilon_0)\left(p/R^3\right)(\hat{a}_R 2\cos\theta + \hat{a}_\theta \sin\theta), \tag{3.67}$$

where the electric dipole moment defined as $\vec{p} = \hat{a}_z qd$ is clear. A comparison of the two fields is shown in the near region in Figure 6.7.

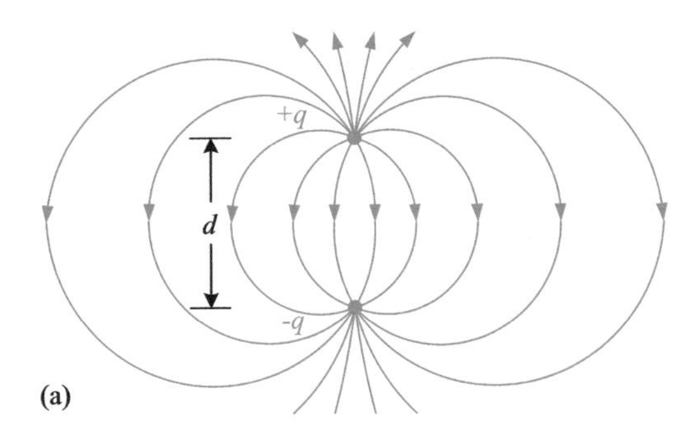

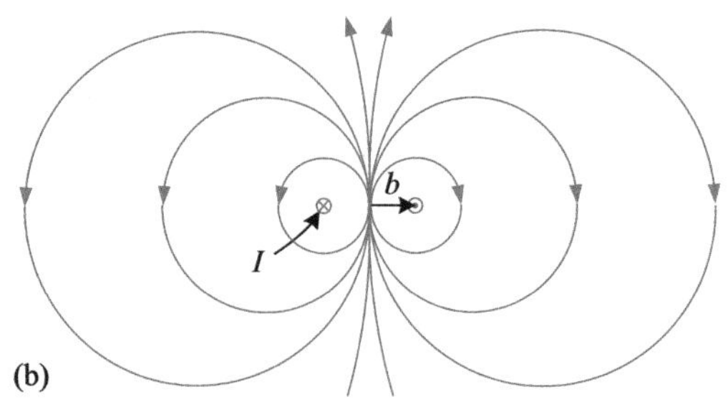

Figure 6.7 (a). Electric field intensity in the near-field region of an electric dipole moment, $\vec{p} = qd\hat{a}_z$. (b). Magnetic flux density in the near-field region of a magnetic dipole moment, $\vec{m} = I\pi b^2\hat{a}_z$, caused by electric current in a loop.

In these two figures, we can see that the electric field intensity and magnetic flux density distribution in the near-region ($R \sim b$, $R \sim d$) is different but that the field intensity distribution in the far-field region ($R \gg b$, $R \gg d$) is similar. In the case of electric dipoles, we normally consider the electric field intensity to be evaluated in the $\lim_{d\to 0}$, and, in the case of magnetic dipoles, we normally consider the magnetic flux density to be evaluated in the $\lim_{b\to 0}$. Thus, in both cases, we normally consider the fields in the far-field region.

The Chinese discovered the effects of magnetic field distributions hundreds of years B.C. by scattering iron filings in the neighborhood of magnetite and arrived at the qualitative distribution given by Equation 6.42, as shown in Figure 6.8.

6.6 ATOMIC MAGNETISM

Because iron is so inexpensive, it is commonly used in the construction of bar magnets. Iron has the good characteristic that it can retain a magnetic N and S pole even after the polarizing field has been removed, but the saturation field is limited

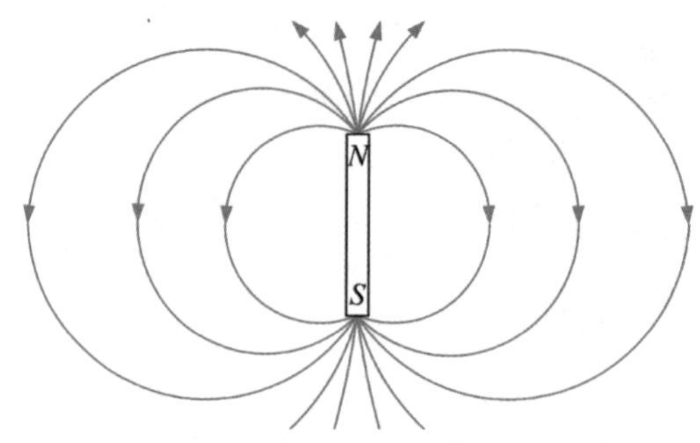

Figure 6.8 Magnetic flux density, $\vec{B}$, in the neighborhood of a bar magnet.

Table 6.3 Relative Magnitude of a Number of Magnetic Field Sources

Source of magnetic field	Typical period	Magnitude
Earth's white noise	Random (1 μs–1 ms)	10^{-6} G
Earth's natural surface field	Continuous	1 G
Iron magnets	Continuous	2 T
Rare earth magnets	Continuous	5 T
Superconducting magnets	Continuous	25 T
Cu+ superconducting magnets	Continuous	45 T
Pulsed magnets	~1 s	65 T
Conventional explosive magnets	~1 μs	100 T

to about 2 T (20 kG [Gauss]). By comparison, the magnetic flux density at the surface of the earth is about 1 G. Table 6.3 gives a few typical fields for a variety of sources:

Sources of Material Magnetism

Material magnetism arises from the behavior of electrons in a pure metal or chemical compound as is determined by their nominal electron structure, as shown in Figure 6.9.

Magnetism Due to Loosely Bound Electrons (in Electronic Energy Bands)

In Figure 6.9, the electronic structure is given for atoms that are widely separated from one another. If the separated atoms are brought closer together, the electrons of a particular atom will begin to interact with the electrons of its nearest neighbors.

Periodic Table of the Elements

1 H 1s^{1}																	2 He 1s^{2}
3 Li [He] 2s^{1}	4 Be [He] 2s^{2}											5 B [He] 2s^{2}1p^{1}	6 C [He] 2s^{2}2p^{2}	7 N [He] 2s^{2}2p^{3}	8 O [He] 2s^{2}2p^{4}	9 F [He] 2s^{2}2p^{5}	10 Ne [He] 2s^{2}2p^{6}
11 Na [Ne] 3s^{1}	12 Mg [Ne] 3s^{2}											13 Al [Ne] 3s^{2}3p^{1}	14 Si [Ne] 3s^{2}3p^{2}	15 P [Ne] 3s^{2}3p^{3}	16 S [Ne] 3s^{2}3p^{4}	17 Cl [Ne] 3s^{2}3p^{5}	18 Ar [Ne] 3s^{2}3p^{6}
19 K [Ar] 4s^{1}	20 Ca [Ar] 4s^{2}	21 Sc [Ar] 3d^{1}4s^{2}	22 Ti [Ar] 3d^{2}4s^{2}	23 V [Ar] 3d^{3}4s^{3}	24 Cr [Ar] 3d^{5}4s^{1}	25 Mn [Ar] 3d^{5}4s^{2}	26 Fe [Ar] 3d^{6}4s^{2}	27 Co [Ar] 3d^{7}4s^{2}	28 Ni [Ar] 3d^{8}4s^{2}	29 Cu [Ar] 3d^{10}4s^{1}	30 Zn [Ar] 3d^{10}4s^{2}	31 Ga [Ar] 3d^{10}4s^{2}4p^{1}	32 Ge [Ar] 3d^{10}4s^{2}4p^{2}	33 As [Ar] 3d^{10}4s^{2}4p^{3}	34 Se [Ar] 3d^{10}4s^{2}4p^{4}	35 Br [Ar] 3d^{10}4s^{2}4p^{5}	36 Kr [Ar] 3d^{10}4s^{2}4p^{6}
37 Rb [Kr] 5s^{1}	38 Sr [Kr] 5s^{2}	39 Y [Kr] 4d^{1}5s^{2}	40 Zr [Kr] 4d^{2}5s^{2}	41 Nb [Kr] 4d^{4}5s^{1}	42 Mo [Kr] 4d^{5}5s^{1}	43 Tc [Kr] 4d^{5}5s^{2}	44 Ru [Kr] 4d^{7}5s^{1}	45 Rh [Kr] 4d^{8}5s^{1}	46 Pd [Kr] 4d^{10}5s^{0}	47 Ag [Kr] 4d^{10}5s^{1}	48 Cd [Kr] 4d^{10}5s^{2}	49 In [Kr] 4d^{10}5s^{2}5p^{1}	50 Sn [Kr] 4d^{10}5s^{2}5p^{2}	51 Sb [Kr] 4d^{10}5s^{2}5p^{3}	52 Te [Kr] 4d^{10}5s^{2}5p^{4}	53 I [Kr] 4d^{10}5s^{2}5p^{5}	54 Xe [Kr] 4d^{10}5s^{2}5p^{6}
55 Cs [Xe] [Xe]6s^{1}	56 Ba [Xe] [Xe]6s^{2}		72 Hf [Xe]4f^{14} 5d^{1}6s^{2}	73 Ta [Xe]4f^{14} 5d^{2}6s^{2}	74 W [Xe]4f^{14} 5d^{3}6s^{2}	75 Re [Xe]4f^{14} 5d^{5}6s^{2}	76 Os [Xe]4f^{14} 5d^{6}6s^{2}	77 Ir [Xe]4f^{14} 5d^{7}6s^{2}	78 Pt [Xe]4f^{14} 5d^{8}6s^{1}	79 Au [Xe]4f^{14} 5d^{10}6s^{1}	80 Hg [Xe]4f^{14} 5d^{10}6s^{2}	81 Tl [Xe]4f^{14} 5d^{10}6s^{2}6p^{1}	82 Pb [Xe]4f^{14} 5d^{10}6s^{2}6p^{2}	83 Bi [Xe]4f^{14} 5d^{10}6s^{2}6p^{3}	84 Po [Xe]4f^{14} 5d^{10}6s^{2}6p^{4}	85 At [Xe]4f^{14} 5d^{10}6s^{2}6p^{5}	86 Rn [Xe]4f^{14} 5d^{10}6s^{2}6p^{6}
87 Fr [Rn] 7s^{1}	88 Ra [Rn] 7s^{2}																

57 La [Xe] 5d^{1}6s^{2}	58 Ce [Xe] 4f^{2}5d^{0}6s^{2}	59 Pr [Xe] 4f^{3}5d^{0}6s^{2}	60 Nd [Xe] 4f^{4}5d^{0}6s^{2}	61 Pm [Xe] 4f^{5}5d^{0}6s^{2}	62 Sm [Xe] 4f^{6}5d^{0}6s^{2}	63 Eu [Xe] 4f^{7}5d^{0}6s^{2}	64 Gd [Xe] 4f^{7}5d^{1}6s^{2}	65 Tb [Xe] 4f^{9}5d^{0}6s^{2}	66 Dy [Xe] 4f^{10}5d^{0}6s^{2}	67 Ho [Xe] 4f^{11}5d^{0}6s^{2}	68 Er [Xe] 4f^{12}5d^{0}6s^{2}	69 Tm [Xe] 4f^{13}5d^{0}6s^{2}	70 Yb [Xe] 4f^{14}5d^{0}6s^{2}	71 Lu [Xe] 4f^{14}5d^{1}6s^{2}
89 Ac [Rn] 6d^{1}7s^{2}	90 Th [Rn] 6d^{2}7s^{2}	91 Pa [Rn] 5f^{3}6d^{1}7s^{2}	92 U [Rn] 5f^{14}6d^{5}7s^{2}	93 Np [Rn] 5f^{5}6d^{1}7s^{2}	94 Pu [Rn] 5f^{6}6d^{0}7s^{2}	95 Am [Rn] 5f^{7}6d^{0}7s^{2}	96 Cm [Rn] 5f^{7}6d^{1}7s^{2}	97 Bk [Rn] 5f^{9}6d^{0}7s^{2}	98 Cf [Rn] 5f^{10}6d^{0}7s^{2}	99 Es [Rn] 5f^{11}6d^{0}7s^{1}	100 Fm [Rn] 5f^{12}6d^{0}7s^{1}	101 Md [Rn] 5f^{13}6d^{0}7s^{1}	102 No [Rn] 5f^{14}6d^{0}7s^{1}	103 Lr [Rn] 5f^{14}6d^{1}7s^{1}

Figure 6.9 Nominal electronic structure of the elements.

In the case of a pure material, the outer electrons (those with the most physical extent from the nucleus such as the *s* electrons) may form an energy band structure like that discussed in Chapter 5. If the atoms come even closer together, the next electrons with physical extent from the nucleus (like the *d* electrons) may also form an energy band structure.

As a practical example, Figure 6.10 shows a typical relationship of the 4s and 3d bands for metallic copper.[iii] As we see from Figure 6.9, an isolated copper atom has one electron in the 4s shell and 10 electrons in the 3d shell (Five with spin-up [↑] and five with spin-down [↓]).

The d bands with spin ↑ and spin ↓ are shown with the same bottom and top energies, and both are fully populated. The relative area of the boxes implies the number of possible states for that original electronic designation (two per atom for the 4s band; about half of which are populated). The fact that the spin ↑ and spin ↓

electronic states in Figure 6.10 are equally populated implies that copper is a non-magnetic material.

As a second example, Figure 6.11 shows the relationship of the 3d and 4s bands for metallic nickel (one less electron per atom than copper). While Figure 6.9 implies that an isolated nickel atom has two electrons in the 4s shell and eight electrons in the 3d shell, some tables list the configuration of nickel metal as having one electron in the 4s shell and nine electrons in the 3d shell, and others list the configuration of nickel metal as having zero electrons in the 4s shell and 10 electrons in the 3d shell. The energy band specification, as shown in Figure 6.11, indicates the inadequacy of using single-energy atomic states for the designation.

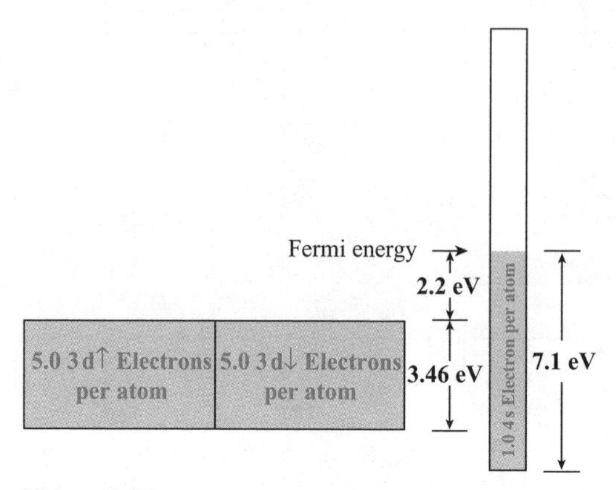

Figure 6.10 Relative energy bands of the 3d and 4s bands (↑ implies spin up) for electrons in metallic copper.

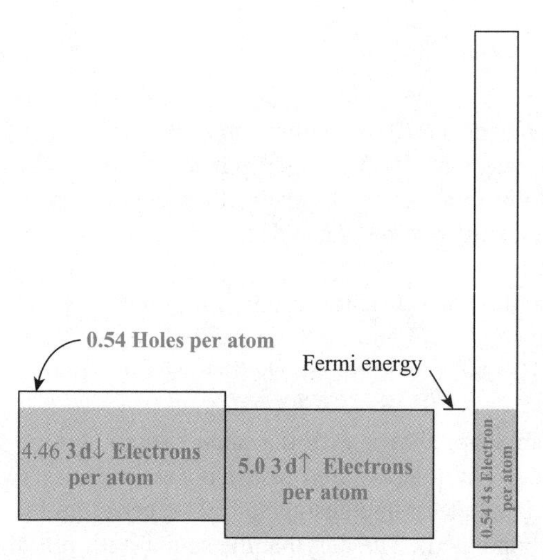

Figure 6.11 Relative energy bands of the 3d and 4s bands for electrons in metallic nickel.

The d bands with spin ↑ and spin ↓ are shown with different bottom and top energies because of an exchange interaction between the two states. Because the d bands are displaced, only one is fully populated, while the other is partially populated with 0.54 electrons per atom in the 4s band and 4.46 electrons per atom in the 3d↓ band.

In this electronic band (itinerant electron) model, the temperature is assumed to be below the Curie temperature of Ni (627 K) and is thus in a ferromagnetic regime. Similar graphs exist for other 3d transition metals Fe and Co. Even this picture is too simple for quantitative analysis because it still needs to take into account a magnetic moment contribution for an orbital electronic motion of electrons from neighboring atoms that are relatively tightly bound to their respective ion core.

For atoms of higher *Z*, such as the rare earth lanthanides or the man-made actinides, the ion cores are small compared with the distance to their nearest neighbors so that even the least tightly bound electrons in the 4f or 5f shell do not interact significantly with the electrons of their nearest neighbors. While the least tightly bound 6s or 7s electrons may form an energy band structure like that discussed above, they are not magnetic and do not contribute significantly to the atomic magnetism. As we will see below, the electrons responsible for magnetism in these elements do not overlap significantly and hence do not have a strong exchange energy between them. These elements thus form ferromagnets (and antiferromagnets) only at very low temperatures. At room temperature, the magnetism is dominantly characterized as paramagnetic or diamagnetic, which is a magnetic alignment similar to the dielectric alignment of electric dipoles in an insulator. However, for a strictly classical atomic system in thermal equilibrium with its surroundings, no magnetism will be created even in an externally applied magnetic field. As we will also see in the next few sections, magnetism depends strongly on quantum mechanics, so much that the two topics are entwined.

Magnetism Due to Tightly Bound Electrons

To determine the source of magnetic fields in f-electron materials, we consider the field produced by an unpaired electron in its orbit about a nucleus, as shown in Figure 6.12:

We can calculate the classical magnetic moment of the electron as

$$\vec{\mu}_{orbital} = iA\hat{a}_z = (\Delta q/\Delta t)A\hat{a}_z = [-e/(2\pi r/v)](\pi r^2)\hat{a}_z = (-evr/2)\hat{a}_z \qquad (6.43)$$

and, because the orbital angular momentum of a mass m_e traveling in a circular orbit of radius, r, at velocity, v, is $\vec{L} = m_e v r \hat{a}_z$,

$$\vec{\mu}_{orbital} = -(e/2m_e)\vec{L}. \qquad (6.44)$$

In quantum mechanics, the orbital angular momentum is a quantized quantity, and the component of angular momentum along the z-axis is

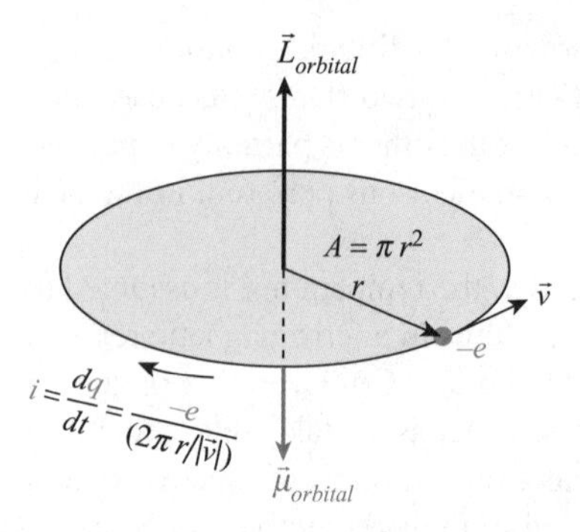

Figure 6.12 Magnetic dipole moment caused by an electron traveling at constant speed, v, in a circular path of radius, r.

$$L_z = m_l(h/2\pi) = m_l\hbar, \tag{6.45}$$

where $m_l = 0, \pm1, \pm2, \pm3, \dots$, and $\hbar$ is Plank's constant. If we express Equation 6.44 in terms of the quantum number m_l, we thus get

$$\mu_{orbital,\,z} = -m_l(e\hbar/4\pi m_e) = -m_l\mu_B, \tag{6.46}$$

where μ_B is a fundamental quantity called the Bohr magneton. Using the known values for e, $\hbar$, and m_e, the Bohr magneton has the value also found at the National Institute for Standards and Technology (NIST) web site http://physics.nist.gov/cgi-bin/cuu/Value?mub|search_for=elecmag_in!

$$\mu_B = 9.27400949(80)\times10^{-24}\ JT^{-1}. \tag{6.47}$$

Spin

Magnetic moments of electrons also arise from their spin, which can be similarly expressed through their spin angular momentum, $\vec{S}$, as

$$\vec{\mu}_{spin} = -(e/m_e)\vec{S}. \tag{6.48}$$

The magnetic flux density produced by the spin of an electron (relative to its spin angular momentum and its spin magnetic dipole moment) is shown in Figure 6.13.

If an electron is located in an external magnetic flux density, $\vec{B}_{external} = B_{external}\hat{a}_z$, neither the orbital angular momentum, $\vec{L}$, nor the spin angular momentum, $\vec{S}$, will necessarily lie in the direction of the external magnetic flux density (the z-direction), as indicated in Figure 6.14.

In general, the total angular momentum of the electron is $\vec{J} = \vec{L} + \vec{S}$, and the total magnetic moment of the electron is $\vec{\mu}_{total} = \vec{\mu}_{orbital} + \vec{\mu}_{spin}$ so that

$$\vec{\mu} = -g\mu_B\vec{J} \tag{6.49}$$

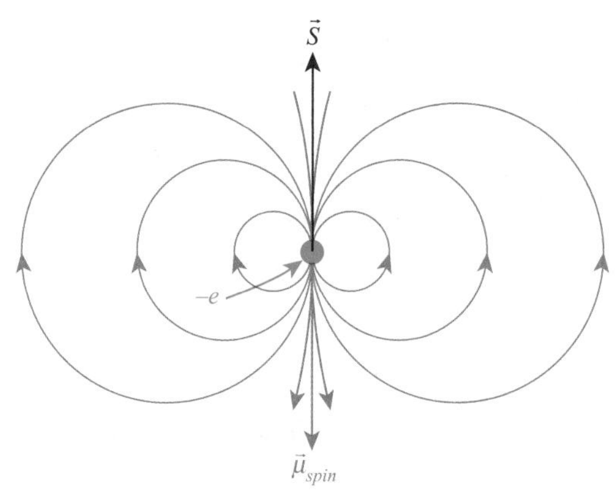

Figure 6.13 Magnetic dipole moment, $\vec{\mu}_{spin}$, of an electron due to its spin angular momentum, $\vec{S}$.

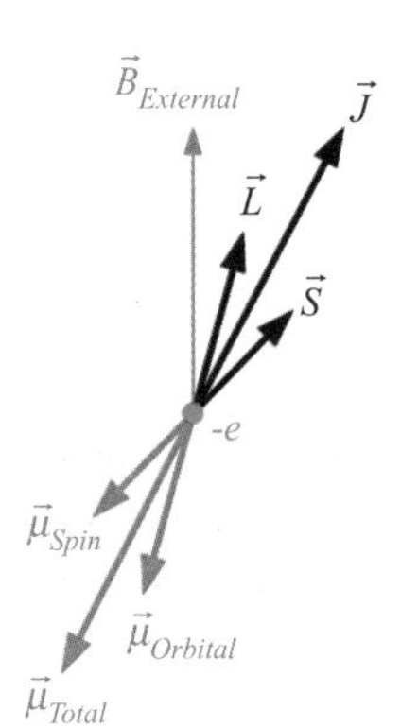

Figure 6.14 Relative directions of the orbital angular momentum, $\vec{L}$, spin angular momentum, $\vec{S}$, and total angular momentum, $\vec{J}$, to an external magnetic flux density in the z-direction.

and, if we take the z-component of the magnetic moment

$$\mu_z = -g\sqrt{J(J+1)}\mu_B, \tag{6.50}$$

where the quantity g is called the Landé g factor. For an electron with only spin (no orbital angular momentum) g = 2.0023, (usually taken as 2.00),[4] and, for a free atom whose electrons have both spin and orbital angular momentum, g, has the value

$$g = 1 + \left[\left(|\vec{J}| + |\vec{S}| - |\vec{L}|\right)/2|\vec{J}|\right] = 1 + [(J(J+1) + S(S+1) - L(L+1))/2J(J+1)]. \tag{6.51}$$

The z-component of the electron's spin can be expressed through its *spin angular quantum number*, m_s (which has values of ±½), so

[4] Quantum physics of the electron, *quantum electrodynamics* (*QED*), reveals that g is slightly greater than 2.00, but most texts ignore this correction factor.

$$\mu_{spin,\,z} = -m_s(eh/2\pi m_e) = \pm\mu_B. \tag{6.52}$$

The z-component of the electron's orbital angular momentum can be expressed through its *orbital angular quantum number*, m_l, (0 for an s-electron with $l = 0$), (−1, 0, 1 for a p-electron with $l = 1$), (−2, −1, 0, 1, 2 for a d-electron with $l = 2$), or (−3, −2, −1, 0, 1, 2, 3 for an f-electron with $l = 2$). The orbital angular momentum of an electron thus depends upon its orbital state (s, p, d, f), and its orbital angular quantum number, m_l, is

$$\mu_{orbital,\,z} = \pm m_l(eh/2\pi m_e) = \pm m_l\mu_B. \tag{6.53}$$

The z-component of the *total* magnetic moment of an individual electron is thus

$$\mu_{total,\,z} = \pm(1 \pm m_l)\mu_B, \tag{6.54}$$

where we need only take the numerical sum because we are expressing the answer in terms of the component of the total angular momentum in the z-direction.

Magnetism Due to Atoms

To find the magnetism due to an isolated atom, we need to sum the total electron magnetic moments overall of the individual electrons in the atom. This would be a large sum for high-atomic number atoms except for the fact that most of the inner shells of electrons in an atom are full; that is, there are as many spin ↑ electrons as spin ↓ electrons and there is no net orbital angular momentum caused by electrons in a full shell. Thus we need only take the sum of the individual electrons over those electrons in an unfilled shell. Because of the Pauli exclusion principle that no two electrons can have the same quantum numbers, there can be only two electrons in an s-shell (one for spin ↑ and one for spin ↓), 6 electrons in a p-shell, 10 electrons in a d-shell, and 14 electrons in an f-shell, consistent with $2(2l + 1)$ electrons in a full shell.

Hund's Rule

Even if we ignore the sum of angular momentum quantum numbers over full shells, the possible number of combinations of spin and orbital angular momentum quantum numbers over an unfilled shell is large. At high temperatures, the thermal energy imparted between atoms in a material will cause a distribution of spin and orbital states to exist in an ensemble of atoms and we would have to take the average overall atoms to find the magnetism due to a material. However, at low temperatures, defined as a lattice kinetic energy small compared with k_BT, the electrons in an unfilled shell will decrease to their lowest energy level or "ground state." Hund was one of the first experimentalists to give a general rule about which states in an atom would exist for atoms in their ground state:

Electrons in the ground state of an atom will arrange themselves to produce

- maximum spin angular momentum ($S_z = \sum m_s$) and then to achieve maximum orbital angular momentum ($L_z = \sum m_l$) consistent with that spin,
- total angular momentum of $J_z = L_z - S_z$ if the shell is half-full or less and $J_z = L_z + S_z$ if the shell is half-full or more.

EXAMPLE

6.3 Let us take the case of uranium in a chemical +3 charge state (with three *5f* electrons in its unfilled shell) or neodymium in a chemical +3 charge state (with three *4f* electons in its unfilled shell). Hund's rule would be satisfied if the three electrons all had their spin ↑; that is, $S_z = \sum m_s = 1/2 + 1/2 + 1/2 = 3/2$. Because the electrons all have the same spin state, m_s, they cannot also have the same orbital state, m_l, so $L_z = \sum m_l = 3 + 2 + 1 = 6$, where we have chosen the maximum value of L_z possible. The total angular momentum will thus be $J_z = L_z - S_z = 6 - 3/2 = 9/2$ because the shell is less than half-full. Thus, we can calculate

$$g = 1 + \left[\left(|\vec{J}| + |\vec{S}| - |\vec{L}|\right)/2|\vec{J}|\right] = 1 + [(J(J+1) + S(S+1) - L(L+1))/2J(J+1)] = -0.727$$

and

$$\mu_{eff} = -g\sqrt{J(J+1)}\mu_B = 0.727\sqrt{(9/2)(11/2)}\mu_B = 3.62\mu_B.$$

A similar calculation has been carried out for each of the atoms in the lathanide (actinide) 4f (5f series) to produce a table of theoretical magnetic moments for those elements in the +3 charge state. Those values are shown in Table 6.4.

The measured effective magnetic moments, μ_{eff}, for the lanthanide and actinide ions in various charge states for several compounds are plotted in Figure 6.15 as a function of their probable ion configuration.

These data tell us a number of things:

1. In its paramagnetic state, the magnetic moment of the lanthanide and actinide elements dysprosium, holmium, californium, and einsteinium in their +3 charge state is 10 Bohr magnetons or greater, as predicted by L-S coupling (above) and Hund's rule. This is the largest magnetic moment ever observed, so these elements could likely be combined with other materials in a ferromagnetic arrangement below their curie temperature to make practical magnets with supermagnet strengths.
2. The *metals* of those same elements above their curie temperature behave paramagnetically as if they were in the +3 charge state because their magnetic moment is the same as that of compounds that are chemically in the +3 charge state. The three conduction electrons per atom are assumed to

Table 6.4 +3 Charge States, Spectroscopic Designation, and Effective Magnetic Moment, μ_{eff}, Given by L-S Coupling and Hund's Rule for the Lanthanides and Actinides

Lanthanide element	Actinide element	+3 Ion configuration	Basic level	L	S	J	g	$-g\sqrt{J(J+1)} = \mu_{eff}(\mu_B)$
Lanthanum	Actinium	f^0	^1S	0	0	0	0	0
Cerium	Thorium	f^1	$^2F_{5/2}$	3	1/2	5/2	6/7	2.54
Praseodymium	Protactinium	f^2	$^3H_4$	5	1	4	4/5	3.58
Neodymium	Uranium	f^3	$^4I_{9/2}$	6	3/2	9/2	8/11	3.62
Promethium	Neptunium	f^4	$^5I_4$	6	2	4	3/5	2.68
Samarium	Plutonium	f^5	$^6H_{5/2}$	5	5/2	5/2	2/7	0.84
Europium	Americium	f^6	$^7F_0$	3	3	0	0	0
Gadolinium	Curium	f^7	$^8S_{7/2}$	0	7/2	7/2	2	7.94
Terbium	Berkelium	f^8	$^7F_6$	3	3	6	3/2	9.72
Dysprosium	Californium	f^9	$^6H_{15/2}$	5	5/2	15/2	4/3	10.63
Holmium	Einsteinium	f^{10}	$^5I_8$	6	2	8	5/4	10.60
Erbium	Fermium	f^{11}	$^4I_{15/2}$	6	3/2	15/2	6/5	9.59
Thulium	Mendelevium	f^{12}	$^3H_6$	5	1	6	7/6	7.57
Ytterbium	Nobelium	f^{13}	$^2F_{7/2}$	3	1/2	7/2	8/7	4.54
Lutetium	Lawrencium	f^{14}	^1S	0	0	0	0	0

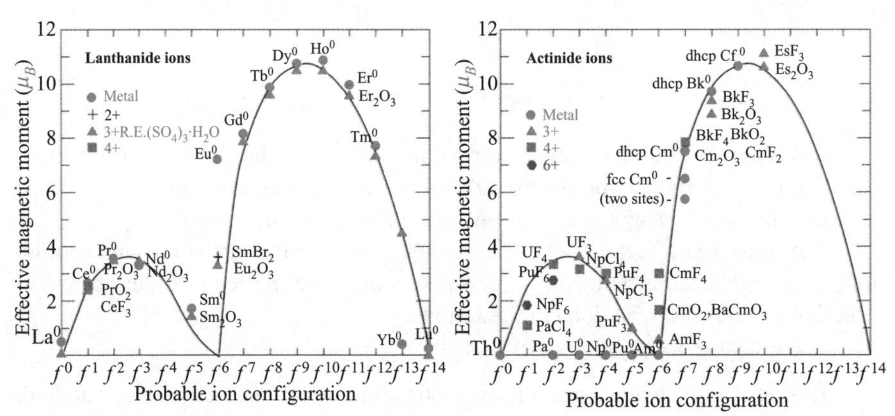

Figure 6.15 Measured effective magnetic dipole moment, μ_{eff} (in μ_b) in the paramagnetic regime for several lanthanide and actinide ions as a function of their probable ion configuration.

exist in a free electron gas or band structure like those of lower Z metals, so their electrical conductivity should be high.

3. The 4f and 5f electrons undergo a small-enough exchange interaction with those of a neighboring atom of the same kind that they act relatively independent of one another at temperatures above their curie temperature in chemical compounds. Even in their metallic form (except for Pa, U, Np, Pu

Table 6.5 Charge States, Number of 3d Electrons, Spectroscopic Designation, and Effective Magnetic Moment Given by L-S Coupling and Hund's Rule for the "Iron Transition Group." Measured Values Are Seen to Be Better than the Value of Predicted Moment with $j = s$ (i.e., When the Orbital Moments Are Quenched)

Ion	Configuration	Basic level	$-g\sqrt{J(J+1)}$	$-g\sqrt{S(S+1)}$	$\mu_{measured}$
Ti^{3+}, V^{4+}	$3d^1$	$^2D_{3/2}$	1.55	1.73	1.8
V^{3+}	$3d^2$	$^3F_2$	1.63	2.83	2.8
Cr^{3+}, V^{2+}	$3d^3$	$^4F_{3/2}$	0.77	3.87	3.8
Mn^{3+}, Cr^{2+}	$3d^4$	$^5D_0$	0	4.90	4.9
Fe^{3+}, Mn^{2+}	$3d^5$	$^6S_{5/2}$	5.92	5.92	5.9
Fe^{2+}	$3d^6$	$^5D_4$	6.70	4.90	5.4
Co^{2+}	$3d^7$	$^4F_{9/2}$	6.63	3.87	4.8
Ni^{2+}	$3d^8$	$^3F_4$	5.59	2.83	3.2
Cu^{2+}	$3d^9$	$^2D_{5/2}$	3.55	1.73	1.9

and Am), the atoms exhibit magnetic behaviors like those of compounds in a +3 charge state, so they must be physically tightly bound to their parent ion so that they have relatively small overlap.

4. The large moments of the lanthanides and actinides must be sufficiently large to overcome crystal field effects provided by the local environment of the crystal structure. Otherwise, we would expect to observe a quenching of the orbital angular momentum of the electronic structure in highly crystalline compounds like the sesqui-oxides.

To complete the magnetic properties of the 3d, 4f, and 5f elements, we show in Table 6.5 the effective magnetic moment of some 3d ions predicted by L-S coupling and Hund's rule and the measured values for those ions.

Finally, Figure 6.16 shows the character of the ferromagnetic and antiferromagnetic state as a function of temperature for the lanthanide and actinide metals. The temperature at which the magnetism changes to the paramagnetic state is the curie temperature. Note that Cm metal occurs in both a double hexagonal close packed (DHCP) and a face centered cubic (FCC) structure.

NOTE In the above tables, we have expressed the "basic level" of the atoms in terms of their so-called "spectroscopic" designation $^{(2S+1)}L_J$ notation.

CONCLUSION We have seen that the magnetic moments of electrons in an atom are caused by their total spin or total orbital angular momentum. For atoms in a lattice, the magnetization of the ensemble will be a vector sum over all of the magnetic dipole moments caused by multiple electrons. Electrons in a full shell (from the periodic table of the elements) contribute no magnetic moment. The remaining

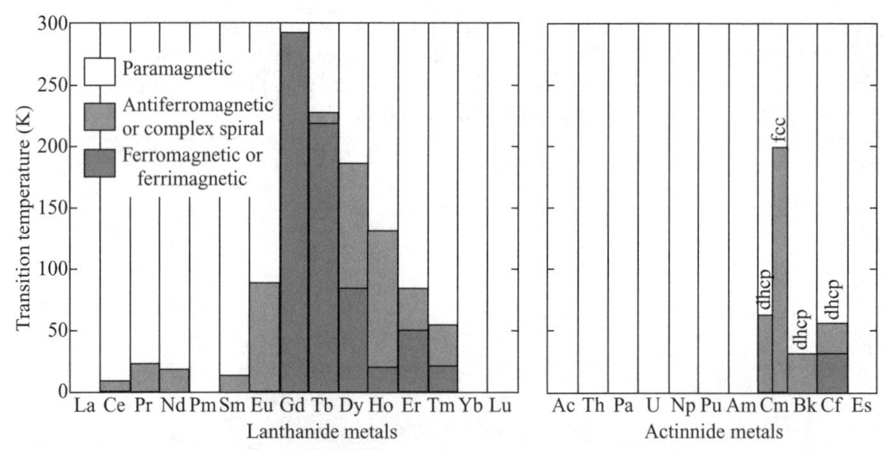

Figure 6.16 Measured magnetic character of the lanthanide and actinide metals as a function of temperature.

unpaired electrons can cause a substantial enhancement of the external magnetic flux density as they align with that field.

6.7 MAGNETIZATION

In the absence of an external magnetic field intensity, the magnetic dipole moments of atoms in most materials (except permanent magnets) have random orientations, resulting in no net magnetic moment. The application of an external magnetic field intensity induces an alignment of the electron spins and their orbital motion that causes an alignment of their magnetic dipole moments into distinct quantized orientations. The thermal collisions between atoms cause a change in the population of atoms in different energy states and result in an average magnetic moment for the ensemble. At low temperatures, low energy levels with magnetic moments parallel to the external magnetic field intensity are dominantly populated. At higher temperatures, higher energy orientations are populated according to a Boltzmann distribution. If the net magnetic alignment is generally opposite to the applied field, the material is said to be *diamagnetic*. If the net magnetic alignment is generally in the same direction as the applied field, the material is said to be *paramagnetic*.

In some materials like iron, cobalt, nickel, gadolinium, dysprosium, and their compounds or alloys, the spins of electrons in an atom can interact strongly at low temperatures with their neighbor atoms so that there is a strong alignment of their magnetic dipoles in spite of the randomizing effect of thermal collisions between neighbors. This alignment can give *ferromagnetic* materials permanent magnetism in the absence of an external field. Sometimes, the interaction of neighboring atoms causes an alternation in direction of the magnetic dipole moments in atomic layers or crystalline structures; materials that exhibit this type of alignment are called *antiferromagnets*, which can also exhibit complicated structures like spiral alignments of their spins. All magnetic structures are temperature dependent, and materials typically revert from strong interatomic magnetic coupling at low temperatures to *paramagnetism* or *diamagnetism* at temperatures above a critical temperature called the curie temperature.

Diamagnetism and Paramagnetism

To understand the partial alignment of magnetic dipole moments in a diamagnetic or paramagnetic material, let us consider the magnetic dipole moment of a single atom to be $\vec{\mu} = \vec{\mu}_{orbital} + \vec{\mu}_{spin}$, as described in Equation 6.49. In diamagnetism and paramagnetism, we assume neighboring atoms interact only slightly, so the main energy of interaction is between the electrons in individual atoms and the external applied magnetic flux. If there are n atoms per unit volume, we can then define a magnetization density vector, $\vec{M}$, as

$$\vec{M} = \lim_{\Delta V \to 0} \sum_{k=1}^{n\Delta V} \vec{\mu}_k / \Delta V, \tag{6.55}$$

which is the volume density of the individual magnetic dipole moments in a material. The magnetic dipole moment $d\vec{\mu}$ of an element volume dv' is thus $d\vec{\mu} = \vec{M}dv'$, and that moment, according to Equation 6.40b, will produce a magnetic vector potential, $d\vec{A}$, at the point $\vec{R}$ due to the magnetic dipole moment at $\vec{R}'$ of

$$d\vec{A} = (\mu_0/4\pi)\left(\vec{M} \times \hat{a}_{R-R'} / \left|\vec{R} - \vec{R}'\right|^2\right) dv', \tag{6.56}$$

where

$$\left|\vec{R} - \vec{R}'\right| = \sqrt{(x-x')^2 + (y-y')^2 + (z-z')^2}$$

and so

$$\vec{A} = \int_{V'} d\vec{A} = (\mu_0/4\pi) \int_{V'} \left(\vec{M} \times \hat{a}_{R-R'} / \left|\vec{R} - \vec{R}'\right|^2\right) dv'. \tag{6.57}$$

Now, by using Equation 3.95, $\vec{\nabla}'(1/|\vec{R} - \vec{R}'|) = \hat{a}_{R-R'}/|\vec{R} - \vec{R}'|^2$

$$\vec{A} = (\mu_0/4\pi) \int_{V'} \vec{M} \times \vec{\nabla}'\left(1/\left|\vec{R} - \vec{R}'\right|\right) dv' \tag{6.58}$$

and using the vector identity

$$\vec{M} \times \vec{\nabla}'\left(1/\left|\vec{R} - \vec{R}'\right|\right) = \left(1/\left|\vec{R} - \vec{R}'\right|\right)\vec{\nabla}' \times \vec{M} - \vec{\nabla}' \times \left(\vec{M}/\left|\vec{R} - \vec{R}'\right|\right), \tag{6.59}$$

we can write

$$\vec{A} = (\mu_0/4\pi) \int_{V'} \left(\vec{\nabla}' \times \vec{M} / \left|\vec{R} - \vec{R}'\right|\right) dv' - (\mu_0/4\pi) \int_{V'} \vec{\nabla}' \times \left(\vec{M}/\left|\vec{R} - \vec{R}'\right|\right) dv'. \tag{6.60}$$

PROBLEM

6.2 Use the divergence theorem $\iiint_{V'} \vec{\nabla} \cdot \vec{A} dv' = \oint\!\!\oint_{S'} \vec{A} \cdot d\vec{s}'$ with $\vec{A} = \vec{F} \times \vec{C}$ where $\vec{C}$ is a constant vector, to show that

$$\int_{V'} \vec{\nabla}' \times \vec{F} dv' = -\oint_{S'} \vec{F} \times d\vec{s}', \tag{6.61}$$

where $\vec{F}$ is any vector field with a continuous first derivative. ***HINT*** See *Schaum's Outlines of Vector Analysis* by Murray R. Spiegel.

Using the identity found in Equation 6.61 in the second integral of Equation 6.60, we can write

$$\vec{A} = (\mu_0/4\pi)\int_{V'}(\vec{\nabla}' \times \vec{M}/|\vec{R} - \vec{R}'|)dv' + (\mu_0/4\pi)\oint_{S'}(\vec{M} \times \hat{a}'/|\vec{R} - \vec{R}'|)\, ds', \tag{6.62}$$

where $\hat{a}_n'$ is the unit outward normal vector from ds', and S' is the surface bounding the volume V'.

6.8 EQUIVALENT SURFACE CURRENT DENSITY

Similar to the case of polarized charges in a dielectric material (Equations 3.100 and 3.101), we can define a volume current density

$$\vec{J}' = \vec{\nabla}' \times \vec{M} \tag{6.63}$$

and a surface current density

$$\vec{J}_S' = \vec{M} \times \hat{a}_n' \tag{6.64}$$

in which case

$$\vec{A} = (\mu_0/4\pi)\int_{V'}(\vec{J}'/|\vec{R} - \vec{R}'|)\, dv' + (\mu_0/4\pi)\oint_{S'}(\vec{J}_S'/|\vec{R} - \vec{R}'|)\, ds'. \tag{6.65}$$

We can interpret this equation with the help of a figure that shows the fields produced by a set of magnetic dipoles that constitute a volume in a material, as shown in Figure 6.17.

Note that adjacent interior currents cancel, while currents in the surface layer of thickness ***b*** do not. We can see from Figure 6.16 that $\vec{J}' = \vec{\nabla}' \times \vec{M} = 0$ interior to the sample but that, on the surface of the sample, we have an equivalent current density $\vec{J}_S' = \vec{M} \times \hat{a}_n' \neq 0$, so Equations 6.62 and 6.65 need consider only the second surface integral to calculate $\vec{A}$ and $\vec{B} = \vec{\nabla} \times \vec{A}$.

6.9 EQUIVALENT MAGNETIC MONOPOLE CHARGE DENSITY

In the current-free region of space, exterior to a sample volume, $\vec{J} = 0$, so from Equation 6.8, $\vec{\nabla} \times \vec{B} = \mu_0 \vec{J} = 0$; that is, the magnetic flux density in free space is curl free. Thus, magnetic flux density $\vec{B}$ can be expressed as the gradient of a scalar field, and we can write

$$\vec{B} = -\mu_0 \vec{\nabla} V, \tag{6.66}$$

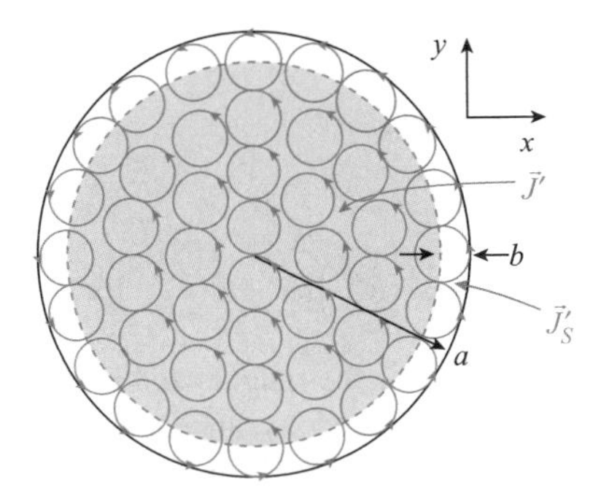

Figure 6.17 Representative cross section of a magnetic material with atomic scale magnetic dipole moments (there would be a much greater number of circles for a real sample) (arrows indicate the direction of current flow).

where V is called the scalar magnetic potential.[5] As in the case for scalar electric potential, V will be a conservative field, so we can conclude that, in free space, for two points P_1 and P_2

$$V_2 - V_1 = -\int_{P_1}^{P_2} (\vec{B}/\mu_0)\cdot d\vec{l} \tag{6.67}$$

independent of the path taken between P_1 and P_2.

If there were magnetic monopole charges with a volume density ρ_V (A/m^2),

$$V = (1/4\pi)\int_{V'} \left(\rho_V / |\vec{R} - \vec{R}'|\right) dv', \tag{6.68}$$

but, because isolated magnetic monopole charges have never been observed experimentally, they must be considered fictitious. Nevertheless, we have seen that, in the far-field region, the magnetic flux density caused by a small bar magnet is the same as that of a magnetic dipole moment. For a bar magnet, we often draw the fictitious north magnetic charge, $+q$, and the south magnetic charge, $-q$, as being separated by a distance $\boldsymbol{d}$ to form an equivalent magnetic dipole moment:

$$\vec{m} = q\vec{d} = (\hat{a}_n IS). \tag{6.69}$$

The scalar magnetic potential V caused by this magnetic dipole moment can then be written as

$$V = \vec{m}\cdot\hat{a}_{R-R'} \Big/ 4\pi|\vec{R} - \vec{R}'|^2. \tag{6.70}$$

If, in terms of a magnetization vector $\vec{M}$ (volume density of magnetic dipole moments), with $d\vec{m} = \vec{M}dv'$, we write

$$dV = \left(\vec{M}\cdot\hat{a}_{R-R'} \Big/ 4\pi|\vec{R} - \vec{R}'|^2\right) dv' \tag{6.71}$$

[5] Maxwell specifically included the scalar magnetic potential (which he called Ω) in his quaternion expressions 1.68. The Heavyside vector field interpretation of Maxwell's quaternions usually ignores this field except as a mathematical convenience.

and

$$V = (1/4\pi)\int_{V'}\left(\vec{M}\cdot\hat{a}_{R-R'}/\left|\vec{R}-\vec{R}'\right|^2\right)dv', \tag{6.72}$$

then V can be expressed as

$$V = (1/4\pi)\oint_{S'}\left(\vec{M}\cdot\hat{a}'/\left|\vec{R}-\vec{R}'\right|^2\right)ds' + (1/4\pi)\int_{V'}\left(-\vec{\nabla}'\cdot\vec{M}/\left|\vec{R}-\vec{R}'\right|^2\right)dv', \tag{6.73}$$

where $\hat{a}_n'$ is the outward normal to the surface element ds' of the magnetized body.

For field calculation, a magnetized body may be replaced by an equivalent (fictitious) magnetization monopole surface charge density ρ_S and an equivalent (fictitious) magnetization monopole volume charge density ρ_V such that

$$\begin{aligned}\rho_S &= \vec{M}\cdot\hat{a}_n \\ \rho_V &= -\vec{\nabla}\cdot\vec{M}\end{aligned} \tag{6.74}$$

We can, for example, consider the cylindrical sample of magnetic material that was shown in Figure 6.12 from the horizontal perspective, as shown in Figure 6.18.

N indicates a north magnetic monopole charge, and S indicates a south magnetic monopole charge. Note that all interior magnetic monopole charges cancel, while charges on the upper and lower surfaces create a magnetic monopole charge density in a half atomic layer thickness.

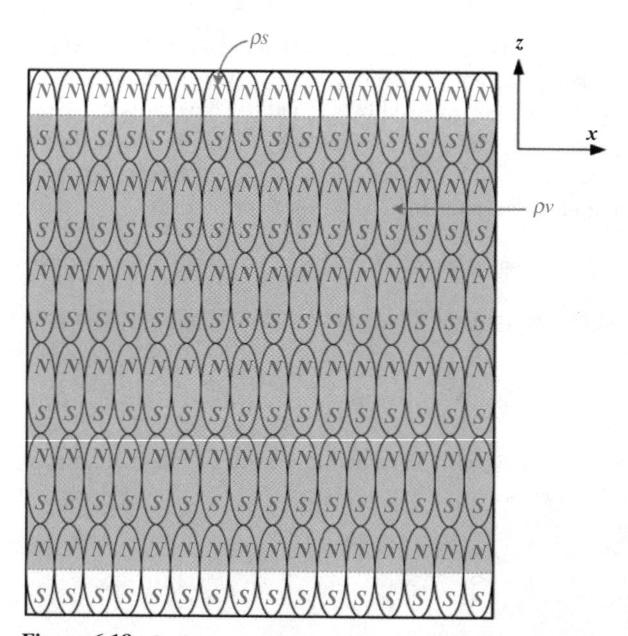

Figure 6.18 Representative cross section of a magnetic material with atomic scale magnetic dipole moments (there would be a much greater number of magnetic dipole moments for a real sample).

Conclusions

We can conclude from the last two sections that the magnetic flux density, $\vec{B}$, for a cylindrical-shaped material may be calculated as shown in Figure 6.19 via the following:

a. A volume integral for the magnetic vector potential over the magnetization $\vec{M}$

$$\vec{B} = \vec{\nabla} \times \vec{A} = \vec{\nabla} \times (\mu_0/4\pi) \int_{V'} \vec{M} \times \vec{\nabla}' \left(1/\left|\vec{R} - \vec{R}'\right|\right) dv' \tag{6.75}$$

b. A surface integral for the magnetic vector potential over the material

$$\vec{B} = \vec{\nabla} \times \vec{A} = \vec{\nabla} \times (\mu_0/4\pi) \oint_{S'} \left(\vec{J}_S'/\left|\vec{R} - \vec{R}'\right|\right) ds' \tag{6.76}$$

with $\vec{J}_S' = \vec{M} \times \hat{a}_\rho$ on the cylindrical surface, or

c. A surface integral for the scalar magnetic potential over the material

$$\vec{B} = -\mu_0 \vec{\nabla} V = -(\mu_0/4\pi) \vec{\nabla} \oint_{S'} \left(\rho_S/\left|\vec{R} - \vec{R}'\right|^2\right) ds' \tag{6.77}$$

with $\rho_S = \vec{M} \cdot \hat{a}_z$ on the top and bottom surfaces

6.10 MAGNETIC FIELD INTENSITY AND PERMEABILITY

When an external magnetic flux density, $\vec{B}$, caused by a free-electric current density, $\vec{J}$, is applied to a material, a partial alignment of the internal magnetic dipole moments occurs according to the atomic interaction between the individual magnetic

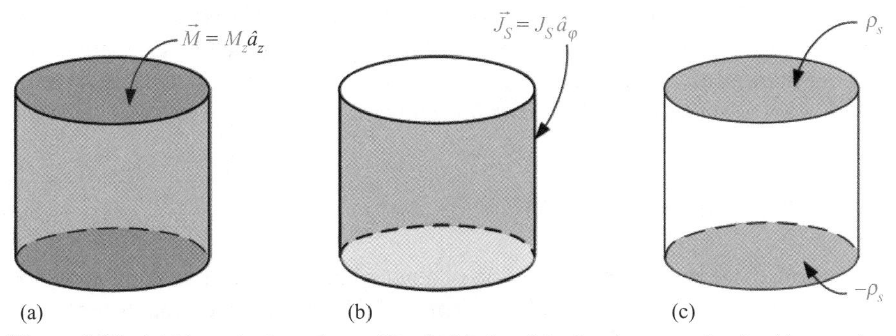

Figure 6.19 (a) Magnetization volume; (b) cylindrical surface electric current density; (c) top and bottom surface magnetic monople charge density for the equivalent integral calculations of magnetic flux density.

moments and the temperature of the sample. In Equation 6.62, we calculated the additional magnetic vector potential due to the induced magnetization of the atomic dipoles via an expression just like Equation 6.21 for the free-electric current density. However, in Equation 6.63, we used an equivalent volume current density, $\vec{\nabla} \times \vec{M}$, to find the magnetic vector potential due to the magnetic dipole moment alignment. If we reason that the sum of the two currents gives the magnetic flux density caused by free currents, $\vec{J}$, and the induced magnetization from a material, we can generally rewrite the point form of Ampere's law (Equation 6.8) for materials as

$$\vec{\nabla} \times (\vec{B}/\mu_0) = \vec{J} + \vec{\nabla} \times \vec{M} \tag{6.78}$$

or

$$\vec{\nabla} \times (\vec{B}/\mu_0 - \vec{M}) = \vec{J}. \tag{6.79}$$

The quantity ($\vec{B}/\mu_0$ – $\vec{M}$) is a fundamental field quantity in magnetism that incorporates not only the applied magnetic flux density but the induced magnetization caused by materials. This quantity is called the magnetic field intensity, $\vec{H}$, where

$$\vec{H} = \vec{B}/\mu_0 - \vec{M} \tag{6.80}$$

so that

$$\vec{\nabla} \times \vec{H} = \vec{J}, \tag{6.81}$$

where the right side is the volume density of free current. The corresponding integral form of Equation 6.81 is

$$\int_S (\vec{\nabla} \times \vec{H}) \cdot d\vec{s} = \int_S \vec{J} \cdot d\vec{s}$$

or

$$\oint_C \vec{H} \cdot d\vec{l} = I \tag{6.82}$$

Equation 6.82 is a more complete form of Ampere's law that implies that "**the circulation of the magnetic field intensity around any closed path is equal to the free electric current flowing through the surface bounded by the path.**" This is a very powerful statement that we will use to calculate magnetic field intensity in space, *even when magnetic materials are present.*

6.11 FERROMAGNETISM

In general, the degree of magnetization of a material will depend upon the direction of the applied external magnetic field intensity. This is especially true of ferromagnets in which there is typically a so-called *easy axis* along which materials may be magnetized that is caused by their crystal structure or grain boundaries that can orient during sample forging or drawing processes. It is also generally the case that sample geometry is important in the magnetization of materials (e.g., long, thin rods are easier to magnetize along their axis of symmetry than in a perpendicular direction). For our understanding of magnetic phenomena, we will first consider a special case in which the magnetic properties of the medium are linear and isotropic.

Special Case

If a magnetic material is linear and isotropic (such as is the case of diamagnetism and paramagnetism),

$$\vec{M} = \chi_m \vec{H}, \tag{6.83}$$

where the constant of proportionality, χ_m, is a dimensionless quantity called the *magnetic susceptibility*. Putting this linear relation back into Equation 6.80, we see

$$\vec{B} = \mu_0(1+\chi_m)\vec{H} = \mu_0\mu_r\vec{H} = \mu\vec{H}, \tag{6.84}$$

where $\mu_r = 1 + \chi_m = \mu/\mu_0$ is a dimensionless quantity called the relative permeability of the medium. $\mu = \mu_0\mu_r$ is usually called the *absolute permeability* (or often just the permeability) of the medium.

Ferromagnetic Applications

The term *susceptibility* is not normally used with ferromagnets because the magnetic flux density is not linearly proportional to the magnetic field intensity. However, the relation $\vec{B} = \mu\vec{H}$ is often used for ferromagnetic materials, with the knowledge that μ is not a constant but depends strongly upon $|\vec{H}|$.

A typical plot of B versus H for a ferromagnetic material is shown in Figure 6.20. The magnetization curves for different iron compounds depend highly on the exact composition and heat treatment, as shown in Figure 6.20, for a number of practical alloys used in motors and transformers. Arrows on each curve indicate the direction of increasing H for the first magnetic induction of B in the material.

The subject of ferromagnetism is complicated by the fact that magnetic moments minimize their energy by forming into magnetic domains (small clumps) that are reluctant to move until they overcome the polycrystalline grain boundaries of the material that "pin" them in place. Once the domains achieve enough field to overcome their "pinning" by the grain boundaries, they orient in the same direction and tend toward saturation where large numbers (and density) of moments then contribute to magnetization. The domains "remember" their orientation because the process is not reversible; a coercive force must be present to coerce the domains to overcome their grain boundary pinning and align in the other direction. Thus, we often see magnetization curves that exhibit a hysteretic behavior as the field intensity is initially increased and then reduced back to zero, as shown in Figure 6.21. Here, a *remnance* of magnetic flux density, B_r, remains induced in the magnetic material (a memory of the ferromagnetic alignment) even after the initial field intensity has been removed. This is the effect that creates permanent magnets. If the magnetic field

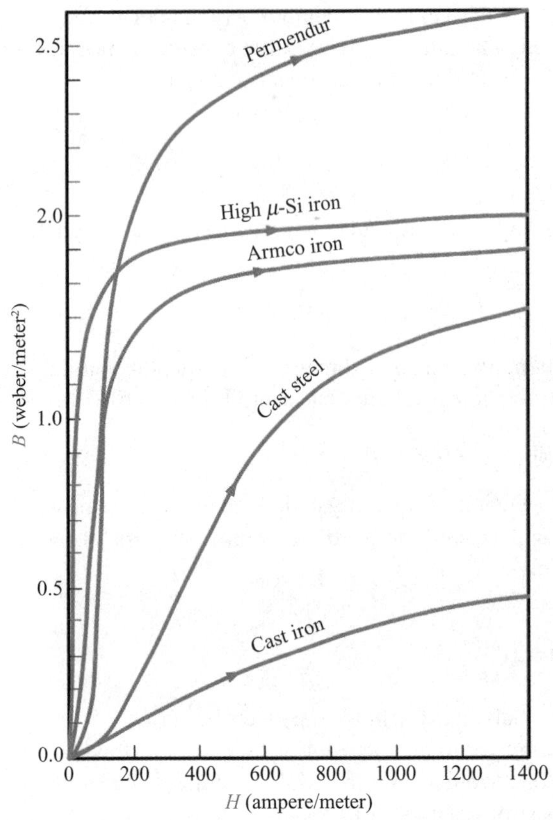

Figure 6.20 Curves of magnetic flux density, b, for several ferromagnetic 3d materials below their curie temperature as a function of externally applied magnetic field intensity, h.

intensity is further reduced to negative values, some of the magnetic domains begin to align in the opposite direction to the point where there is no longer any net induced magnetic flux density in the material; the applied field for which this occurs is called the *coercive* field intensity (sometimes called the *coercive force*), H_c. Continuing to decrease the magnetic field intensity will result in a saturation of the domains in the opposite direction from their initial saturation direction. After that, increasing and decreasing the applied magnetic field intensity cause the magnetization to follow the external curve, as shown in Figure 6.21 for Alnico V iron. In practical applications, there will likely be multiple *turns* of the inducing field wires that proportionately increase the applied field intensity, so the horizontal axis includes units of ampere-*turns* per meter.

Analyses of magnetic circuits are often based on Ampere's law (Equations 6.81 and 6.82) in which $\vec{\nabla} \times \vec{H} = \vec{J}$ or $\oint_C \vec{H} \cdot d\vec{l} = I$. For example, if a closed path C is

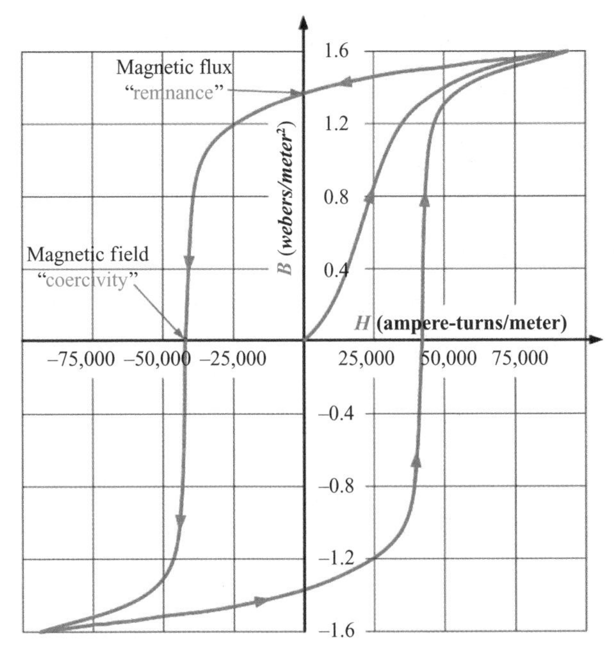

Figure 6.21 Hysteresis curve for ferromagnetic alnico V iron.

chosen to enclose N turns of a winding carrying a current, I, in a toroidal configuration such as that shown in Figure 6.22, we have

$$\oint_C \vec{H} \cdot d\vec{l} = NI. \tag{6.85}$$

The toroidal geometry is useful because, except for some small leakage flux, the magnetic flux is primarily contained within the donut-shaped torus, and it is relatively constant as a function of the poloidal radius (the small radius of the donut body). For this reason, the geometry is employed in a Tokomak confinement scheme for plasmas used in fusion reactors. The geometry is often also used in the construction of high-magnetic field magnets used in chemistry or physics labs where an air gap (also shown in Figure 6.22) is created in the torus so that samples may be inserted to test their magnetic properties.

Using Ampere's law for this configuration with the path C taken to be a circle of radius r_0, as shown in Figure 6.22, we can see that the magnetic field intensity $\vec{H}$ will lie in a direction parallel to C at all points and will be a constant over this path. Thus,

$$\oint_C \vec{H} \cdot d\vec{l} = |\vec{H}| 2\pi r_0 = NI \text{ or } |\vec{H}| = NI/2\pi r_0. \tag{6.86}$$

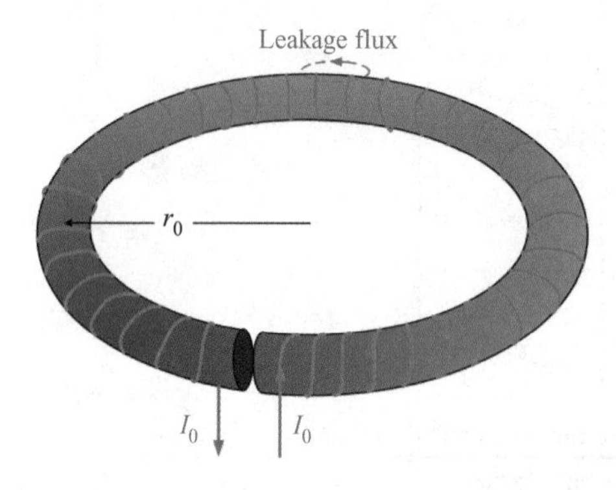

Figure 6.22 Current carrying coil on a toroid form with an air gap.

Note that the magnetic field intensity is the same for any toroidal material, but the magnetic flux density depends upon the magnetic permeability of the torus as

$$|\vec{B}| = \mu_r \mu_0 NI/2\pi r_0. \tag{6.87}$$

If the toroidal form is a diamagnetic or paramagnetic (e.g., copper), then $\mu_r \approx 1$, and we can see that the largest magnetic flux density that can be obtained in the air gap (which the magnetic flux extends across) depends upon the number of turns, N, the current, I, and is inversely proportional to r_0. The number of turns, N, is often alternately written as $N = N_l 2\pi r_0$, where N_l is the number of turns per unit length. In this form, for a given number of turns per unit length, Equation 6.83 becomes simply $|\vec{B}| = \mu_0 N_l I$.

If the toroidal form is a ferromagnet (e.g., iron), then $\mu_r \approx 10^4$ unless the magnetic domains all lie in the direction of the applied external magnetic field intensity; in this case, we say the material is saturated because there is no more magnetization to be had from more magnetic field intensity except that applied by the current. The engineering associated with ferromagnetism is closer to *hammer-and-tongs* forgery than science because the relative permeability and the saturation field of iron depend strongly on its impurities and polycrystalline structure or other alloy materials. The science of ferromagnetic materials is closer to an art because the exchange mechanism between material atoms is so complex. Historically, engineers have tested many compounds and alloys of materials in an attempt to find the largest value of μ_r for applied magnetic field intensities before they begin to saturate. Modern applications include the use of magnetic *chokes* that operate in the reversible field region where the inductance ($\Phi = BA = LI$) is high until they experience a large current to induce saturation of the domains (where the effective inductance is relatively lower due to a lower value of dB/dH). In radar applications, the high to low inductance transition can act as a fast, high-power switch to turn on high current at a rapid rate; for example, for 10^6 V, current can rise as fast as 10^3 A in 10 ns.

6.12 BOUNDARY CONDITIONS FOR MAGNETIC FIELDS

Normal Direction

Similarly as in the process we used for electric fields, we can see from Figure 6.23 $\oint_S \vec{B}\cdot d\vec{s} = 0$ and in the limit as $\Delta h \to 0$ we have $\vec{B}_1 \cdot A_1 \hat{a}_{n2} + \vec{B}_2 \cdot A_2 \hat{a}_{n1} = 0$.

Considering $A_1 = A_2$ and $\hat{a}_{n2} = -\hat{a}_{n1}$, we finally have

$$B_{1n} = B_{2n}. \tag{6.88}$$

For linear media, $\vec{B}_1 = \mu_1 \vec{H}_1$, and $\vec{B}_2 = \mu_2 \vec{H}_2$, Equation 6.88 becomes

$$\mu_1 H_{1n} = \mu_2 H_{2n}. \tag{6.89}$$

Tangential Direction

Starting from Ampere's equation,

$$\oint_C \vec{H}\cdot d\vec{l} = I_{enclosed}. \tag{6.90}$$

Choosing the integral path, as shown in Figure 6.24, in the limit as $\Delta h \to 0$, we have

$$\oint_C \vec{H}\cdot d\vec{l} = \vec{H}_1 \cdot w\hat{a}_{t,1} + \vec{H}_2 \cdot (-w\hat{a}_{t,1}) = J_l w$$
$$H_{1t} - H_{2t} = J_l \tag{6.91a}$$

or

$$\hat{a}_{n2} \times \left(\vec{H}_1 - \vec{H}_2\right) = \vec{J}_l, \tag{6.91b}$$

where J_l is the surface current density normal to the contour C.

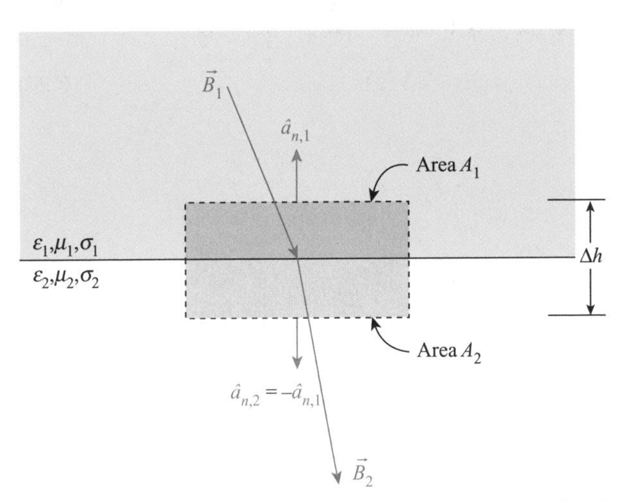

Figure 6.23 Gaussian surface in the form of a pillbox extending through the interface between two materials.

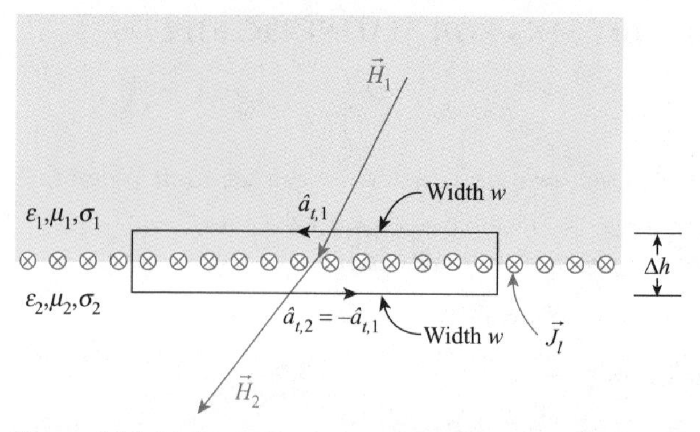

Figure 6.24 Path c used in the integral of magnetic field intensity that penetrates the interface between two materials.

6.13 INDUCTANCE AND INDUCTORS

Consider two neighboring closed loops, C_1 and C_2, bounding surfaces S_1 and S_2, respectively, as shown in Figure 6.25. If a current I_1 flows in C_1, a magnetic field $\vec{B}_1$ will be created.[6] Let us designate a mutual flux Φ_{12}.

$$\Phi_{12} = \int_{S_2} \vec{B}_1 \cdot d\vec{s}_2 \tag{6.92}$$

From the Biot–Savart law, Equation 6.32, we know that $\vec{B}_1$ is directly proportional to I_1; hence, Φ_{12} is also proportional to I_1

$$\Phi_{12} = L_{12} I_1, \tag{6.93}$$

where L_{12} is called the mutual inductance between loops C_1 and C_2, with units Henry (H) = Wb/A. If C_2 has N_2 turns, the flux linkage Λ_{12} due to Φ_{12} is

$$\Lambda_{12} = N_2 \Phi_{12} (\text{Wb}). \tag{6.94}$$

Equation 6.94 yields

$$\Lambda_{12} = L_{12} I_1$$

or

$$L_{12} = \frac{\Lambda_{12}}{I_1} = \frac{N_2}{I_1} \int_{S_2} \vec{B}_1 \cdot d\vec{s}_2. \tag{6.95}$$

The mutual inductance between two circuits is then the magnetic flux linkage with one circuit per unit current in the other.

[6] Lenz showed that a time-varying I_l (and, therefore, a time-varying Φ_{12}) will produce an induced electromotive force or voltage in C_2, as we shall see in Chapter 7 when we consider time-variable magnetic flux density.

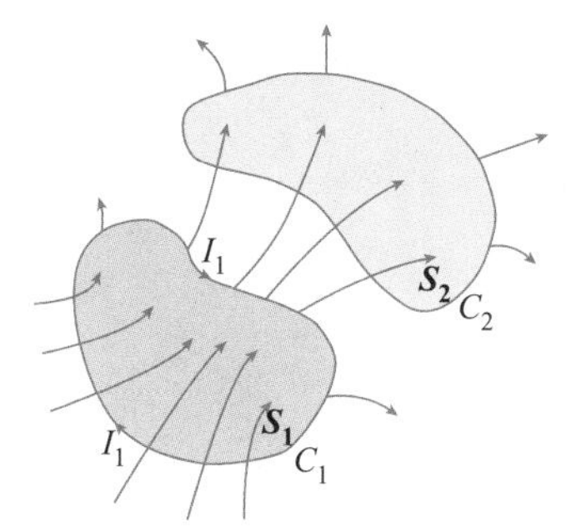

Figure 6.25 Two magnetically flux coupled loops.

A more general definition for L_{12} is

$$L_{12} = \frac{d\Lambda_{12}}{dI_1} \tag{6.96}$$

The total flux linkage with C_1 caused by I_1 is

$$\Lambda_{11} = N_1\Phi_{11} > N_1\Phi_{12}. \tag{6.97}$$

The self-inductance of loop C_1 is defined as the magnetic flux linkage per unit current in the loop itself; that is,

$$L_{11} = \frac{\Lambda_{11}}{I_1} = N_1 \int_{S_1} \left(\vec{B}_1/I_1\right) \cdot d\vec{s}_1 \tag{6.98}$$

for a linear medium, in general

$$L_{11} = \frac{d\Lambda_{11}}{dI_1}. \tag{6.99}$$

EXAMPLE

6.4 If an air-filled coaxial transmission line (Figure 6.26) has a solid inner conductor of radius, a, and a very thin outer conductor of inner radius, b, determine the inductance per unit length of the line.

SOLUTION See Example 6.1.

1. Inside the inner conductor

$$(0 \le r \le a):$$

$$\vec{B}_1 = \hat{a}_\phi B_{\phi 1} = \hat{a}_\phi \frac{\mu_0 r}{2\pi a^2} I$$

2. Outside the inner conductor:

$$(a \le r \le b):$$

$$\vec{B}_2 = \hat{a}_\phi B_{\phi 2} = \hat{a}_\phi \frac{\mu_0}{2\pi r} I$$

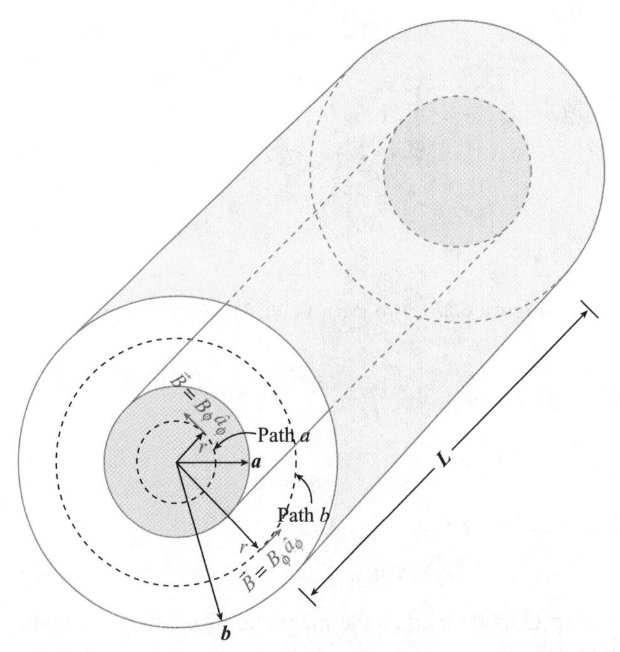

Figure 6.26 Perspective view of an air-filled coaxial transmission line.

Now, consider an angular ring in the inner conductor between radii r and $r + dr$. The current in a unit length of this annular ring is linked by the magnetic flux that can be obtained by integrating magnetic flux densities.

$$d\Phi' = \int_r^a B_{\phi 1} dr + \int_a^b B_{\phi 2} dr = \left(\mu_0 I/2\pi a^2\right)\int_r^a r dr + (\mu_0 I/2\pi)\int_a^b (1/r)\,dr$$
$$= \left(\mu_0 I/4\pi a^2\right)\left(a^2 - r^2\right) + (\mu_0 I/2\pi)\ln\frac{b}{a}$$

The current in the annular ring is only a fraction of the total current $I(2\pi r dr/\pi a^2 = 2rdr/a^2)$, and the flux linkage for this annular ring is

$$d\Lambda' = \left(2rdr/a^2\right)d\Phi'.$$

The total magnetic flux linkage per unit length is

$$\Lambda' = \int_0^a d\Lambda' = \left(\mu_0 I/\pi a^2\right)\left[\left(1/2a^2\right)\int_0^a \left(a^2 - r^2\right) r dr + \ln(b/a)\int_0^a r dr\right]$$

$$\Lambda' = (\mu_0 I/2\pi)[1/4 + \ln(b/a)]$$

The inductance of a unit length of the coaxial transmission line is therefore

$$L' = \Lambda'/I = (\mu_0/2\pi)[1/4 + \ln(b/a)] = (\mu_0/8\pi) + (\mu_0/2\pi)\ln(b/a).$$

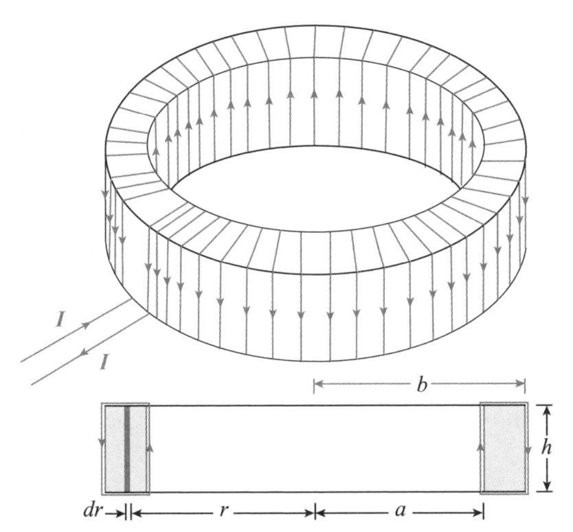

Figure 6.27 A closely wound toroidal coil with a rectangular cross section.

EXAMPLE

6.5 N turns of wire are tightly wound on a toroidal frame of rectangular cross section with dimensions as shown in Figure 6.27. Assuming the permeability of the medium to be μ_0, find the self-inductance of the toroidal coil.

SOLUTION

Assuming a current I in the conducting wire,

$$\vec{B} = B_\phi \hat{a}_\phi$$

and

$$d\vec{l} = \hat{a}_\phi r d\phi$$

so

$$\oint_C \vec{B} \cdot d\vec{l} = 2\pi r B_\phi = \mu_0 NI$$

so

$$B_\phi = \mu_0 NI/2\pi r$$

so that

$$\Phi = \int_S \vec{B} \cdot d\vec{s} = \int_S \left(\frac{\mu_0 NI}{2\pi r} \hat{a}_\phi \right) \cdot (\hat{a}_\phi h dr) = \frac{\mu_0 NIh}{2\pi r} \int_a^b \frac{dr}{r} = (\mu_0 NIh/2\pi r) \ln(b/a).$$

The magnetic flux linkage is

$$\Lambda = N\Phi = \left(\mu_0 N^2 Ih/2\pi r\right) \ln(b/a)$$

so that we obtain

$$L = \Lambda/I = \left(\mu_0 N^2 h/2\pi r\right) \ln(b/a)(H).$$

6.14 TORQUE AND ENERGY

Torque on a Magnetic Dipole

The torque on a rectangular loop of wire of length a and width b carrying current I is shown in Figure 6.28.

The forces on the four individual sides of the rectangular loop are shown as F_1, F_2, F_3, and F_4; the two forces on sides of 2 and 4 cancel one another, while the two forces on sides 1 and 3 cause a torque:

$$\vec{\tau} = \vec{\mu} \times \vec{B} \tag{6.100}$$

that tends to align the magnetic moment of the loop with the external magnetic field intensity.

PROBLEM

6.3 Show that the torque on a *circular loop* of radius b (area πb^2) carrying current I gives the same torque as 6.100.

NOTE Equation 6.100 is the same form as that for torque that tends to align an electric dipole, $\vec{p}$, in an external electric field intensity, $\vec{E}$.

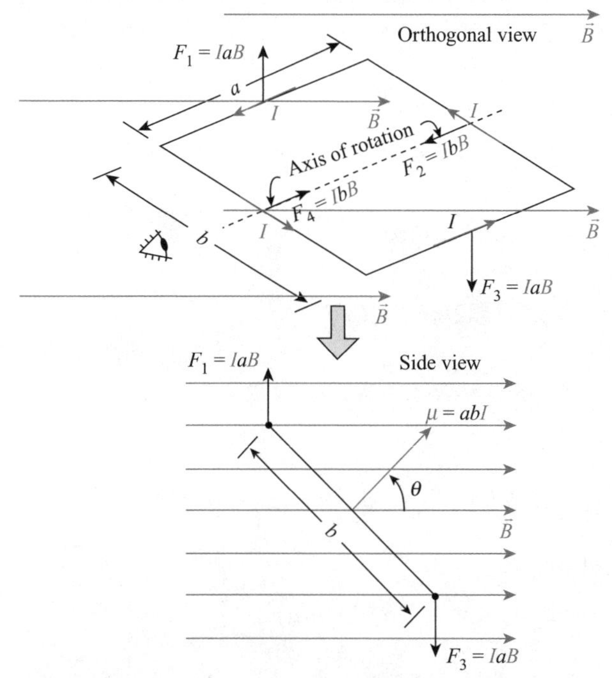

Figure 6.28 Rectangular loop of wire carrying current I in an external magnetic flux density B. The loop is free to rotate about an axis, as shown in the top isometric view; below that is a side view.

The **potential energy** of a magnetic dipole moment, $\vec{\mu}$, that makes angle θ with respect to an external magnetic flux density, $\vec{B}$, is

$$U_{\parallel}(\theta) = \int_0^{\theta} \tau(\theta')\,d\theta' = \mu B(1-\cos\theta). \tag{6.101}$$

Because the zero of potential energy is arbitrary, we can alternately express the potential energy relative to the angle $\theta = \pi/2$ as

$$U_{\perp}(\theta) = \int_{\pi/2}^{\theta} \tau(\theta')\,d\theta' = -\mu B\cos\theta = -\vec{\mu}\cdot\vec{B} \tag{6.102}$$

Equation 6.102 is the form that Zeeman used to express the potential energy of different magnetic states in an external magnetic flux density; that is, if $\vec{\mu}$ is aligned with $\vec{B}$, its potential energy is $-\mu B$; if opposed, its potential energy is μB. The energy level representation of this equation is shown in Figure 6.29 for a magnetic dipole moment with spin quantum number $m_s = ½$ or $m_s = -½$, $\mu_z = -gJ\mu_B = -2.003\ S\mu_B$ (as would be the case for a hydrogen atom electron with only spin angular momentum).

In the special case that a system has only two levels the equilibrium populations, N_1 and N_2 are given[iv] by the Boltzmann distribution

$$\frac{N_1}{N} = \frac{e^{(\mu B/k_BT)}}{e^{(\mu B/k_BT)} + e^{(-\mu B/k_BT)}} \text{ and } \frac{N_2}{N} = \frac{e^{(-\mu B/k_BT)}}{e^{(\mu B/k_BT)} + e^{(-\mu B/k_BT)}}. \tag{6.103}$$

Here, N_1 is the population of the lower level, N_2 is the population of the upper level, and $N = N_1 + N_2$. The fractional population of the magnetic moments of the upper and lower states is plotted in Figure 6.30.

The resultant magnetization for N atoms per unit volume is

$$M = (N_1 - N_2)\mu = N\mu \frac{e^{(\mu B/k_BT)} - e^{(-\mu B/k_BT)}}{e^{(\mu B/k_BT)} + e^{(-\mu B/k_BT)}} \approx N\mu_B \tanh\left(\frac{\mu_B B}{k_BT}\right), \tag{6.104}$$

where the approximate symbol comes from $\mu = g(1/2)\mu_B$ and $g \approx 2$. This function is shown plotted in Figure 6.31.

Note that the comparison of the average magnetic dipole moment, $\langle\mu\rangle$, plotted as a function of $(\mu_B B/k_BT)$ in Figure 6.30 to that of the average electric dipole moment, $\langle p\rangle$, plotted as a function of (p_0E/k_BT) in Figure 5.10 of The Foundations of Signal Integrity[v]. Main differences areas follows:

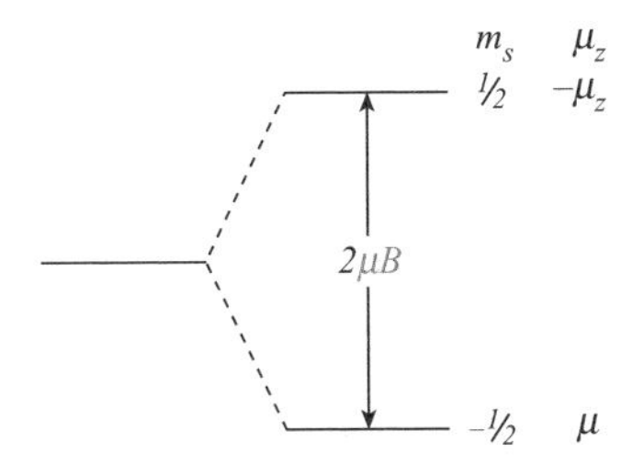

Figure 6.29 Energy level "Zeeman" splitting for one electron in a magnetic flux density B directed along the z axis.

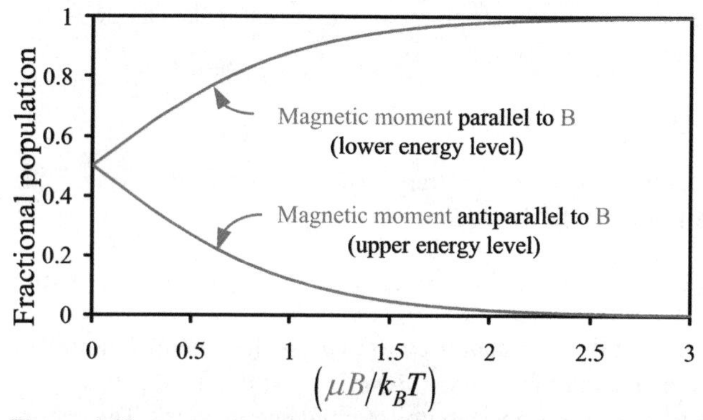

Figure 6.30 Relative populations of electrons with total angular momentum quantum numbers ±½ in a magnetic flux density, B, in thermal equilibrium at temperature t.

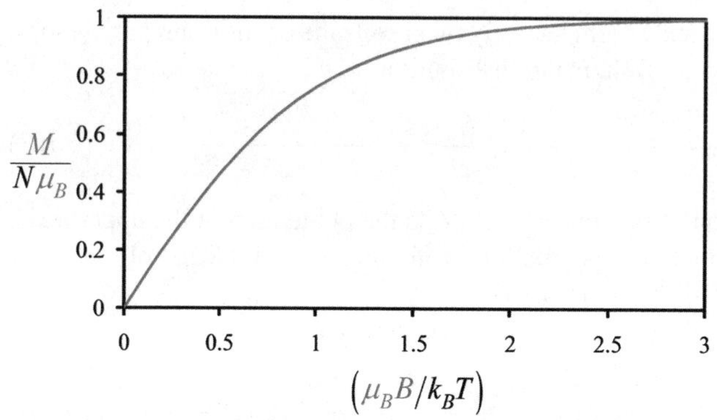

Figure 6.31 Average magnetic dipole moment per atom, $\langle \mu \rangle = M/N\mu_B$, plotted as a function of $(\mu_B B/k_B T)$.

1. The orientation of electric dipoles is continuous with angle, while magnetic dipoles are either parallel or antiparallel to the external fields.
2. $p_0 E$ is typically much smaller than $\mu_B B$.

For the more general case of an atom with angular momentum quantum number J, there will be $2J + 1$ equally spaced energy levels, with a magnetization given in terms of the Brillouin function, B_J:

$$M = NgJ\mu_B B_J\left(\frac{\mu_B B}{k_B T}\right), \tag{6.105}$$

where

$$B_J\left(\frac{\mu_B B}{k_B T}\right) = \frac{2J+1}{J}\coth\left(\frac{g(2J+1)}{2}\frac{\mu_B B}{k_B T}\right) - \frac{1}{2J}\coth\left(\frac{\mu_B B}{k_B T}\right) \tag{6.106}$$

Summary Conclusions

1. Magnetic forces are a result of electric monopole charges, q, and the vector cross product of their velocity, $\vec{v}$, and an externally applied magnetic flux density, $\vec{B}$. Maxwell's equations for divergence and curl are similar to their electric analogs if we assume that there is no magnetic monopole charge, q.
2. Electric dipole fields and magnetic dipole fields follow the same far-field basic geometric distributions if we take the electric dipole moment to be $\vec{p} = q\vec{d}$ and the magnetic dipole moment to be $\vec{m} = i\pi b^2 \hat{a}_\perp$.
3. Magnetic flux density, $\vec{B}$, can be calculated through the use of Ampere's equation if the current distribution is known and circuit self and mutual inductances, L_{11} and L_{12}, can be calculated from the ratio of the magnetic flux to that current by $L = \Phi/I$.
4. If electronic orbitals of atoms cause outer electrons to be disassociated from their parent ions, band structure formalism is required to determine magnetic effects.
5. Material magnetism is primarily a result of quantum effects of electronic states in atoms. The orbital and spin angular momentum of electrons yields a net vector magnetic moment whose z-component can be described in terms of the projection of the magnetic moment along the applied field direction. For atoms with unfilled shells, the vector sum of all such electronic states, consistent with the Pauli Exclusion Principle and Hund's rule, can be added to find effective atomic magnetic moments for isolated ions in chemical compounds.
6. At temperatures in the paramagnetic regime, the population of possible quantum states follows a Boltzmann distribution. The net magnetization depends on the sum of the proportion of those states individual magnetic moments.
7. If the crystal structure of a chemical compound is tight enough to cause the states of extended atomic orbitals, for example, in many 3d ions, to be quenched by their neighbors, the magnetic moments will be caused primarily by the atom's net spin quantum number.
8. In the paramagnetic regime, the magnetic vector potential of a specimen may be calculated by using three equivalent mathematical formalisms: direct calculation due to the vector sum of individual magnetic dipole moments, equivalent macroscopic surface currents, and a surface distribution of magnetic monopoles.
9. For tightly bound electrons, for example, in many 4f and 5f atoms, the crystal structure permits the total angular momentum to determine the net magnetic moments.
10. Below a curie temperature, the exchange interaction of atomic orbitals with their neighbors can overcome the Boltzmann thermal energy to form local

magnetic ordering. In a few cases, antiferromagnetic ordering occurs at a higher temperature, and ferromagnetism occurs at lower temperature.

11. Magnetism of metals and alloys is a result of magnetic domains that align in directions determined by the physical geometry, polycrystalline grain boundaries, and impurities. Once pinning energy is exceeded, domains can align to form permanent magnets.

ENDNOTES

i. Paul G. Huray, *The Foundations of Signal Integrity* (Hoboken, NJ: John Wiley & Sons, 2009), Chapter 4.

ii. http://arxiv.org/abs/hep-ex/0302011.

iii. Charles Kittel, *Introduction to Solid State Physics*, 7th ed. (Hoboken, NJ: John Wiley & Sons, 1996), 451.

iv. Ibid., 421.

v. Huray, *The Foundations of Signal Integrity*, Figure 5.10.

Chapter 7

Time-Varying Fields

LEARNING OBJECTIVES

- Use Maxwell's equations to derive the macroscopic relations between time-varying electromagnetic field quantities $\vec{E}$, $\vec{D}$, $\vec{H}$, and $\vec{B}$
- Develop the concept of time-dependent scalar electric potential V and magnetic vector potential $\vec{A}$ and relate them to the electromagnetic field quantities $\vec{E}$, $\vec{D}$, $\vec{H}$, and $\vec{B}$
- Derive time dependent boundary conditions that apply to $\vec{E}$, $\vec{D}$, $\vec{H}$, and $\vec{B}$
- Review separation of variable and Green's functions techniques and employ them to find causal solutions to Maxwell's equations using time-dependent scalar and vector fields
- Review the use and characteristics of the electromagnetic frequency spectrum
- Show that time-dependent magnetic fields in a dielectric medium penetrate conductor boundaries and slowly propagate into the conductor inducing electric fields in both regions thereby creating "surface impedance, Z_s"
- Explore the field interpretation of memristance and electric vector potential
- Compare field definitions of the fully symmetric Maxwell's equations with circuit definitions of resistance, inductance capacitance, and memristance

INTRODUCTION

In constructing an electro*static* model in the first chapters of this text, an electric field intensity vector field, $\vec{E}$, and an electric flux density (electric displacement) field, $\vec{D}$ have been given a definition that satisfy the relations

$$\vec{\nabla} \times \vec{E} = 0 \tag{7.1}$$

$$\vec{\nabla} \cdot \vec{D} = \rho_V. \tag{7.2}$$

For linear, isotropic (homogeneous) media, the macroscopic electric field quantities $\vec{E}$ and $\vec{D}$ are proportional to one another through the constitutive relation

$$\vec{D} = \varepsilon \vec{E}. \tag{7.3}$$

For the magneto*static* model, we have also defined the magnetic field intensity vector field, $\vec{H}$, and a magnetic flux density field, $\vec{B}$, that satisfy another set of

Maxwell's Equations, by Paul G. Huray

Table 7.1 Fundamental *Static* Relations

Fundamental relations	Electro*static* model	Magneto*static* model
Governing equations	$\vec{\nabla}\times\vec{E}=0$ $\vec{\nabla}\cdot\vec{D}=\rho_V$	$\vec{\nabla}\times\vec{H}=\vec{J}$ $\vec{\nabla}\cdot\vec{B}=0$
Constitutive relations (linear, isotropic media)	$\vec{D}=\varepsilon\vec{E}$	$\vec{B}=\mu\vec{H}$

macroscopic relations:

$$\vec{\nabla}\times\vec{H}=\vec{J} \tag{7.4}$$

$$\vec{\nabla}\cdot\vec{B}=0. \tag{7.5}$$

For linear, isotropic (homogeneous) media, the magnetic field quantities are proportional to one another through the constitutive relation

$$\vec{B}=\mu\vec{H}. \tag{7.6}$$

As was pointed out in Chapter 6, the latter approximation is poor for microscopic analysis of a solid material (especially a single crystal) where Equations 7.3 and 7.6 would need to take into account directions relative to the crystal structure and hence would, in general, be described via a tensor relationship. In addition, on a microscopic level in a solid, fields will be strongly influenced by the orbital and spin angular behavior of electrons associated with material atoms and those in the conduction band. Equations 7.3 and 7.6 will be altered by those electrons in the presence of electric and magnetic fields and they can in turn produce enormous fields at the nuclei of atoms so that Equations 7.3 and 7.6 are not even good approximations.

On a macroscopic scale of millions of atoms, the field quantities $\vec{E}$, $\vec{D}$, $\vec{H}$, and $\vec{B}$ can be averaged over the volume element so that Equations 7.3 and 7.6 begin to represent average fields in the media. In a liquid, the approximation is better, and, in a gas or vacuum, the constitutive relations become very good approximations. In the following sections, only average fields on a macroscopic scale will be considered, and, because many of the applications will be to field quantities propagating in a liquid or gas, our approximations will be considered acceptable. With these qualifications, the fundamental relations of electro*static* and magneto*static* fields are summarized in Table 7.1.

In these *static* cases, we observe that electric fields and magnetic fields form separate, independent pairs.

7.1 FARADAY'S LAW OF INDUCTION

Michael Faraday, in 1831, discovered experimentally that a current was induced in a conducting loop when the magnetic flux linking the loop changed with time. The quantitative relationship between the induced *emf* and the rate of charge flux linkage based on experimental observation is known as Faraday's experimental law.

$$emf \propto -\partial\left(\iint_S \vec{B}\cdot d\vec{s}\right)/\partial t = -\partial\Phi/\partial t, \quad (7.7)$$

where Φ is called the magnetic flux passing through the surface S. Thus, a fundamental postulate for electromagnetic induction between electric and magnetic quantities was made

$$\nabla\times\vec{E} = -\partial\vec{B}/\partial t \quad (7.8)$$

because applying Stokes's theorem yields

$$\iint_S \vec{\nabla}\times\vec{E}\cdot d\vec{s} = \oint_C \vec{E}\cdot d\vec{l} = emf = -\iint_S (\partial\vec{B}/\partial t)\cdot d\vec{s} = -\partial\Phi/\partial t. \quad (7.9)$$

According to Faraday's experimental law, Equation 7.9 should be valid for any surface S with bounding contour C, whether or not a physical circuit exists around C. Note that the sign in Faraday's law is arbitrary. It was chosen to be negative because an increasing magnetic flux with time causes an *emf* that acts to resist the imposed flux; that is, the **current induced in the loop creates its own magnetic flux density in the opposite direction to the one applied**; most texts call this response **Lenz's law**.

7.2 E&M EQUATIONS BEFORE MAXWELL

Differential Form of Maxwell's Equations

Including Faraday's law, the fundamental equations for electromagnetic fields, prior to Maxwell's addition of a displacement current, were thus given by

$$\vec{\nabla}\times\vec{E} = -\partial\vec{B}/\partial t \quad (7.8)$$

$$\vec{\nabla}\cdot\vec{D} = \rho_V \quad (7.10)$$

$$\vec{\nabla}\times\vec{H} = \vec{J} \quad (7.11)$$

$$\vec{\nabla}\cdot\vec{B} = 0. \quad (7.12)$$

If we also apply the principle of conservation of charge to these equations, we found the continuity equation, (Equation 2.58)

$$\vec{\nabla}\cdot\vec{J} = -\partial\rho_V/\partial t \quad (7.13)$$

or

$$\iiint_V \vec{\nabla}\cdot\vec{J}dv = \oiint_S \vec{J}\cdot d\vec{s} = -\partial\iiint_V \rho_V dV/\partial t = -\partial Q/\partial t,$$

which is contradictory to Equation 7.11 because we can show by direct differentiation that $\vec{\nabla}\cdot(\vec{\nabla}\times\vec{H}) = 0$, and, therefore, according to Equation 7.11,

$$\vec{\nabla}\cdot\left(\vec{\nabla}\times\vec{H}\right) = 0 = \vec{\nabla}\cdot\vec{J}. \quad (7.14)$$

Except for the very special case of time-independent, free charge density, Equations 7.13 and 7.14 come to two different conclusions. Is one of these equations wrong? Did we miss something?

This conundrum needed resolution in 1860 and was resolved by revising Ampere's Equation 7.11 by adding another time-dependent term that related magnetic and electric field quantities and that was symmetric with Faraday's Equation 7.09:

$$\vec{\nabla} \times \vec{H} = \vec{J} + \partial \vec{D} / \partial t. \tag{7.15}$$

The term $\partial \vec{D} / \partial t$ was called *displacement current density* and introduced by James Clerk Maxwell (1831–1879).

7.3 MAXWELL'S DISPLACEMENT CURRENT

When the symmetric differential equations and their integral equivalents was completed by Maxwell by adding a *displacement current* to Ampere's law, it was the leap in thinking that made James Clerk Maxwell the greatest electromagnetic theorist of all time! Because of the historical importance of this step, we will consider the logic in some detail to assure ourselves that it makes sense. The displacement current is the only term that was deduced and not taken from an experimental law.

At the time Maxwell made the addition, the concept almost violated common sense because it stated that a current existed between two terminals of a simple parallel plate capacitor, as shown in Figure 7.1, even if the conductivity in the intervening medium was zero.

To make his analysis, Maxwell used a conundrum. For the positive terminal of a capacitor to be charged, a current, I, must flow. However, thinking before the 1860s said that no current could flow between the $+Q$ and $-Q$ terminals if the space between them was a vacuum, as shown in Figure 7.1.

Thus,

a. if we consider a circular surface S_1 (bounded by path C), Ampere's law states that the magnetic field on C is $\oint_C \vec{B} \cdot d\vec{l} = \mu_0 I_{inside\ C}$ and because S_1 has current I flowing through it, $|\vec{B}| \neq 0$ on C; and

b. if we consider the bullet-shaped surface S_2 (also bounded by path C) and use 1860's "common sense" that no current flows through this surface, we conclude that $\oint_C \vec{B} \cdot d\vec{l} = 0$ and, thus, $|\vec{B}| = 0$ on C.

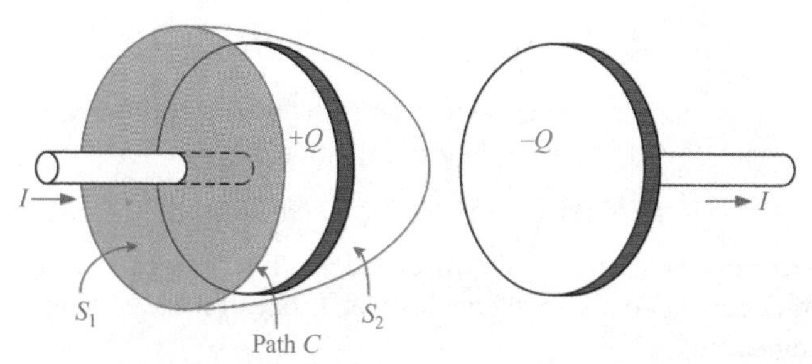

Figure 7.1 Conceptual view of displacement current in a simple parallel plate capacitor.

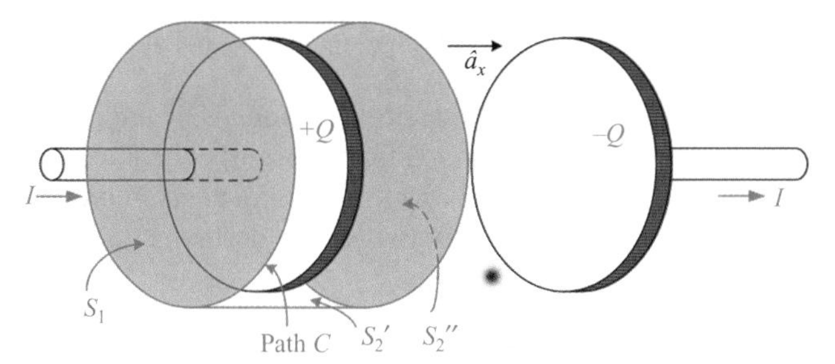

Figure 7.2 Conceptual view of displacement current in a simple parallel plate capacitor with a modified surface.

The two statements are contradictory, so Maxwell concluded that either Ampere's law was wrong or something was missing. He reasoned from Faraday's law that, because electric fields are induced by changing magnetic flux density, magnetic fields should be induced by changing electric flux density. To quantitate this symmetry, he redrew the closed Figure 7.2.

Maxwell noted that the electric field produced between the two charged parallel plates is $\vec{E} = \hat{a}_x \rho_s / \varepsilon_0$, where ρ_s is the charge per unit area. If we ignore the fringe fields at the edge of the plates and say that $\vec{E} = 0$ everywhere except between the two plates, then $\Phi = \oiint_{S_1+S_2'+S_2''} \vec{E} \cdot d\vec{s} = \oiint_{S_2''} \vec{E} \cdot d\vec{s} = \vec{E} \cdot \vec{A} = (Q/A)/\varepsilon_0 A = Q/\varepsilon_0$, where A is the area of the $+Q$ terminal. Thus,

$$dQ/dt = \varepsilon_0 \, d\Phi/dt. \tag{7.16a}$$

Maxwell reasoned that, if Q changes with time, a "current" is caused by the displacement of charge to the $+Q$ terminal; that is,

$$I_d \equiv dQ/dt = \varepsilon_0 \, d\Phi/dt \tag{7.16b}$$

may be thought of as a *displacement current* that passes through S_2'', and Ampere's law must be modified to include this current; that is,

$$\oint_C \vec{B} \cdot d\vec{l} = \mu_0 I_{inside\ C} + \mu_0 I_d. \tag{7.17}$$

Maxwell concluded that magnetic fields are caused by conduction currents **and** are induced by changing electric fields.

Using Stokes's theorem, he concluded the differential form

$$\oint_C \vec{B} \cdot d\vec{l} = \oiint_S \left(\vec{\nabla} \times \vec{B}\right) \cdot d\vec{s} = \mu_0 \int_{S_1} \vec{J} \cdot d\vec{s} + \mu_0 \varepsilon_0 d \iint_{S_2''} \vec{E} \cdot d\vec{s}/dt \tag{7.18}$$

or

$$\vec{\nabla} \times \vec{B} = \mu_0 \vec{J} + \mu_0 \varepsilon_0 \, d\vec{E}/dt. \tag{7.19}$$

Finally, we obtain the point form (or differential form) of Maxwell's equations:

$$\vec{\nabla} \times \vec{E} = -\partial \vec{B}/\partial t \tag{7.20a}$$

$$\vec{\nabla} \times \vec{H} = \vec{J} + \partial \vec{D}/\partial t \tag{7.20b}$$

$$\vec{\nabla} \cdot \vec{D} = \rho_V \tag{7.20c}$$

$$\vec{\nabla} \cdot \vec{B} = 0. \tag{7.20d}$$

Note that ρ_V is the volume density of free charges and $\vec{J}$ is the density of free currents.

Although Maxwell's equations consist of four vector equations, they are not all independent. Two divergence equations can be derived from two curl equations. The four fundamental field vectors $\vec{E}$, $\vec{D}$, $\vec{H}$, $\vec{B}$, each having three components, represent 12 unknowns, which, in principle, can be solved by two vector curl equations and two vector constitutive relations.

7.4 INTEGRAL FORM OF MAXWELL'S EQUATIONS

For solving problems with finite objects of specified shapes and boundaries, it is convenient to convert Maxwell's equations into their integral forms.

$$\oint_C \vec{E}\cdot d\vec{l} = -\iint_S (\partial\vec{B}/\partial t)\cdot d\vec{s} \quad \left(\iint_S \vec{\nabla}\times\vec{E}\cdot d\vec{s} = \oint_C \vec{E}\cdot d\vec{l}\right) \tag{7.21a}$$

$$\oint_C \vec{H}\cdot d\vec{l} = I + \iint_S (\partial\vec{D}/\partial t)\cdot d\vec{s} \quad \left(I = \iint_S \vec{J}\cdot d\vec{s}\right) \tag{7.21b}$$

$$\oiint_S \vec{D}\cdot d\vec{s} = Q \quad \left(\iiint_V \vec{\nabla}\cdot\vec{D}dv = \oiint_S \vec{D}\cdot d\vec{s}\right) \tag{7.21c}$$

$$\oint_S \vec{B}\cdot d\vec{s} = 0 \quad \left(Q = \iiint_V \rho_V \cdot dv\right) \tag{7.21d}$$

Maxwell's equations are summarized in Table 7.2.

Maxwell's equations represent the vector field quantities:

$\vec{E}$ = Electric field intensity (volts/meter).

$\vec{B}$ = Magnetic flux density (weber/meter^2 or tesla)

$\vec{H}$ = Magnetic field intensity (ampere/meter)

$\vec{D}$ = Electric flux density (coulombs/meter^2)

$\vec{J}$ = Electric current density (ampere/meter^2)

ρ_V = Electric charge density (coulomb/meter^3)

Units of the field quantities in SI units are shown in parentheses.

Table 7.2 Maxwell's Equations

Differential form	Integral form	Name
$\vec{\nabla}\times\vec{E} = -\partial\vec{B}/\partial t$	$\oint_C \vec{E}\cdot d\vec{l} = -\iint_S (\partial\vec{B}/\partial t)\cdot d\vec{s}$	Faraday's law
$\vec{\nabla}\times\vec{H} = \vec{J} + \partial\vec{D}/\partial t$	$\oint_C \vec{H}\cdot d\vec{l} = I + \iint_S (\partial\vec{D}/\partial t)\cdot d\vec{s}$	Ampere's law
$\vec{\nabla}\cdot\vec{D} = \rho_V$	$\oiint_S \vec{D}\cdot d\vec{s} = Q$	Gauss's law for electric charge
$\vec{\nabla}\cdot\vec{B} = 0$	$\oiint_S \vec{B}\cdot d\vec{s} = 0$	Gauss's law for magnetic charge

7.5 MAGNETIC VECTOR POTENTIAL

We have shown in Chapter 3 that the divergence of the curl of **any** vector field is identically zero; that is,

$$\vec{\nabla}\cdot\left(\vec{\nabla}\times\vec{A}\right)=0 \quad \text{for any vector field } \vec{A}. \tag{7.22}$$

Because the fourth of Maxwell's equations states that $\vec{B}$ is solenoidal, as given by Equation 7.20d ($\vec{\nabla}\cdot\vec{B} = 0$), we can thus assume that $\vec{B}$ may be written in terms of another vector field, $\vec{A}$, that we will call the **magnetic vector potential**:

$$\vec{B}=\vec{\nabla}\times\vec{A}. \tag{7.23}$$

NOTE We can see from Equation 7.23 that, given a magnetic flux density, $\vec{B}$, there will be an infinite number of vector fields, $\vec{A}$, that can satisfy the identity; for example, adding a constant to $\vec{A}$ will also satisfy Equation 7.23. This means that, to specify a *unique* definition of the vector field, $\vec{A}$, we will need to make an additional restriction on $\vec{A}$. The additional restriction is called a gauge and it is arbitrary. Coulomb and Lorenz made two different restrictions, as we will see below.

Substituting Equation 7.23 into Faraday's Law (Equation 7.20a) ($\vec{\nabla}\times\vec{E} = -\partial\vec{B}/\partial t$), we can write

$$\vec{\nabla}\times\vec{E}=-\partial\left(\vec{\nabla}\times\vec{A}\right)/\partial t$$

or

$$\vec{\nabla}\times\left(\vec{E}+\partial\vec{A}/\partial t\right)=0. \tag{7.24}$$

We have also shown in Chapter 3 that $\vec{\nabla}\times(-\vec{\nabla}V) = 0$ for *any* scalar field. Thus, because the curl of the vector field shown in parenthesis in Equation 7.24 is zero (i.e., it is irrotational), then that field can be written as the negative gradient of another scalar field, V, that we will call the **electric scalar potential**:

$$\vec{E}+\partial\vec{A}/\partial t=-\vec{\nabla}V$$

or

$$\vec{E}=-\vec{\nabla}V-\partial\vec{A}/\partial t. \tag{7.25}$$

NOTE We can see from Equation 7.25 that the electric field intensity, $\vec{E}$, can be written in terms of the electric scalar potential, V, and the magnetic vector potential, $\vec{A}$. As long as these two potentials are unique, the electric field intensity will also be unique.[1]

In the special case of static (time-independent) fields and potentials, $\partial\vec{A}/\partial t = 0$, and we can see that Equation 7.25 reduces to $\vec{E} = -\vec{\nabla}V$, as we found in Chapter 6 for static fields.

[1] Nature always finds unique electric field intensities. If the electric field intensity as defined by Equation 7.25 is multiple valued, then there is a problem with our mathematical definition of electric scalar potential or magnetic vector potential, and it must be resolved. Lorenz defined (in Equation 7.28 below) a further restraint on the magnetic vector potential that resolved multiple values for the electric field intensity.

In the more general case of time-varying fields and potentials, we can substitute Equations 7.23 and 7.25 into Ampere's law, $\vec{\nabla} \times \vec{H} = \vec{J} + \partial \vec{D}/\partial t$ (Equation 7.20b), along with $\vec{B} = \mu \vec{H}$ and $\vec{D} = \varepsilon \vec{E}$ for homogeneous media, to yield

$$\vec{\nabla} \times \vec{B} = \mu \vec{J} + \mu \, \partial \vec{D}/\partial t$$

or

$$\vec{\nabla} \times \left(\vec{\nabla} \times \vec{A}\right) = \mu \vec{J} + \mu\varepsilon \, \partial \vec{E}/\partial t$$

or

$$\vec{\nabla} \times \vec{\nabla} \times \vec{A} = \mu \vec{J} + \mu\varepsilon \partial\left(-\vec{\nabla} V - \partial \vec{A}/\partial t\right)/\partial t. \tag{7.26}$$

Now, using the identity $\vec{\nabla} \times \vec{\nabla} \times \vec{A} = \vec{\nabla}(\vec{\nabla} \cdot \vec{A}) - \vec{\nabla}^2 \vec{A}$ in Equation 7.26, we see

$$\vec{\nabla}\left(\vec{\nabla} \cdot \vec{A}\right) - \vec{\nabla}^2 \vec{A} = \mu \vec{J} - \vec{\nabla}\left(\mu\varepsilon \, \partial V/\partial t\right) - \mu\varepsilon \, \partial^2 \vec{A}/\partial t^2$$

or

$$\vec{\nabla}^2 \vec{A} - \mu\varepsilon \, \partial^2 \vec{A}/\partial t^2 = -\mu \vec{J} + \vec{\nabla}\left(\vec{\nabla} \cdot \vec{A} + \mu\varepsilon \, \partial V/\partial t\right). \tag{7.27}$$

Now, the definition of a **unique** vector field $\vec{A}$ requires an additional restriction or gauge. One way to provide this restriction is to specify its divergence; that is, although the curl of $\vec{A}$ is designated $\vec{B}$, we are still liberty to choose the divergence of $\vec{A}$. Lorenz made the choice[i]

$$\vec{\nabla} \cdot \vec{A} + \mu\varepsilon \, \partial V/\partial t = 0, \tag{7.28}$$

which is now called the **Lorenz gauge** for potentials. The rationale for that choice is clear because it reduces Equation 7.27 to a second-order, linear, inhomogeneous partial differential equation (PDE)

$$\vec{\nabla}^2 \vec{A} - \mu\varepsilon \, \partial^2 \vec{A}/\partial t^2 = -\mu \vec{J}, \tag{7.29}$$

which is called the inhomogeneous **wave equation** for the *magnetic vector potential*.

NOTE Equation 7.29 does not include the *electric scalar potential*, V, the electric field intensity, $\vec{E}$, or the magnetic field intensity, $\vec{H}$; only the electric current density, $\vec{J}$, need be known to solve for $\vec{A}$. Through the choice of the Lorenz gauge, Equation 7.28, we have been able to separate the terms in the four coupled Maxwell's differential Equations 7.20 into a single, PDE.

We can find a corresponding wave equation for the electric scalar potential, V, by substituting Equation 7.25 into Gauss's law, $\vec{\nabla} \cdot \vec{D} = \rho_V$ (Equation 7.20c):

$$\vec{\nabla} \cdot \vec{E} = \rho_V/\varepsilon \Rightarrow \vec{\nabla} \cdot \left(\vec{\nabla} V + \partial \vec{A}/\partial t\right) = -\rho_V/\varepsilon, \tag{7.30}$$

which leads to

$$\vec{\nabla}^2 V + \partial\left(\vec{\nabla} \cdot \vec{A}\right)/\partial t = -\rho_V/\varepsilon. \tag{7.31}$$

Using the Lorenz gauge, Equation 7.28 ($\vec{\nabla} \cdot \vec{A} + \mu\varepsilon \partial V/\partial t = 0$), we see that the *electric scalar potential*, V, also satisfies an inhomogeneous wave equation

$$\vec{\nabla}^2 V - \mu\varepsilon \, \partial^2 V/\partial t^2 = -\rho_V/\varepsilon. \tag{7.32}$$

NOTE Equation 7.32 does not include the *magnetic vector potential*, $\vec{A}$, the electric field intensity, $\vec{E}$, or the magnetic field intensity, $\vec{H}$; only the charge density, ρ_V, need be known to solve a single, PDE to find the electric scalar potential, V.

CONCLUSION We can separate the x, y, and z components of Equation 7.29 and solve each component equation independent of the others; that is, in Cartesian coordinates, Equation 7.29 is actually three wave equations (one for each of the components of $\vec{A}$). All three of these equations and Equation 7.32 are in the form of the same inhomogeneous wave equation and are independent of one another. Thus, given the electric charge density, ρ_V, and vector electric current density, $\vec{J}$, we can solve the inhomogeneous wave equation (subject to boundary conditions specified by a particular application) to find the potentials V and $\vec{A}$ from which we can then find all of the components of the electric field intensity and magnetic field intensity.

The equations for V and $\vec{A}$ form a set of four equations equivalent in all respects to Maxwell's equations (and they are subject to the restriction of the Lorenz gauge). However, unlike Maxwell's equations, these four inhomogeneous PDEs are independent of one another, so they will be much easier to solve. The four equations in Cartesian coordinates are called the time-dependent inhomogeneous **potential wave equations** and are all of the same form:

$$\vec{\nabla}^2 V - \mu\varepsilon\partial^2 V/\partial t^2 = -\rho_V/\varepsilon \tag{7.33a}$$

$$\vec{\nabla}^2 A_x - \mu\varepsilon\partial^2 A_x/\partial t^2 = -\mu J_x \tag{7.33b}$$

$$\vec{\nabla}^2 A_y - \mu\varepsilon\partial^2 A_y/\partial t^2 = -\mu J_y \tag{7.33c}$$

$$\vec{\nabla}^2 A_z - \mu\varepsilon\partial^2 A_z/\partial t^2 = -\mu J_z. \tag{7.33d}$$

Coulomb Gauge

Coulomb placed an alternate restriction on $\vec{A}$ by choosing $\vec{\nabla}\cdot\vec{A} = 0$. This choice reduces Equations 7.31 and 7.27, respectively, to

$$\vec{\nabla}^2 V = -\rho_V/\varepsilon \tag{7.34a}$$

$$\vec{\nabla}^2 \vec{A} - \mu\varepsilon\partial^2 \vec{A}/\partial t^2 = -\mu\vec{J} + \mu\varepsilon\vec{\nabla}(\partial V/\partial t). \tag{7.34b}$$

This choice makes the solution for scalar electric potential the same as that for the *static* potential, which we have already solved in Chapter 4. In principle, we could thus put the solution to Equation 7.34a into 7.34b to find the solution for *magnetic vector potential*; however, we see that this makes Equations 7.34 dependent on one another. We can see the Coulomb gauge produces a different *scalar electric potential* and *magnetic vector potential* from the Lorenz gauge, so we must state which gauge we are using in solutions that employ potentials.

Gauge Invariance

Maxwell's original quaternion equations used the gradient of a scalar magnetic vector potential $\vec{H} = -\vec{\nabla}\Omega$ (Equation 1.68) to calculate magneto*static* field intensity and included the equation of continuity as fundamental. This made the *magnetic vector potential* $\vec{A}$ a physically meaningful field (not a mathematical convenience like we have just described). However, the original equations included a magnetic mass m that was part of the Lorenz force equation, as shown in Equation 1.59. In addition, the mass m could be calculated according to Equation 1.67 as $m = S.\vec{\nabla}\vec{M}$, which takes the scalar part of the four vector $\vec{M}$ to yield an equation similar to Gauss's Law for electric charge density, as shown in Equation 1.66, $e = S.\vec{\nabla}\vec{D}$. This became the basis for the P.A.M. Dirac formulation of a magnetic charge density and the possible existence of magnetic monopoles (instantatons) in 1931. Although no one has found an isolated magnetic monopole, we have seen in Chapter 6 that it is possible to think of a magnetic monopole layer and a magnetic monopole current as a mathematical convenience in which case we arrive at the same answers for electromagnetic field quantities. Note also that, if we compare Equation 7.25 ($\vec{E} = -\vec{\nabla}V - \partial\vec{A}/\partial t$) with Equation 1.58 ($\vec{E} = V.v\vec{B} - \partial\vec{A}/\partial t - \vec{\nabla}\varphi$), the term $V.v\vec{B}$ (the vector part of $v\vec{B}$) is missing. Thus, $\vec{E}_{motion} = \mu(\vec{v} \times \vec{H})$ is the one term that appears to be discarded. Hertz interpreted the velocity, v, as the (absolute) motion of charges relative to the luminiferous ether; however, if v is interpreted as relative velocity between charges, then the Maxwell Heavyside equations are defined for the case $v = 0$ (i.e., test charges do not move in the observer's reference frame). The first Einstein postulate that, in a uniform moving system all physical laws take their simplest form, independent of the velocity, requires that observers always measure the undamped wave equation (e.g., Equation 7.29) with no terms that pertain to the first time derivative.

Maxwell agreed with Faraday that there existed a medium characterized by polarization and strain, through which radiation propagated from one local region to another local region. Instead of force residing in the medium, Maxwell adopted the Faraday force field concept that there was a distinction between quantity and intensity. Magnetic intensity was represented by the line integral described in Chapter 6 and referred to the magnetic polarization of the medium and magnetic quantity was represented by a surface integral that referred to the magnetic induction in the medium. In all cases, a medium was required and was the basis for electromagnetic phenomena, that is, the electromagnetic field did not exist except as a state of the propagating medium in which there were no electrical conduction currents (only displacement currents) and in which there were no electrical or magnetic sources. Rather than electricity producing a disturbance in a medium, the presence of electricity *was* the disturbance. Maxwell thus thought of electric current as a moving system with forces communicating the motion from one part of the system to another; equations of motion were defined only locally.

7.6 SOLUTION OF THE TIME-DEPENDENT INHOMOGENEOUS POTENTIAL WAVE EQUATIONS

To solve a time-dependent inhomogeneous PDE (wave equation) of the form

$$\vec{\nabla}^2\psi - \mu\varepsilon\,\partial^2\psi/\partial t^2 = f(\vec{x}, t), \tag{7.35a}$$

we need to find a particular solution to the inhomogeneous PDE (called the particular solution, $\psi_{\text{particular}}$). To solve the full equation, we will need a solution (called the homogeneous solution, $\psi_{\text{homogeneous}}$) to the homogeneous PDE

$$\vec{\nabla}^2\psi - \mu\varepsilon\,\partial^2\psi/\partial t^2 = 0. \tag{7.35b}$$

Then, the general solution to the problem will be

$$\psi = \psi_{\text{particular}} + \psi_{\text{homogeneous}}. \tag{7.36}$$

NOTE The student is asked to refer to mathematical books like *Schaum's Outlines on Differential Equations* by Richard Bronson or *Mathematical Methods for Physicists* by George Arfken for techniques to solve these equations. These and other authors conclude that it is possible to show that

a. a particular solution to the inhomogeneous PDE 7.35a is unique; that is, if you find a solution to Equation 7.35a by any technique, you need not search further because you have found the only solution; and

b. there are, in general, two linearly independent solutions to the second-order homogeneous PDE 7.35b, so the most general solution to this equation is a linear combination of the two. Any arbitrary coefficients in the final problem solution 7.36 will need to be chosen to satisfy the boundary conditions of the particular problem.

NOTE In the *special case* of a *source-free* problem ($\rho_V = 0$ and $\vec{J} = 0$), for example, in a nonconducting medium like free space, we can see that all four of the PDEs in Equation 7.33 are homogeneous PDEs. Thus, the solution to these equations *in this special case* need not contain a particular solution, and the most general solution will be a linear combination of the two linearly independent solutions to the homogeneous PDEs. Even in the general case of the inhomogeneous PDEs, we will need the solutions to the homogeneous PDEs in our answer, so we begin with the solution to the homogeneous PDEs.

NOTE Many students and professionals become so caught up in the detailed solutions of these equations that they lose sight of what they are trying to accomplish. Entire math and physics course sequences are designed to justify a solution that is unique and complete, so it is not surprising that a state of misunderstanding or confusion results, and sometimes students decide they are not capable of doing this kind of work. It is thus recommended by the author that the student first read to the end of this chapter (without trying to understand the detailed solutions) to get a clear picture of where we are headed.

7.7 ELECTRIC AND MAGNETIC FIELD EQUATIONS FOR *SOURCE-FREE* PROBLEMS

For the *special case* of *source-free* problems (i.e., $\rho_V = 0$ and $\vec{J} = 0$), we can see that Maxwell's equations reduce to the form shown in Table 7.3:

So, if we take the curl of Faraday's law,

$$\vec{\nabla} \times \vec{\nabla} \times \vec{E} = -\vec{\nabla} \times \partial \vec{B}/\partial t$$

or

$$\vec{\nabla}(\vec{\nabla} \cdot \vec{E}) - \vec{\nabla}^2 \vec{E} = -\mu \partial(\vec{\nabla} \times \vec{H})/\partial t. \tag{7.37}$$

Now, substituting Gauss's law ($\vec{\nabla} \cdot \vec{E} = 0$) and Ampere's Law into Equation 7.37, we see

$$\vec{\nabla}^2 \vec{E} - \mu\varepsilon \partial^2 \vec{E}/\partial t^2 = 0. \tag{7.38}$$

Likewise, taking the curl of Ampere's law,

$$\vec{\nabla} \times \vec{\nabla} \times \vec{H} = \vec{\nabla} \times \partial \vec{D}/\partial t$$

or

$$\vec{\nabla}(\vec{\nabla} \cdot \vec{H}) - \vec{\nabla}^2 \vec{H} = \varepsilon \partial(\vec{\nabla} \times \vec{D})/\partial t. \tag{7.39}$$

And using the fourth of Maxwell's equations ($\vec{\nabla} \cdot \vec{H} = 0$) with Faraday's law, we see

$$\vec{\nabla}^2 \vec{H} - \mu\varepsilon \partial^2 \vec{H}/\partial t^2 = 0. \tag{7.40}$$

CONCLUSION In source-free space, V, all of the components of $\vec{A}$, all of the components of $\vec{E}$, and all of the components of $\vec{H}$ satisfy the homogeneous wave equation.

NOTE We often label $\mu\varepsilon = 1/u_p^2$ and $\mu_0\varepsilon_0 = 1/c^2$ in the wave equation.

Table 7.3 Maxwell's Equations for *Source-Free* Problems

Differential form	Integral form	Name of law
$\vec{\nabla} \times \vec{E} = -\partial\vec{B}/\partial t$	$\oint_C \vec{E} \cdot d\vec{l} = -\iint_S (\partial\vec{B}/\partial t) \cdot d\vec{s}$	Faraday's law
$\vec{\nabla} \times \vec{H} = \partial\vec{D}/\partial t$	$\oint_C \vec{H} \cdot d\vec{l} = \iint_S (\partial\vec{D}/\partial t) \cdot d\vec{s}$	Ampere's law
$\vec{\nabla} \cdot \vec{D} = 0$	$\oint_S \vec{D} \cdot d\vec{s} = 0$	Gauss's law
$\vec{\nabla} \cdot \vec{B} = 0$	$\oint_S \vec{B} \cdot d\vec{s} = 0$	No isolated magnetic charge

7.8 SOLUTIONS FOR THE HOMOGENEOUS WAVE EQUATION

Time Dependence

The homogeneous wave equation lends itself well to a solution using the *separation of variables* technique. In this technique we *assume* that the solution to the PDE, $\psi(\vec{x}, t)$, can be written in the form of a product, $\psi(\vec{x}, t) = \psi_S(\vec{x})T(t)$, where $\psi_S(\vec{x})$ is a function of only the spatial variables and $T(t)$ is a function of only the time variable. Putting this product form into Equation 7.36, we see

$$\vec{\nabla}^2\psi_S(\vec{x})T(t) - \mu\varepsilon\,\partial^2[\psi_S(\vec{x})T(t)]/\partial t^2 = 0. \tag{7.41}$$

Now we recognize that the operator $\vec{\nabla}^2$ operates only on the spatial part of the product and that the operator $\partial^2/\partial t^2$ operates only on the time part of the product so that

$$T(t)\vec{\nabla}^2\psi_S(\vec{x}) = \mu\varepsilon\psi_S(\vec{x})\partial^2 T(t)/\partial t^2, \tag{7.42}$$

and, if we then divide Equation 7.42 through by the product $\psi_S(\vec{x})T(t)$, we get

$$\frac{\vec{\nabla}^2\psi_S(\vec{x})}{\psi_S(\vec{x})} = \mu\varepsilon\frac{\partial^2 T(t)/\partial t^2}{T(t)}. \tag{7.43}$$

In Equation 7.43, we see that the left-hand side of the equation depends only on spatial variables and that the right-hand side of the equation depends only on time. We can argue that there is no way that these two quantities can equal one another except in the case that they both equal a constant (the same constant). We are free to name this constant, so let us call it $-k^2$ (recognizing that this constant might turn out to be a complex number).

Then,

$$\frac{\vec{\nabla}^2\psi_S(\vec{x})}{\psi_S(\vec{x})} = -k^2 \quad\text{and}\quad \mu\varepsilon\frac{\partial^2 T(t)/\partial t^2}{T(t)} = -k^2. \tag{7.44}$$

The beauty of these two equations is that they are easy to solve because they reduce to

$$\vec{\nabla}^2\psi_S(\vec{x}) + k^2\psi_S(\vec{x}) = 0 \quad\text{and}\quad \partial^2 T(t)/\partial t^2 = -\left(k^2/\mu\varepsilon\right)T(t) \tag{7.45}$$

The first of these equations is called the scalar Helmholtz equation, and the second is the equation for a harmonic oscillator with solution

$$T(t) = A\sin\omega t + B\cos\omega t = Ce^{j\omega t} + De^{-j\omega t} = \begin{Bmatrix}\sin\omega t\\ \cos\omega t\end{Bmatrix} = \begin{Bmatrix}e^{j\omega t}\\ e^{-j\omega t}\end{Bmatrix} \tag{7.46}$$

with

$$\omega = k/\sqrt{\mu\varepsilon}.$$

This solution[2] is a linear combination of the two linearly independent functions $\sin\omega t$ and $\cos\omega t$ or alternately of the linearly independent functions $e^{j\omega t}$ and $e^{-j\omega t}$

[2] Some authors use the matrix symbol { } to mean take a linear combination of the contents and some use a matrix symbol ( ) to mean the same thing; both are equally common are interchangeable in this text.

(the choice of form is usually obvious when we examine the boundary conditions for the problem). The matrix form of the answer on the right of Equation 7.46 is shorthand for the same linear combinations.

We have now found the time-dependent part of the product $\psi(\vec{x}, t) = \psi_S(\vec{x})T(t)$ and we can find the spatial part of the product by solving the scalar Helmholtz equation

$$\vec{\nabla}^2 \psi_S(\vec{x}) + k^2 \psi_S(\vec{x}) = 0 \tag{7.47}$$

subject to the condition that $k^2 = \omega^2 \mu \varepsilon$, and we will write the general answer as

$$\psi(\vec{x}, t) = \psi_S(\vec{x}) \begin{Bmatrix} e^{j\omega t} \\ e^{-j\omega t} \end{Bmatrix}. \tag{7.48}$$

The solution to Equation 7.47 will depend on the coordinate system because $\vec{\nabla}^2$ is different in Cartesian, cylindrical, or spherical coordinates. The choice of coordinate system is usually apparent from the boundary conditions given in the problem; for example, if the quantity $\psi_S(\vec{x})$ or $\partial \psi_S(\vec{x})/\partial n$ is specified on the surface of a rectangular box, the obvious choice of coordinate system is Cartesian.

NOTE There are very special problems in which the boundary conditions are specified on the surface of an elliptical shape or a toriodal shape. For those problems, other more complex coordinate systems might be employed with their concurrent forms for the Laplacian, $\vec{\nabla}^2$.

For most real-world problems, the boundary conditions are very hard to specify in any simple coordinate system, and we will need to revert to a numerical solution of the Helmholtz equation. Many commercial computer codes[ii] have been written for these solutions such as High Frequency Structure Simulator (HFSS) by the Ansoft Corporation of Pittsburg, PA.

NOTE The choice of coordinate systems cannot change the answer to the problem. In nature, there is only one unique answer to the problem so our choice of coordinate system is only a convenience to being able to state the boundary conditions in a simple way.

Spatial Dependence

The spatial dependence of the homogeneous wave equation is found by solving the scalar[3] Helmholtz equation $\vec{\nabla}^2 \psi_S(\vec{x}) + k^2 \psi_S(\vec{x}) = 0$.

For *Cartesian coordinates*, we can express the Helmholtz equation as

$$\partial^2 \psi_S(\vec{x})/\partial x^2 + \partial^2 \psi_S(\vec{x})/\partial y^2 + \partial^2 \psi_S(\vec{x})/\partial z^2 + k^2 \psi_S(\vec{x}) = 0. \tag{7.49}$$

[3] In Cartesian coordinates, the **vector** Helmholtz equation separates into three independent scalar equations. In cylindrical or spherical coordinates, this is not the case; the topic is considered in Huray, *The Foundations of Signal Integrity,* Chapter 6.

If we again employ the separation of variables technique by assuming that the spatial function can be written as the product of functions, $\psi_S(\vec{x}) = X(x)\,Y(y)\,Z(z)$, then we can substitute this into Equation 7.49 and divide by $\psi_S(\vec{x})$ to obtain

$$\frac{\partial^2 X(x)/\partial x^2}{X(x)} + \frac{\partial^2 Y(y)/\partial y^2}{Y(y)} + \frac{\partial^2 Z(z)/\partial z^2}{Z(z)} = -k^2. \tag{7.50}$$

Again, we can see that all of the x dependence is in the first term, all of the y dependence is in the second term, and all of the z dependence is in the third term. The only way these terms could cancel one another for all values of x, y, and z is that the terms individually, at most, be equal to a constant. For convenience, let us call the three constants $-k_x^2$, $-k_y^2$, and $-k_z^2$, where any one or all of the constants could potentially be a complex number. This choice renders Equation 7.50 into the three separate equations

$$\partial^2 X(x)/\partial x^2 = -k_x^2 X(x) \tag{7.51a}$$
$$\partial^2 Y(y)/\partial y^2 = -k_y^2 Y(y) \tag{7.51b}$$
$$\partial^2 Z(z)/\partial z^2 = -k_z^2 Z(z) \tag{7.51c}$$

with the condition that

$$k_x^2 + k_y^2 + k_z^2 = k^2. \tag{7.51d}$$

By separating variables, the partial derivatives now operate only on functions that contain the derivative variable, so they can be replaced by ordinary derivatives. The solutions to these differential equations are all the same in Cartesian coordinates:

$$X(x) = \begin{pmatrix} \sin k_x x \\ \cos k_x x \end{pmatrix} = \begin{pmatrix} e^{jk_x x} \\ e^{-jk_x x} \end{pmatrix} \tag{7.52a}$$

$$Y(y) = \begin{pmatrix} \sin k_y y \\ \cos k_y y \end{pmatrix} = \begin{pmatrix} e^{jk_y y} \\ e^{-jk_y y} \end{pmatrix} \tag{7.52b}$$

$$Z(z) = \begin{pmatrix} \sin k_z z \\ \cos k_z z \end{pmatrix} = \begin{pmatrix} e^{jk_z z} \\ e^{-jk_z z} \end{pmatrix}. \tag{7.52c}$$

The choice of form of the linear combination is often obvious because it is easy to specify the boundary conditions in a particular problem. For example, in Cartesian coordinates with free-space boundary conditions, it is convenient to use the vector definition $\vec{k} = k_x\hat{a}_x + k_y\hat{a}_y + k_z\hat{a}_z$ to specify a direction of propagation so that $\vec{k}\cdot\vec{x} = k_x x + k_y y + k_z z$ and the most general solution to the homogeneous wave equation can be simply written as

$$\psi(\vec{x}, t) = e^{j\vec{k}\cdot\vec{x}} \begin{Bmatrix} e^{j\omega t} \\ e^{-j\omega t} \end{Bmatrix} \quad \text{with } k^2 = \omega^2\mu\varepsilon. \tag{7.53}$$

This solution is a plane wave traveling in the $\pm\vec{k}$-direction, depending on the choice of the sign in the exponent of the time-dependent term. The constraint $k^2 = \omega^2\mu\varepsilon$ will place very important restrictions on the kind of solution that can exist in specific boundary value problems (e.g., rectangular waveguides).

In other applications (e.g., potential interior to a rectangular cavity with fixed wall potentials), it may be more convenient to use the solution form

$$\psi(\vec{x},t)=\begin{pmatrix}\sin k_x x\\ \cos k_x x\end{pmatrix}\begin{pmatrix}\sin k_y y\\ \cos k_y y\end{pmatrix}\begin{pmatrix}\sin k_z z\\ \cos k_z z\end{pmatrix}\begin{pmatrix}\sin \omega t\\ \cos \omega t\end{pmatrix} \quad \text{with } \omega^2\mu\varepsilon = k_x^2+k_y^2+k_z^2. \tag{7.54}$$

For **cylindrical coordinates**, we can express the *scalar* Helmholtz equation as

$$\frac{1}{\rho}\frac{\partial}{\partial\rho}\left(\rho\frac{\partial\psi_S(\vec{x})}{\partial\rho}\right)+\frac{1}{\rho^2}\frac{\partial^2\psi_S(\vec{x})}{\partial\phi^2}+\frac{\partial^2\psi_S(\vec{x})}{\partial z^2}+k^2\psi_S(\vec{x})=0. \tag{7.55}$$

If we again employ the separation of variables technique by assuming that the spatial function can be written as the product of functions, $\psi_S(\vec{x}) = \mathrm{P}(\rho)\Phi(\phi)Z(z)$, then we can substitute this into Equation 7.55 and divide by $\psi_S(\vec{x})$ to obtain

$$\frac{1}{\mathrm{P}\rho}\frac{\partial}{\partial\rho}\left(\rho\frac{\partial \mathrm{P}}{\partial\rho}\right)+\frac{1}{\Phi\rho^2}\frac{\partial^2\Phi}{\partial\phi^2}+\frac{1}{Z}\frac{\partial^2 Z}{\partial z^2}=-k^2. \tag{7.56}$$

NOTE In Equation 7.56, we have dropped the arguments of the $\mathrm{P}(\rho)$, $\Phi(\phi)$, $Z(z)$ functions for brevity but can see that the partial differentials may be exchanged by ordinary differentials. Here, we can recognize that all of the z dependence occurs in the last term on the left-hand side (LHS) so we can conclude

$$\frac{1}{Z}\frac{d^2Z}{dz^2}=-a^2 \tag{7.57a}$$

and Equation 7.56 becomes

$$\frac{\rho}{\mathrm{P}}\frac{\partial}{\partial\rho}\left(\rho\frac{\partial \mathrm{P}}{\partial\rho}\right)+\frac{1}{\Phi}\frac{\partial^2\Phi}{\partial\phi^2}-a^2\rho^2=-k^2\rho^2$$

from which we can conclude

$$\frac{1}{\Phi}\frac{\partial^2\Phi}{\partial\phi^2}=-m^2 \tag{7.57b}$$

and

$$\frac{\rho}{\mathrm{P}}\frac{\partial}{\partial\rho}\left(\rho\frac{\partial P}{\partial\rho}\right)-m^2-a^2\rho^2=-k^2\rho^2$$

or with $\gamma^2 = k^2 - a^2$

$$\rho\frac{\partial}{\partial\rho}\left(\rho\frac{\partial \mathrm{P}}{\partial\rho}\right)+(\gamma^2\rho^2-m^2)\mathrm{P}=0. \tag{7.57c}$$

The solution to Equation 7.57c is a linear combination of a Bessel function, $J_m(\gamma\rho)$, and a Neumann function, $N_m(\gamma\rho)$, so

$$\psi(\vec{x},t)=\begin{pmatrix}J_m(\gamma\rho)\\ N_m(\gamma\rho)\end{pmatrix}\begin{pmatrix}\sin m\phi\\ \cos m\phi\end{pmatrix}\begin{pmatrix}\sin az\\ \cos az\end{pmatrix}\begin{pmatrix}e^{-j\omega t}\\ e^{j\omega t}\end{pmatrix}. \tag{7.58}$$

The linear combination of Bessel and Neumann functions can have coefficients that describe a real and an imaginary part, in which case the answer can be expressed as a linear combination of a Hankel function of the first kind, $H_m^{(1)}(\gamma\rho) =$

$J_m(\gamma\rho) + jN_m(\gamma\rho)$ and a Hankel function of the second kind, $H_m^{(2)}(\gamma\rho) = J_m(\gamma\rho) - jN_m(\gamma\rho)$. These functions are often used for problems in which there is a known incoming or outgoing cylindrical wave (e.g., for antenna problems).

In the event that the argument is negative (i.e., when $\gamma^2 = k^2 - a^2$ is negative), the solution can be written as a linear combination of a modified or "hyperbolic" Bessel function, $I_m(\gamma\rho) = e^{-j\pi m/2}J_m(j\gamma\rho)$ and a McDonald function, $K_m(\gamma\rho) = (j\pi/2)e^{j\pi m/2}H_m^{(1)}(\gamma\rho)$. These functions occur in potential problems associated with cylindrical superconductors.

In some circumstances (e.g., for fields inside a good conductor), the argument of the radial function can be expressed in terms of $\sqrt{j} = e^{j\pi/4}$ (called a phase ¼ number) in which case we may write the real and imaginary part of the modified Bessel function as $I_m(\sqrt{j}\gamma\rho) = Ber_m(\gamma\rho) + jBei_m(\gamma\rho)$ and the real and imaginary part of the McDonald function as $K_m(\gamma\rho\sqrt{j}) = Ker_m(\gamma\rho) + jKei_m(\gamma\rho)$.

Because these various forms of the solution of the Helmholtz equation are linear combinations with coefficients that may be complex numbers, we are at liberty to choose the form of a solution that most easily satisfies our given boundary conditions.

For **spherical coordinates**, we can express the *scalar* Helmholtz equation as

$$\frac{1}{R^2}\frac{\partial}{\partial R}\left(R^2\frac{\partial\psi_S(\vec{x})}{\partial R}\right)+\frac{1}{R^2\sin\theta}\frac{\partial}{\partial\theta}\left(\sin\theta\frac{\partial\psi_S(\vec{x})}{\partial\theta}\right)+\frac{1}{R^2\sin^2\theta}\frac{\partial^2\psi_S(\vec{x})}{\partial\phi^2}+k^2\psi_S(\vec{x})=0. \tag{7.59}$$

If we again employ the separation of variables technique by assuming that the spatial function can be written as the product of functions, $\psi_S(\vec{x}) = \Gamma(R)\Theta(\theta)\Phi(\phi)$, then we can substitute this into Equation 7.59 and divide by $\psi_S(\vec{x})$ to obtain

$$\frac{1}{\Gamma R^2}\frac{\partial}{\partial R}\left(R^2\frac{\partial\Gamma}{\partial R}\right)+\frac{1}{\Theta R^2\sin\theta}\frac{\partial}{\partial\theta}\left(\sin\theta\frac{\partial\Theta}{\partial\theta}\right)+\frac{1}{\Phi R^2\sin^2\theta}\frac{\partial^2\Phi}{\partial\phi^2}=-k^2. \tag{7.60}$$

NOTE In Equation 7.60, we have dropped the arguments of the $\Gamma(R)$, $\Theta(\theta)$, and $\Phi(\phi)$ functions for brevity but can see that the partial differentials may be exchanged by ordinary differentials. Here, we can also multiply through by $R^2\sin^2\theta$ to recognize that all of the ϕ dependence occurs in the last term on the LHS so we can conclude

$$d^2\Phi/d\varphi^2 = -m^2\Phi \tag{7.61a}$$

and Equation 7.60 becomes

$$\frac{1}{\Gamma R^2}\frac{\partial}{\partial R}\left(R^2\frac{\partial\Gamma}{\partial R}\right)+\frac{1}{\Theta R^2\sin\theta}\frac{\partial}{\partial\theta}\left(\sin\theta\frac{\partial\Theta}{\partial\theta}\right)-\frac{m^2}{R^2\sin^2\theta}=-k^2.$$

Multiplying through by R^2, we can conclude

$$\frac{1}{\Gamma}\frac{\partial}{\partial R}\left(R^2\frac{\partial\Gamma}{\partial R}\right)+\frac{1}{\Theta\sin\theta}\frac{\partial}{\partial\theta}\left(\sin\theta\frac{\partial\Theta}{\partial\theta}\right)-\frac{m^2}{\sin^2\theta}=-k^2R^2.$$

And rearranging terms on the LHS and right-hand side (RHS),

$$\frac{1}{\Gamma}\frac{\partial}{\partial R}\left(R^2\frac{\partial \Gamma}{\partial R}\right)+k^2R^2=\frac{-1}{\Theta\sin\theta}\frac{\partial}{\partial\theta}\left(\sin\theta\frac{\partial\Theta}{\partial\theta}\right)+\frac{m^2}{\sin^2\theta}.$$

We see that the variables are again separated, so we can set them equal to an arbitrary constant; in this case, $-l(l+1)$, so that

$$\frac{1}{\sin\theta}\frac{d}{d\theta}\left(\sin\theta\frac{d\Theta}{d\theta}\right)+l(l+1)\Theta-\frac{m^2\Theta}{\sin^2\theta}=0 \tag{7.61b}$$

and

$$\frac{1}{\Gamma}\frac{\partial}{\partial R}\left(R^2\frac{\partial \Gamma}{\partial R}\right)+k^2R^2-l(l+1)=0$$

or

$$\frac{1}{R^2}\frac{\partial}{\partial R}\left(R^2\frac{\partial \Gamma}{\partial R}\right)+k^2\Gamma-\frac{l(l+1)\Gamma}{R^2}=0. \tag{7.61c}$$

The solution to Equation 7.61b is the associated Legendre function of the first kind, $P_l^m(\cos\theta)$, and the associated Legendre function of the second kind, $Q_l^m(\cos\theta)$. For $k \neq 0$, the solution to Equation 7.61c is the spherical Bessel function, $j_l(kR)$, and the spherical Neumann function, $\eta_l(kR)$. We can thus write the solution to the homogeneous wave equation in spherical coordinates as

$$\psi(\vec{x},t)=\begin{pmatrix} j_l(kR)\\ \eta_l(kR)\end{pmatrix}\begin{pmatrix} P_l^m(\cos\theta)\\ Q_l^m(\cos\theta)\end{pmatrix}\begin{pmatrix} e^{jm\phi}\\ e^{-jm\phi}\end{pmatrix}\begin{pmatrix} e^{j\omega t}\\ e^{-j\omega t}\end{pmatrix}. \tag{7.62}$$

The linear combination of spherical Bessel and Neumann functions can have coefficients that describe a real and an imaginary part, in which case the answer can be expressed as a linear combination of a spherical Hankel function of the first kind, $h_m^{(1)}(kR) = j_m(kR) + j\eta_m(kR)$ and a spherical Hankel function of the second kind, $h_m^{(2)}(kR) = j_m(kR) - j\eta_m(kR)$. These functions are often used for problems in which there is a known incoming or outgoing spherical wave (e.g., scattering from a spherical shape, as are discussed in Huray, *The Foundations of Signal Integrity*, Chapters 6 and 7.

In a few applications, it is desired to write the solution as a linear combination of a modified spherical Bessel function, $i_m(kR)=\sqrt{\pi/2kR}I_{m+1/2}(kR)$, and a modified spherical McDonald function, $k_m(kR)=\sqrt{2/\pi kR}K_{m+1/2}(kR)$. These functions occur in potential problems associated with spherical conductors.

In some circumstances, we write the real and imaginary part of the modified spherical Bessel function as $i_m(kR) = ber_m(kR) + jbei_m(kR)$ and the real and imaginary part of the spherical McDonald function as $k_m(kR) = ker_m(kR) + jkei_m(kR)$.

As in the case of cylindrical coordinates, these various forms of the solution of the *scalar* Helmholtz equation are linear combinations with coefficients that may be complex numbers, so we are at liberty to choose the form of a solution that most easily satisfies our given boundary conditions.

NOTE For $k = 0$ (the **special case** when $\omega = 0$, i.e., the ***static*** case), Equation 7.59c is simply

$$\frac{1}{R^2}\frac{\partial}{\partial R}\left(R^2\frac{\partial \Gamma}{\partial R}\right)-\frac{l(l+1)\Gamma}{R^2}=0, \tag{7.63}$$

which has two linearly independent solutions R^l and $R^{-(l+1)}$ so that

$$\Gamma(R)=\begin{Bmatrix} R^l \\ R^{-(l+1)} \end{Bmatrix}. \tag{7.64}$$

The total solution for the *static* case can be written as

$$\psi(\vec{x},t)=\begin{pmatrix} R^l \\ R^{-(l+1)} \end{pmatrix}\begin{pmatrix} P_l^m(\cos\theta) \\ Q_l^m(\cos\theta) \end{pmatrix}\begin{pmatrix} e^{jm\phi} \\ e^{-jm\phi} \end{pmatrix} \tag{7.65}$$

because both of the time-dependent exponentials are $e^{j0t} = 1$.

7.9 PARTICULAR SOLUTION FOR THE INHOMOGENEOUS WAVE EQUATION

Wave Equations with Sources

We now consider the particular solution of the inhomogeneous wave equations:

$$\vec{\nabla}^2\psi-\mu\varepsilon\partial^2\psi/\partial t^2=f(\vec{x},t) \text{ where} \tag{7.66a}$$

$$f(\vec{x},t)=-\rho(\vec{x},t)/\varepsilon \text{ when } \psi(\vec{x},t)=V(\vec{x},t) \text{ and} \tag{7.66b}$$

$$f(\vec{x},t)=-\mu J_i(\vec{x},t) \text{ when } \psi(\vec{x},t)=A_i(\vec{x},t) \tag{7.66c}$$

for each of the $\boldsymbol{i}$ components of the *magnetic vector potential* in Cartesian coordinates. Equation 7.66c thus represents three equations.

As noted previously, any technique that provides *a* solution provides *the* solution because the particular solution is unique. Some authors (e.g., Matthews and Walker) often use an informed guess technique, and others (e.g., Jackson) use a formal Green's function technique to obtain an answer, as we discussed in section 4.3 for a static case. We provide the latter technique here for completeness. The Green's function technique is optional in the sense that the student may go straight to the answer below and try that answer in Equations 7.66 to see that it satisfies the PDEs. Using this result as their informed guess, the students can conclude that they have found *an* answer and thus have found *the* mathematical solution to the problem. The only remaining issue will be to determine what physical characteristics the answer implies.

Green's Function Technique

For time-varying fields we can find the solution for an inhomogeneous PDE by first taking its Fourier transform with respect to the variable t. We will use the Fourier integral representations:[4]

[4] Justified by the last paragraph of Appendix C because $e^{-j\omega t}/\sqrt{2\pi}$ are solutions of a Sturm–Liouville differential equation.

$$\psi(\vec{x},t)=(1/2\pi)\int_{-\infty}^{\infty}\psi(\vec{x},\omega)e^{-j\omega t}d\omega \tag{7.67a}$$

$$f(\vec{x},t)=(1/2\pi)\int_{-\infty}^{\infty}f(\vec{x},\omega)e^{-j\omega t}d\omega \tag{7.67b}$$

and their inverse transformations

$$\psi(\vec{x},\omega)=\int_{-\infty}^{\infty}\psi(\vec{x},t)e^{j\omega t}dt \tag{7.67c}$$

$$f(\vec{x},\omega)=\int_{-\infty}^{\infty}f(\vec{x},t)e^{j\omega t}dt. \tag{7.67d}$$

Substituting these representations back into Equation 7.66a, we find that the Fourier transforms satisfy the inhomogeneous Helmholtz equation:

$$\left(\vec{\nabla}^2+k^2\right)\psi(\vec{x},\omega)=f(\vec{x},\omega) \tag{7.68}$$

for each value of ω. We will later consider the restriction $\omega=k/\sqrt{\mu\varepsilon}$ to see that it imposes a causality condition on the solution.

Two important relations between the Fourier transforms are that

$$(1/2\pi)\int_{-\infty}^{\infty}e^{j(\omega-\omega')t}dt=\delta(\omega-\omega') \text{ (completeness relation)} \tag{7.69a}$$

$$(1/2\pi)\int_{-\infty}^{\infty}e^{j\omega(t-t')}d\omega=\delta(t-t') \text{ (orthogonality relation),} \tag{7.69b}$$

where $\delta(x-a)$ is the one-dimensional Dirac delta function.

NOTE The one-dimensional delta function is defined by its properties:

1. $\delta(x-a)=0$ for $x\neq a$
2. $\int_{-\infty}^{\infty}\delta(x-a)dx=1$
3. $\int_{-\infty}^{\infty}f(x)\delta(x-a)dx=f(a)$
4. $\int_{-\infty}^{\infty}f(x)\delta'(x-a)dx=-f'(a)$

One picture of the one-dimensional delta function is that it is in the shape of a Gaussian distribution centered about its mean at $\boldsymbol{a}$ and with a vanishingly small standard deviation.

NOTE It is also possible to define a three-dimensional delta function in Cartesian coordinates for which

1. $\delta(\vec{x}-\vec{x}')=\delta(x-x')\,\delta(y-y')\,\delta(z-z')$
2. $\iiint_{\Delta V}\delta(\vec{x}-\vec{x}')d^3x=\begin{cases}1 & \text{if } \Delta V \text{ contains } \vec{x}=\vec{x}' \\ 0 & \text{if } \Delta V \text{ does not contain } \vec{x}=\vec{x}'\end{cases}$

Fourier reasoned that we could solve Equation 7.68 by taking its transform with respect to time and then each of the spatial variables, x, y, and z, put them into Equation 7.69, find an algebraic solution for the transform, and then take its inverse Fourier transforms to find the answer. We might call this the Fourier technique, but it suffers from the fact that we need to solve Equation 7.68 for each possible charge or current density distribution, $f(\vec{x}, \omega)$.

In 1824, George Green claimed that, if we solve the equation

$$\left(\vec{\nabla}^2 - \mu\varepsilon\,\partial^2/\partial t^2\right)G(\vec{x}, t; \vec{x}', t') = \delta(\vec{x} - \vec{x}')\delta(t - t') \tag{7.70}$$

then (in infinite space with no boundary surfaces) the solution of Equation 7.66a will be

$$\psi(\vec{x}, t) = \int\!\!\iiint G(\vec{x}, t; \vec{x}', t')\, f(\vec{x}', t')\, d^3x'dt'. \tag{7.71}$$

Proof

Let us assume that we have found $G(\vec{x}, t; \vec{x}', t')$ that satisfies Equation 7.70. If George Green's claim is correct, then $\psi(\vec{x}, t)$ formed by Equation 7.71 should satisfy Equation 7.66a.

To see if this is true, we will put $\psi(\vec{x}, t) = \int\!\!\iiint G(\vec{x}, t; \vec{x}', t')\, f(\vec{x}', t')\, d^3x'dt'$ into $\vec{\nabla}^2\psi - \mu\varepsilon\partial^2\psi/\partial t^2$ to see if it produces $f(\vec{x}, t)$.

Check

$$\vec{\nabla}^2 \int\!\!\iiint G(\vec{x}, t; \vec{x}', t') f(\vec{x}', t') d^3x'dt' - \mu\varepsilon\,\partial^2 \int\!\!\iiint G(\vec{x}, t; \vec{x}', t') f(\vec{x}', t') d^3x'dt' \Big/ \partial t^2$$
$$= \int\!\!\iiint \vec{\nabla}^2 G(\vec{x}, t; \vec{x}', t') f(\vec{x}', t') d^3x'dt' - \mu\varepsilon \int\!\!\iiint \partial^2 G(\vec{x}, t; \vec{x}', t')/\partial t^2\, f(\vec{x}', t')$$
$$d^3x'dt' = \int\!\!\iiint \left[\vec{\nabla}^2 G(\vec{x}, t; \vec{x}', t') - \mu\varepsilon\,\partial^2 G(\vec{x}, t; \vec{x}', t')/\partial t^2\right] f(\vec{x}', t') d^3x'dt'$$
$$= \int\!\!\iiint [\delta(\vec{x} - \vec{x}')\delta(t - t')] f(\vec{x}', t') d^3x'dt' = f(\vec{x}, t)$$

Thus, it is proven.

Of course, the Green's function $G(\vec{x}, t; \vec{x}', t')$ will have to satisfy appropriate boundary conditions demanded by physical considerations and causality. $G(\vec{x}, t; \vec{x}', t')$ is often called the *point source response function* because it is the answer (response) at the point $\vec{x}$ and time t to a point source that *was* created at the point $\vec{x}'$ and time t'; i.e., we can see that Equation 7.66 is the same as Equation 7.70 in the special case that $f(\vec{x}, t) = \delta(\vec{x} - \vec{x}')\delta(t - t')$. The emphasis on the word *was* is used to emphasize that the point source has to be created at time t' before the time t that we seek the answer; that is, causality demands that we obtain a response only ***after*** the point source originates. Furthermore, we can conclude that the response at the point $\vec{x}$ and time t must provide enough time for the potential or vector potential point source at the point $\vec{x}'$ and time t' to reach the response point traveling at the speed of light in the medium, $c = 1/\sqrt{\mu\varepsilon}$. This condition is called the ***retarded*** response.

The power of using the Green's technique is that we can solve Equation 7.70 (in whatever coordinate system we choose), and the answer is independent of the source distribution $f(\vec{x}, t)$ in our particular application. That is, once we have found an answer, $G(\vec{x}, t; \vec{x}', t')$, to Equation 7.70, we will then put it into Equation 7.71 along with $f(\vec{x}', t')$ to get an answer, $\psi(\vec{x}, t)$, for a particular source distribution $f(\vec{x}, t)$.

To solve Equation 7.70, we can insert the four dimensional Fourier transform of the Green's function, $g(\vec{k}, \omega)$, on the LHS of the equation and the four dimensional delta function representation on the RHS of the equation as follows:

$$G(\vec{x}, t; \vec{x}', t') = \iiint d^3k \int d\omega g(\vec{k}, \omega) e^{j\vec{k}\cdot(\vec{x}-\vec{x}')} e^{-j\omega(t-t')} \tag{7.72}$$

$$\delta(\vec{x}-\vec{x}')\delta(t-t') = 1/(2\pi)^4 \iiint d^3k \int d\omega e^{j\vec{k}\cdot(\vec{x}-\vec{x}')} e^{-j\omega(t-t')} \tag{7.73}$$

The result is a simple algebraic equation:

$$g(\vec{k}, \omega) = \left[1/(2\pi)^4\right]\left(k^2 - \mu\varepsilon\omega^2\right)^{-1} = \left[1/(2\pi)^4\right]\left(k^2 - \omega^2/c^2\right)^{-1}. \tag{7.74}$$

PROBLEM

7.1 Show that Equation 7.74 results from putting Equations 7.72 and 7.73 into Equation 7.70.

We can substitute Equation 7.74 back into Equation 7.72 to find the Green's function but we must recognize that, in carrying out the integral over ω, there is a singularity in the integrand at the points $\omega = \pm kc$. We can use a Cauchy integral in the complex ω-plane to evaluate the answer using the blue path C shown in Figure 7.3:

Formally, the integral $\int_C \frac{-c^2}{(2\pi)^4} \frac{e^{j\vec{k}\cdot(\vec{x}-\vec{x}')} e^{-j\omega(t-t')}}{(\omega - kc)(\omega + kc)} d\omega$ is not defined because there are two singularities **on** the path C. This mathematical problem can be overcome by displacing the singularities below the path C by adding an infinitesimal term $j\alpha$ to each term in the denominator and taking the limit as $\alpha \to 0$ (α being a positive real

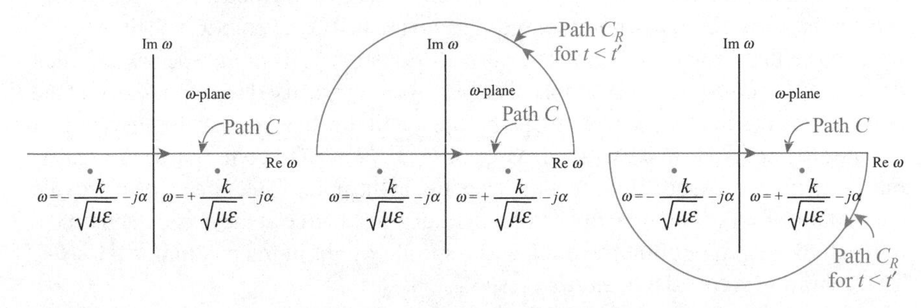

Figure 7.3 (Left) Path C for integration over variable ω because $\int_{-\infty}^{\infty} > f(\omega)d\omega = \int_C f(\omega)d\omega$; (middle) closed contour path C plus C_R for evaluating the ω integral using the method of residues for the case $t < t'$; (right) closed contour path C plus C_R for evaluating the ω integral using the method of residues for the case $t > t'$.

number). The reason for displacement below the path C is that the result satisfies the physical solution as opposed to a displacement above the path C.

According to the Residue Theorem for integrals in the complex ω-plane,

$$G(\vec{x},t;\vec{x}',t')=\iiint d^3k\int d\omega g(\vec{k},\omega)e^{j\vec{k}\cdot(\vec{x}-\vec{x}')}e^{-j\omega(t-t')}$$

with

$$g(\vec{k},\omega)=-\lim_{\alpha\to 0}\frac{c^2}{(2\pi)^4}\frac{1}{(\omega-kc-j\alpha)(\omega+kc-j\alpha)}$$

has two poles of order 1, so, for $t>t'$, the solution is

$$G(\vec{x},t;\vec{x}',t')=\iiint d^3k\frac{c^2 2\pi j}{(2\pi)^4}e^{j\vec{k}\cdot(\vec{x}-\vec{x}')}\left\{\frac{e^{-jkc(t-t')}}{2kc}-\frac{e^{jkc(t-t')}}{2kc}\right\} \tag{7.75}$$

and, for $t<t'$, the solution is 0. Thus,

$$G(\vec{x},t;\vec{x}',t')=\left[c/(2\pi)^3\right]\iiint d^3k e^{j\vec{k}\cdot(\vec{x}-\vec{x}')}\frac{\sin kc(t-t')}{k}, \tag{7.76}$$

PROBLEM

7.2 Show that the integrand goes to zero for points on the semicircle (red Path C_R) in Figure 7.3b for $t<t'$ and for points on the semicircle (red Path C_R) in Figure 7.3c for $t>t'$.

We can take the integral over d^3k by first integrating over angles:

$$G(\vec{x},t;\vec{x}',t')=(c/4\pi^2)\int_0^\infty dk\int_0^\pi d\theta\frac{e^{jkR\cos\theta}}{k}\sin kc(t-t')\sin\theta d\theta,$$

where

$$R=|\vec{x}-\vec{x}'|$$

or

$$G(\vec{x},t;\vec{x}',t')=(c/4\pi^2)\int_0^\infty dk\int_{-1}^1 d\theta\frac{e^{jkR\cos\theta}}{k}\sin kc(t-t')d(\cos\theta)$$

$$G(\vec{x},t;\vec{x}',t')=(c/2\pi^2 R)\int_0^\infty dk\sin kR\sin kc(t-t'). \tag{7.77}$$

Now, because the integrand is even in k, the integral can be written over the interval, $-\infty<k<\infty$ with a change of variable, $x=kc$:

$$G(\vec{x},t;\vec{x}',t')=(-1/8\pi^2 R)\int_{-\infty}^\infty dx\left[e^{j((t-t')-R/c)x}-e^{j[(t-t')+R/c]x}\right] \tag{7.78}$$

and from Equation 7.69b, we can see the integrals are just Dirac delta functions:

$$G(\vec{x},t;\vec{x}',t')=(-1/4\pi R)[\delta((t-t')-R/c)-\delta((t-t')+R/c)] \tag{7.79}$$

However, because $t>t'$, the argument of the second delta function is never 0, so

$$G(\vec{x},t;\vec{x}',t')=(-1/4\pi R)\delta((t-t')-R/c)$$

or

$$G(\vec{x},t;\vec{x}',t')=(-1/4\pi|\vec{x}-\vec{x}'|)\delta((t-t')-|\vec{x}-\vec{x}'|/c). \quad (7.80)$$

This Green's function is called the ***retarded Green's function*** because it exhibits causal behavior associated with the propagation of a wave source to a response location; that is, an effect observed at a point $\vec{x}$ due to a source at a point $\vec{x}'$ and time t' will not occur until the wave has had time to propagate the distance $|\vec{x} - \vec{x}'|$ traveling at speed $c = 1/\sqrt{\mu\varepsilon}$.

Finally, we can use Equation 7.74 to find the solution to the inhomogeneous wave equation in the absence of boundary conditions as

$$\psi(\vec{x},t)=-\int\iiint\frac{\delta((t-t')-|\vec{x}-\vec{x}'|/c)}{4\pi|\vec{x}-\vec{x}'|}f(\vec{x}',t')d^3x'dt'. \quad (7.81)$$

The integration over dt' can be performed to yield the "retarded solution"

$$\psi(\vec{x},t)=-\iiint\frac{[f(\vec{x}',t')]_{retarded}}{4\pi|\vec{x}-\vec{x}'|}d^3x'. \quad (7.82)$$

The potential due to a charge distribution over a volume V' is then

$$V(R,t)=(1/4\pi\varepsilon)\iiint_{V'}\frac{\rho(t-R/c)}{R}d^3x' \quad (7.83)$$

called the *retarded scalar electric potential*, which indicates that the scalar electric potential at (R, t) depends on the value of charge at an earlier time $(t - R/c)$.

Similarly, we can obtain the retarded magnetic vector potential

$$\vec{A}(R,t)=(\mu/4\pi)\iiint_{V'}\frac{\vec{J}(t-R/c)}{R}d^3x'. \quad (7.84)$$

Time-Harmonic Solutions

If we have the *special case* of

$$\rho(\vec{x},t)=\rho_S(\vec{x})e^{j\omega t} \quad \text{and} \quad \vec{J}(\vec{x},t)=\vec{J}_S(\vec{x})e^{j\omega t},$$

then

$$\rho(\vec{x},t_{retarded})=\rho_S(\vec{x})e^{-jk|\vec{x}-\vec{x}'|}e^{j\omega t} \quad \text{and} \quad \vec{J}(\vec{x},t)=\vec{J}_S(\vec{x})e^{-jk|\vec{x}-\vec{x}''|}e^{j\omega t}.$$

In that case, Equations 7.83 and 7.84 reduce to the *special case* answers

$$V(\vec{x},\vec{x}',t)=(1/4\pi\varepsilon)\iiint_{V'}\rho_S(\vec{x}')\left(e^{-jk|\vec{x}-\vec{x}'|}/|\vec{x}-\vec{x}'|\right)d^3x'e^{j\omega t} \quad (7.85)$$

$$\vec{A}(\vec{x},\vec{x}',t)=(\mu/4\pi)\iiint_{V'}\vec{J}_S(\vec{x}')\left(e^{-jk|\vec{x}-\vec{x}'|}/|\vec{x}-\vec{x}'|\right)d^3x'e^{j\omega t}. \quad (7.86)$$

NOTE As compared with Equations 7.83 and 7.84, Equations 7.85 and 7.86 contain the factor $e^{-jk|\vec{x}-\vec{x}'|}$, which takes into account the retarded time.

7.10 TIME-HARMONIC FIELDS

Special Case of Time-Harmonic Electromagnetics

In this section, we summarize the special case of *time-harmonic* (steady-state sinusoidal) field relationships. If we can express

$$\vec{E}(x, y, x, t) = \mathrm{Re}\left[\vec{E}_S(x, y, z)e^{j\omega t}\right], \tag{7.87}$$

where $\vec{E}_S(x, y, z)$ is a purely space-dependent, vector phasor that contains information on direction, magnitude, and phase, the *time-harmonic* Maxwell's equations can thus be written in terms of phasors:

$$\vec{\nabla} \times \vec{E}_S = -j\omega\mu\vec{H}_S \quad (\partial/\partial t \to j\omega) \tag{7.88a}$$
$$\vec{\nabla} \times \vec{H}_S = \vec{J}_S + j\omega\varepsilon\vec{E}_S \tag{7.88b}$$
$$\vec{\nabla} \cdot \vec{E}_S = \rho_S/\varepsilon \tag{7.88c}$$
$$\vec{\nabla} \cdot \vec{B}_S = 0. \tag{7.88d}$$

From the last section, the time-harmonic wave equations for scalar potential and vector potentials, respectively, become

$$\vec{\nabla}^2 V_S + k^2 V_S = -\rho_S(\vec{x})/\varepsilon \tag{7.89}$$
$$\begin{aligned}\vec{\nabla}^2 \vec{A}_S + k^2 \vec{A}_S &= -\mu\vec{J}_S \\ \partial^2/\partial t^2 \to (j\omega)^2 &= -\omega^2,\end{aligned} \tag{7.90}$$

where

$$k = \omega/c \tag{7.91}$$

is called the wave number.

Lorentz's condition is

$$\vec{\nabla} \cdot \vec{A}_S + j\omega\mu\varepsilon V_S = 0, \tag{7.92}$$

and the phasor solutions of potentials for the inhomogeneous equations are

$$V_S(R) = (1/4\pi\varepsilon)\int_{V'} \rho_S\left[e^{-jkR}/R\right]dv' \tag{7.93}$$
$$\vec{A}_S(R) = (\mu/4\pi)\int_{V'} \vec{J}_S\left[e^{-jkR}/R\right]dv', \tag{7.94}$$

which give the retarded scalar and vector potentials due to time-harmonic sources.

Very Special Case of Time-Harmonic, Source-Free Fields in Simple Media

In a simple, nonconducting, *source-free* medium characterized by ($\vec{J} = 0$, $\rho_V = 0$, $\sigma = 0$), the *time-harmonic* ($\partial/\partial t \to j\omega$) Maxwell's equations become

$$\vec{\nabla}\times\vec{E}_S = -j\omega\mu\vec{H}_S \quad (7.95a)$$
$$\vec{\nabla}\times \underline{H}_S = j\omega\varepsilon\vec{E}_S \quad (7.95b)$$
$$\vec{\nabla}\cdot\vec{E}_S = 0 \quad (7.95c)$$
$$\vec{\nabla}\cdot\vec{B}_S = 0, \quad (7.95d)$$

and the vector wave equations become

$$\vec{\nabla}^2\vec{E}_S + k^2\vec{E}_S = 0 \quad (7.96a)$$
$$\vec{\nabla}^2\vec{H}_S + k^2\vec{H}_S = 0. \quad (7.96b)$$

For this very special case, there are no sources, so there are no particular solutions, and the most general solutions to problems are linear combinations of the two linearly independent solutions of the homogeneous equations.

7.11 ELECTROMAGNETIC SPECTRUM

We note that Maxwell's and wave equations impose no limit on frequency of the waves and that all electromagnetic waves propagate at any frequency with the same velocity (Figure 7.4).

7.12 ELECTROMAGNETIC BOUNDARY CONDITIONS

By using Gauss's law (integrating over the surface of a small pillbox), as seen in Figure 7.5a, and Stokes theorem (integrating around the perimeter of a small rectangle), as seen in Figure 7.5b, the boundary conditions between two media for time-varying fields are derived as

$$\oiint_{Pill-Box}\vec{D}\cdot d\vec{A} = q_{inside} \text{ and taking } \lim_{\Delta h\to 0}\vec{D}_1\cdot\vec{A}_1 + \vec{D}_2\cdot\vec{A}_2 = \Sigma_{e,S}A$$
$$\text{or } \left(\vec{D}_2 - \vec{D}_1\right)\cdot\hat{a}_x = \Sigma_{e,S}\ (D_{2n} - D_{1n}) = \Sigma_{e,S} \quad (7.97a)$$

$$\iint_{\text{Rectangle}}\left(\vec{\nabla}\times\vec{E}\right)\cdot d\vec{s} = \oint_C \vec{E}\cdot d\vec{l} = -\frac{\partial}{\partial t}\oiint_{\text{Rectangle}}\vec{B}\cdot d\vec{s} \text{ and taking}$$
$$\lim_{\Delta h\to 0}\left(\vec{E}_2 - \vec{E}_1\right)\cdot w\hat{a}_y = 0 \text{ or } E_{1t} = E_{2t} \quad (7.97b)$$

$$\oiint_{Pill-Box}\vec{B}\cdot d\vec{A} = 0 \text{ and taking } \lim_{\Delta h\to 0}\mathrm{B}_1\cdot\vec{A}_1 + \vec{B}_2\cdot\vec{A}_2 = \left(\vec{B}_2 - \vec{B}_1\right)\cdot\hat{a}_x A = 0 \text{ or } B_{1n} = B_{2n} \quad (7.97c)$$

$$\iint_{\text{Rectangle}}\left(\vec{\nabla}\times\vec{H}\right)\cdot d\vec{s} = \oint_C \vec{H}\cdot d\vec{l} = \oiint_{\text{Rectangle}}\vec{J}\cdot d\vec{s} + \partial\oiint_{\text{Rectangle}}\vec{B}\cdot d\vec{s}\Big/\partial t$$

and taking $\lim_{\Delta h\to 0}\hat{a}_x\times\left(\vec{H}_2 - \vec{H}_1\right)w = \vec{J}_{inside}$

$$(H_{2t} - H_{1t}) = J_{inside}/w = J_l. \quad (7.97d)$$

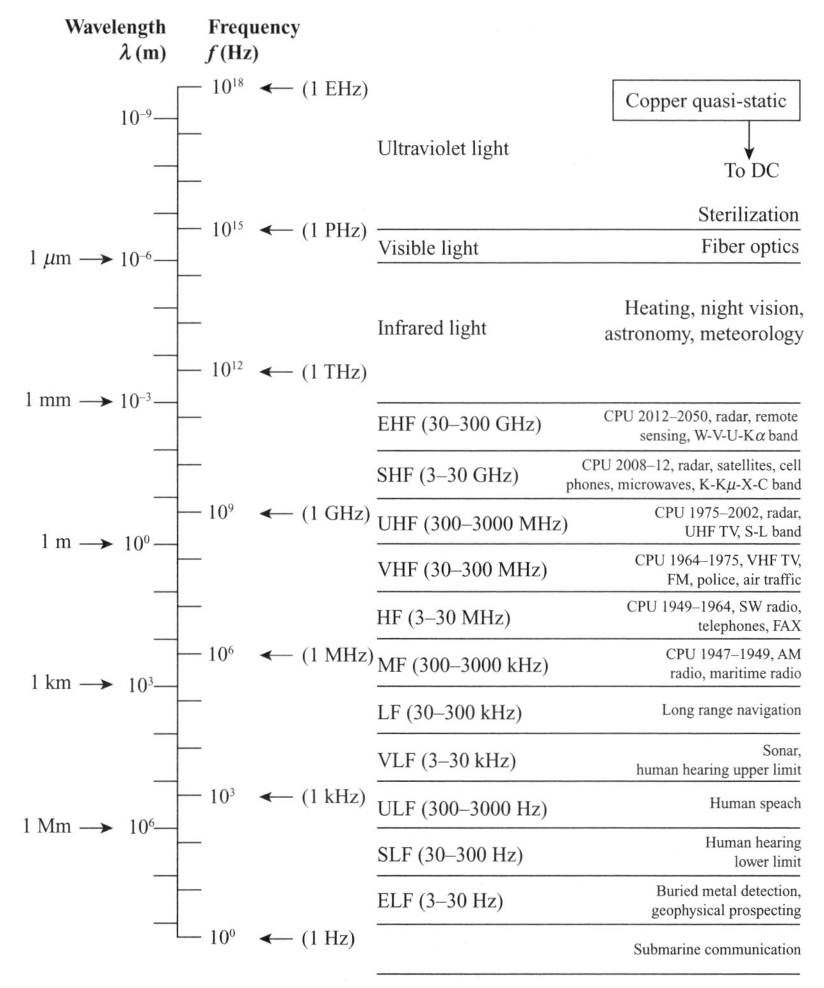

Figure 7.4 Spectrum of electromagnetic waves.

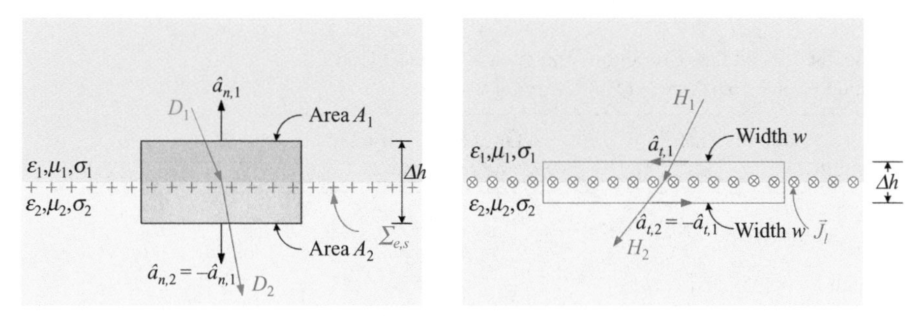

Figure 7.5 (Left) Pillbox of area A and thickness Δh used in the surface integral for Gauss's Law; (right) rectangular path of width w and thickness Δh used in the Stokes theorem integral around a closed path. $\hat{a}_z$ is into the page.

Interface between a Dielectric and a Perfect Electric Conductor

For a perfect conductor (having infinite conductivity, $\sigma \to \infty$), the boundary conditions can be described as in Table 7.4.

The direction of the fields and their corresponding sources are shown in Figure 7.5 and J_l is the current per unit length.

NOTE In *The Foundations of Signal Integrity*,[iii] we consider an electric field intensity E_{2n} as it propagates in the z-direction with velocity $c_2 = 1/\sqrt{\mu_2 \varepsilon_2}$. In that case, the current per unit length will be produced by the motion of the surface charge density $\Sigma_{e,S}$ and $J_l = \Sigma_{e,S} c_2$. Thus, the magnetic field intensity created by the moving charge density will be $H_{2t} = \Sigma_{e,S} / \sqrt{\mu_2 \varepsilon_2}$. The ratio of the normal component of the electric field intensity, E_{2n}, to the tangential component of the magnetic field intensity in medium 2, H_{2t}, will thus be

$$E_{2n}/H_{2t} = (\Sigma_{e,S}/\varepsilon_2)/\Sigma_{e,S}/\sqrt{\mu_2 \varepsilon_2} = \sqrt{\mu_2/\varepsilon_2}. \tag{7.98}$$

This ratio has the units of volts/amps (ohms) and is called the *wave impedance of medium 2*.

Interface between a Dielectric and a Good Electric Conductor

For a good conductor (e.g., for copper, $\sigma_{Cu} = 5.8 \times 10^7 \, \Omega^{-1} \text{m}^{-1}$) that has finite conductivity, the solutions are more complicated because magnetic field intensity can penetrate the conductor in an exponentially decaying fashion. For copper, we will restrict our consideration of time-harmonic fields to frequencies below 10^{18} Hz because we have seen in example 5.1 that conduction electrons in copper quickly reduce electric fields (with a time constant of $\tau_{copper} = 1.5 \times 10^{-19}$ s) to zero. This frequency range[5] is called the "Quasistatic" regime because $\partial \vec{D}_1/\partial t \approx 0$ inside the conductor (just as it would for the static case). Thus, Ampere's law in the quasistatic regime is $\vec{\nabla} \times \vec{H}_1 \approx \vec{J}_1$. Multiplying through by μ_1 and using the identity for magnetic flux density in terms of magnetic vector potential and the current density in terms of the electric field intensity,

Table 7.4 Boundary Conditions Between a Perfect Electric Conductor and a Dielectric (Time-Varying Case)

On the conducting side (medium 1)	On the dielectric side (medium 2)
$E_{1t} = 0$	$E_{2t} = 0$
$H_{1t} = 0$	$H_{2t} = J_l$
$D_{1n} = 0$	$D_{2n} = \Sigma_{e,S}$
$B_{1n} = 0$	$B_{2n} = 0$

[5] Some engineers refer to the frequency range $0 \le f \le 10^{15}$ Hz as "DC to daylight."

$$\vec{\nabla}\times\vec{B}_1 \approx \mu_1\vec{J}_1 \text{ or } \vec{\nabla}\times\vec{\nabla}\times\vec{A}_1 \approx \mu_1\sigma_1\vec{E}_1 \quad \text{or}$$
$$\vec{\nabla}\left(\vec{\nabla}\cdot\vec{A}_1\right)-\vec{\nabla}^2\vec{A}_1 = \mu_1\sigma_1\vec{E}_1. \tag{7.99}$$

Now, using the Lorentz gauge, from Equation 7.28, $\vec{\nabla}\cdot\vec{A}_1 + \mu_1\varepsilon_1\partial V_1/\partial t = 0$, and $\vec{E}_1 = -\vec{\nabla}V_1 - \partial\vec{A}_1/\partial t$ from Equation 7.25, with the scalar electric potential $\vec{\nabla}V_1 \approx 0$ and $\partial V_1/\partial t \approx 0$ (the quasistatic approximation), we deduce

$$\vec{\nabla}^2\vec{A}_1 \approx \mu_1\sigma_1\,\partial\vec{A}_1/\partial t. \tag{7.100}$$

This equation, called the *diffusion equation*, determines how the magnetic vector potential in the conductor, $\vec{A}_1$, propagates with time, and it is fundamentally different from the wave equation because of the *first* derivative with respect to time.

For time-harmonic fields, $\vec{A}_1(\vec{x}, t) = \vec{A}_1(\vec{x})e^{j\omega t}$, we can replace the derivative with respect to time with $j\omega$ to see that

$$\vec{\nabla}^2\vec{A}_1(\vec{x}) = j\omega\mu_1\sigma_1\vec{A}_1(\vec{x}). \tag{7.101}$$

For a large, flat boundary between medium 1 and medium 2 in Figure 7.5, we would expect no variation of $\vec{A}_1$ in the *y*- or the *z*-direction, so Equation 7.101 would simplify to

$$\partial^2\vec{A}_1/\partial x^2 = j\omega\mu_1\sigma_1\vec{A}_1, \tag{7.102}$$

which has solution $\vec{A}_1(x) = \vec{A}_0e^{\pm(1+j)x/\delta}$ if $\delta = \sqrt{2/\omega\mu_1\sigma_1}$. We thus conclude that $\vec{A}_1$ propagates in medium 1 according to

$$\vec{A}_1(x) = \vec{A}_0e^{x/\delta}e^{j(x/\delta+\omega t)}. \tag{7.103}$$

Here, we have chosen the positive exponent of *e* under the assumption that, for large negative values of *x* (penetration *into* the conductor), the magnetic vector potential will decay exponentially rather than grow to infinity. Note that the propagation of the magnetic vector potential is in the –*x*-direction. By considering a point on the wave for which $(x/\delta + \omega t)$ is a constant, $u_p = dx/dt = -\omega\delta = -\sqrt{2\omega/\mu_1\sigma_1} = -\sqrt{2}/\sqrt{\mu_1\varepsilon_1}\sqrt{\omega\varepsilon_1/\sigma_1}$. Thus, we can see that the magnetic vector potential is decaying exponentially with x/δ and propagating at the speed of light in medium 1, $1/\sqrt{\mu_1\varepsilon_1}$, divided by the quantity $\sqrt{\sigma_1/2\omega\varepsilon_1}$. We choose to express the ratio this way because $(\sigma_1/\omega\varepsilon_1) \gg 1$ is the definition of a *good* conductor. **δ is called the skin depth** because it is the depth to which *x* must go to decrease the surface quantity by e^{-1}. We can compute $\delta = \sqrt{2/\mu_1\omega\sigma_1}$ for copper using $\mu_1 \approx \mu_0 = 4\pi \times 10^{-7}\,\Omega\text{s/m}$ and $\sigma_{Cu} = 5.8 \times 10^7\,\Omega^{-1}\text{m}^{-1}$, so that $\delta = 0.066/\sqrt{f}\,\text{m}$ if *f* is measured in Hz. For $f = 60\,\text{Hz}$, $\delta = 0.85\,\text{cm}$. For $f = 10^{10}\,\text{Hz}$, $\delta = 0.66\,\mu\text{m}$.

We also see that the field is propagating into the conductor at a phase velocity **much less** than the speed of light in medium 2, so the field is diffusing slowly into the conductor. For a copper conductor at room temperature, we can take $\varepsilon_1 = \varepsilon_0 = (1/36\pi) \times 10^{-9}(\text{s}/\Omega\text{m})$ then $(\sigma_1/\omega\varepsilon_1) = 1.04 \times 10^{18}/f$ so that $u_p = \left(c/7.2\times10^8\right)\sqrt{f}$, where *f* is measured in Hz. For $f = 60\,\text{Hz}$, the speed of the propagation is only *3.2* m/s! For $f = 10^{10}\,\text{Hz}$, the speed of propagation is 42 km/s, which is faster but well below the speed of light. Thus, at $f = 60\,\text{Hz}$, the time for the initial field intensity at $x = 0$ to

Table 7.5 Values of Various Copper Properties as a Function of Frequency

f (Hz)	$\frac{\sigma}{\omega\varepsilon_0}$	u_p (m/s)	δ (m)	Δt_δ (s)	$\omega\Delta t_\delta$ (rad)
60	1.74×10^{16}	3.22	8.53×10^{-3}	2.65×10^{-3}	1
10^3	1.04×10^{15}	13.2	2.09×10^{-3}	1.59×10^{-4}	1
10^6	1.04×10^{12}	416	66.1×10^{-6}	1.59×10^{-7}	1
10^9	1.04×10^9	13.2×10^3	2.09×10^{-6}	1.59×10^{-10}	1
10^{10}	1.04×10^8	41.6×10^3	0.66×10^{-6}	1.59×10^{-11}	1
10^{11}	1.04×10^7	132×10^3	0.21×10^{-6}	1.59×10^{-12}	1
10^{12}	1.04×10^6	4.16×10^5	66.1×10^{-9}	1.59×10^{-13}	1

From the last column we see that the field at a depth of 1 δ is 1 rad different than the field at the surface for all frequencies; that is, the field at a depth of 1 δ is caused by the field that occurred at the surface 57° before the increment of time, Δt_δ.

reach one skin depth is $\Delta t_\delta = \delta/u_p = (0.158/f)s = 2.6$ ms. We call the currents that are produced by this diffusion *eddy currents* and we use them for heating copper pans on an induction stove.

Values of various quantities for copper at different frequencies are given in Table 7.5.

PROBLEMS

7.3 Using a geological pole reversal time of 10^6 years for the earth's magnetic field intensity, estimate the conductivity of molten iron and the permeability of iron to calculate the skin depth in the earth's mantle and the time for field intensity to penetrate one skin depth.

7.4 A microwave oven heats a bowl of soup (roughly the salinity of seawater) by induction at a frequency $f = 2.4 \times 10^9$ Hz. Using the permittivity of seawater, explain why the soup boils around the edge of the bowl instead of in the center.

7.5 If you put a piglet in a weak microwave oven, how far does the heating extend into its muscle?

Field Penetration into a *Good* Conductor

We have shown in Equations 7.86 and 7.94 that the direction of the *magnetic vector potential*, $\vec{A}_1$, in medium 1 is in the same direction as the current density, $\vec{J}_1$. However, we must be very careful to recognize that the electric current density in a conductor can be[iv] written as the sum of two terms:

$$\vec{J}_1 = \vec{J}_{1,l} + \vec{J}_{1,t}, \tag{7.104}$$

where $\vec{A}_{1,l}$ is called the *longitudinal current density* (or irrotational current density because $\vec{\nabla} \times \vec{J}_{1,l} = 0$) and $\vec{J}_{1,t}$ is called the *transverse current density* (or solenoidal current density because $\vec{\nabla} \cdot \vec{J}_{1,t} = 0$). The longitudinal current density found in the Coulomb gauge Equation 7.34b is $\varepsilon\vec{\nabla}(\partial V/\partial t) = \vec{J}_{1,l}$. This substitution cancels the longitudinal part of the current density in Equation 7.34b and yields

$$\vec{\nabla}^2 \vec{A}_1 - \mu\varepsilon \partial^2 \vec{A}_1 / \partial t^2 = -\mu \vec{J}_{1,t}. \tag{7.105}$$

Here, the longitudinal wave is moving normal to the conductor interface at a relatively low phase velocity, as shown above. The quasistatic approximation permits a fast charge relaxation (relative to the phase velocity) in this direction and cancels out most of the longitudinal current density (except at very high frequencies). By comparison, the fields tangential to the conductor interface are moving at a high phase velocity because of their continuity with the fields in medium 2 (the dielectric), where $u_{2,p} = c/\sqrt{\varepsilon_2}$. These displacements are so fast that the quasistatic approximation does not cancel the current density tangential to the conductor interface, so $\vec{J}_{1,t}$ is finite. Furthermore, we have argued that, for the quasistatic case (frequencies below 10^{18} *Hz*) Ampere's equation is $\vec{\nabla} \times \vec{H}_1 \approx \vec{J}_1$ and $\vec{J}_1 = \sigma_1 \vec{E}_1$, so we should be able to find the relative directions of the electromagnetic fields. Let us begin by arguing that a very good electric conductor should be an approximation to a perfect electric conductor, the difference being that tangential fields penetrate with an exponential decay into the surface of the good conductor.

For a perfect conductor, Table 7.4 shows that magnetic field intensity, $\vec{H}_2$, just outside of a conducting boundary is tangent to the surface and that electric flux density, $\vec{D}_2$, in medium 2 is normal to the boundary (i.e., in the $\hat{a}_x$-direction in Figure 7.5). Let us choose $\vec{H}_2(\vec{x}, t) = H_{\|} e^{j\omega t} \hat{a}_y$ for the magnetic field intensity outside a good conductor (consistent with the field outside a perfect conductor). We now know that $\vec{A}_1$ penetrates into the conductor and, thus, that $\vec{J}_1$ penetrates into the conductor, so we do not expect the magnetic field intensity to be discontinuous at the boundary as it was in the case of a perfect conductor. We can also make use of the fact that the spatial variations of the fields normal to the surface are more rapid than the variations parallel to the surface (especially with the constant value of external field with the y and z variables as chosen above). We can thus take $\vec{\nabla} \approx \hat{a}_x \partial/\partial x$ so that $\nabla \times \vec{H}_1 \approx -\hat{a}_y \partial H_{1,z}/\partial x|_{x=0-} + \hat{a}_z \partial H_{1,y}/\partial x|_{x=0-} \approx \vec{J}_1 \approx \sigma_1 \vec{E}_1$ just inside the conductor. But we have chosen the magnetic field intensity just outside the conductor (at $x = 0+$) to have only a y component, so we expect the field just inside (at $x = 0-$) to have a mainly y component. Thus, we expect the second term (in the z-direction) to be the dominant term that shows that $\vec{E}_1$ is in the z-direction and that $\vec{H}_1$ is in the y-direction just below the surface. Because of the charge density on the boundary, the electric field intensity is discontinuous across the boundary; that is, just outside the good conductor, $\vec{E}_2$ is mainly normal to the surface (in the x-direction), while, just inside the good conductor, $\vec{E}_1$ is mainly parallel (in the z-direction). By comparison, the magnetic field intensity is continuous across the boundary. We can write the continuity condition

$$\hat{a}_x \times \left(\vec{H}_2 - \vec{H}_1\right)\Big|_{x=0} = 0 \tag{7.106}$$

and we can use the expressions above to write

$$\vec{E}_1 = (1/\sigma_1)\vec{\nabla} \times \vec{H}_1 \approx (1/\sigma_1)(\partial H_{1y}/\partial x)\hat{a}_z \quad (7.107a)$$

and

$$\vec{H}_1 = (1/j\omega\mu_1)\vec{\nabla} \times \vec{E}_1 \approx (1/j\omega\mu_1)(\partial E_{1z}/\partial x)\hat{a}_y. \quad (7.107b)$$

We can also take another derivative of Equation 7.107a with respect to the x variable and use Equation 7.107b to substitute for $\partial E_{1z}/\partial x$ to write

$$(1/\sigma_1)(\partial^2 H_{1y}/\partial x^2) = j\omega\mu_1 H_{1y}$$

or

$$\partial^2 H_{1y}/\partial x^2 = j\omega\sigma_1\mu_1 H_{1y} = [(1+j)/\delta]^2 H_{1y} \quad (7.108)$$

where

$$\delta = \sqrt{2/\omega\mu_1\sigma_1}$$

The solution to this second-order, ordinary differential equation with boundary conditions is

$$\vec{H}_1(\vec{x}, t) = H_{1y}(0)e^{\left(\frac{1+j}{\delta}\right)x} e^{j\omega t}\,\hat{a}_y \quad (7.109a)$$

and using Equation 7.107a,

$$\vec{E}_1(\vec{x}, t) = [(1+j)/\sigma_1\delta]H_{1y}(0)e^{\frac{(1+j)}{\delta}x} e^{j\omega t}\,\hat{a}_z, \quad (7.109b)$$

where the variable x is a negative number in medium 1. In the figure below, we thus substitute $\xi = -x$. We can see from Equation 7.109b that the surface impedance Z_s is given by

$$Z_s = E_{1z}(0)/H_{1y}(0) = [(1+j)/\sigma\delta], \quad (7.110)$$

and the boundary conditions at $x = 0$ can be described as in Table 7.6 .

The magnitude of these fields are shown as a function of $x = -\xi$ in Figure 7.6. Figure 7.6 is shown in this orientation to mimic the physical arrangement of electromagnetic plane waves that propagate in a dielectric (medium 2) under a conducting trace (medium 1). The transverse electromagnetic waves that propagate in FR-4 are thus into the page, as seen in this rear view of a cross section of a copper trace above the propagating medium. The graph of relative amplitude is to be read by rotating the figure by 90 degrees in a clockwise manner. Thus, we see that

Table 7.6 Boundary Conditions Between a *Good* Electric Conductor and a Dielectric (Time-Varying Case)

On the conducting side (medium 1)	On the dielectric side (medium 2)
$E_{1t} = E_{1z}(0) = [(1+j)/\sigma_1\delta]H_{1y}(0)$	$E_{2t} = E_{2z}(0) \approx E_{1z}(0)$
$H_{1t}(0) = H_{\parallel}$	$H_{2t} = H_{\parallel}$
$D_{1n} \approx 0$	$D_{2n} = \Sigma_{e,S}$
$B_{1n} \approx 0$	$B_{2n} \approx 0$

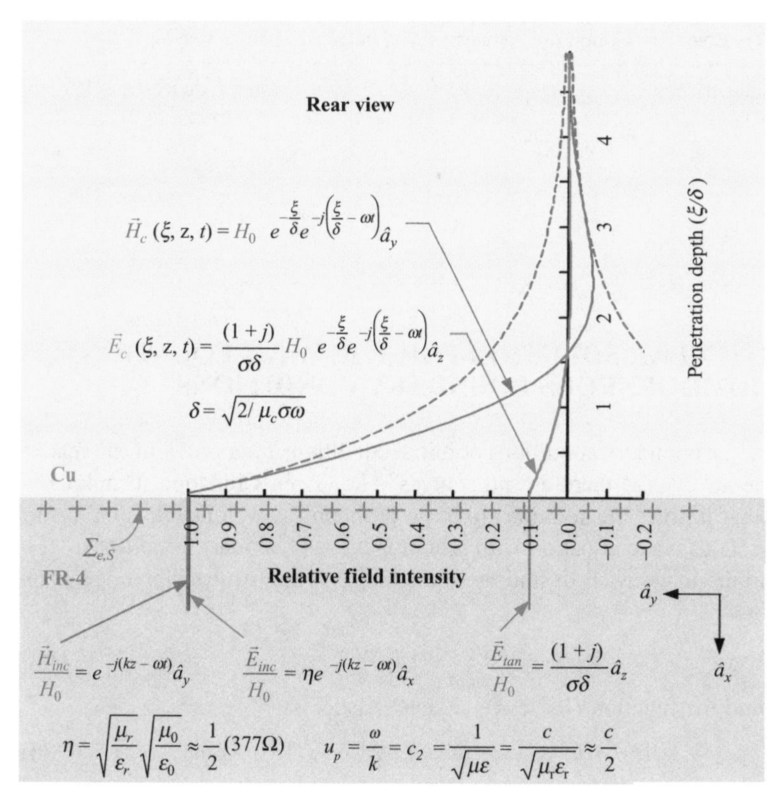

Figure 7.6 Variation of the electric and magnetic field intensity as a function of the penetration depth, $\xi = -x$, in the region near the boundary of a dielectric and a good conductor.

the magnitude of the electric and magnetic field intensities in medium 2 is relatively constant with the variable x, whereas the magnitude of the electric and magnetic field intensities in medium 1 (Cu) falls in a co-sinusoidal manner limited by an exponentially decaying function with the variable $\xi = -x$.

Interface between Two Lossless Linear Media

A lossless linear medium can be specified by a permittivity, ε, and a permeability, μ, with $\sigma = 0$. There are no free charges and no surface currents at the interfaces, so the tangential components of the electric and magnetic field intensity are continuous at the boundary between the two media as are the normal components of the electric and magnetic flux density, as shown in Equations 7.111 and Table 7.7.

$$E_{1t} = E_{2t} \Leftrightarrow D_{1t}/D_{2t} = \varepsilon_1/\varepsilon_2 \tag{7.111a}$$

$$H_{1t} = H_{2t} \Leftrightarrow B_{1t}/B_{2t} = \mu_1/\mu_2 \tag{7.111b}$$

$$D_{1n} = D_{2n} \Leftrightarrow \varepsilon_1 E_{1n} = \varepsilon_2 E_{2n} \tag{7.111c}$$

$$B_{1n} = B_{2n} \Leftrightarrow \mu_1 H_{1n} = \mu_2 H_{2n}. \tag{7.111d}$$

Table 7.7 Boundary Conditions between Two Dielectrics (Time-Varying Case)

On the dielectric side (medium 1)	On the dielectric side (medium 2)
$E_{1t} = E_{\parallel}$	$E_{2t} = E_{\parallel}$
$H_{1t} = H_{\parallel}$	$H_{2t} = H_{\parallel}$
$D_{1n} = D_{\perp}$	$D_{2n} = D_{\perp}$
$B_{1n} = B_{\perp}$	$B_{2n} = B_{\perp}$

7.13 PARTICULAR SOLUTION FOR THE WAVE EQUATION WITH INHOMOGENEOUS BOUNDARY CONDITIONS

In some cases, the boundary conditions or initial conditions for a problem can render it inhomogeneous even if there are no sources. The Green's function technique is so powerful that it gives the solutions to these problems as well as the solutions to the inhomogeneous wave equation with inhomogeneous boundary conditions.

For example, if we wish to find the particular solution of the inhomogeneous wave equations

$$\vec{\nabla}^2\psi - \mu\varepsilon\partial^2\psi/\partial t^2 = f(\vec{x},t) \tag{7.66a}$$

and we can find the function $G(\vec{x}, t; \vec{x}', t')$ that satisfies

$$\left(\vec{\nabla}^2 - \mu\varepsilon\partial^2/\partial t^2\right)G(\vec{x},t;\vec{x}',t') = \delta(\vec{x}-\vec{x}')\delta(t-t') \tag{7.70}$$

as we did in section 7.10, then we will require as part of causality that $G(\vec{x}, t; \vec{x}', t')$ must be zero if $t < t' + |\vec{x}-\vec{x}'|\sqrt{\mu\varepsilon}$; that is, the response at the position $\vec{x}$ at time t will be zero unless the source that occurred at position $\vec{x}'$ at time t' has had time to move the distance $|\vec{x} - \vec{x}'|$, traveling at speed of light $c_2 = 1/\sqrt{\mu\varepsilon}$ in that medium.

Some texts like to use the argument that responses are symmetric and that we can exchange the primed for the unprimed variables to argue that, inversely, the source that occurs at point $\vec{x}$ at time t will cause the response $G(\vec{x}', t'; \vec{x}, t)$ at point $\vec{x}'$ at time t' to be zero if

$$t' < t + |\vec{x}-\vec{x}'|/c. \tag{7.112}$$

They accomplish this symmetry by a "causality requirement"

$$G(\vec{x},t;\vec{x}',t') = G(\vec{x}',-t';\vec{x},-t). \tag{7.113}$$

The "causality requirement" is useful when we derive the time-dependent Green's theorem.

Time-Dependent Green's Theorem

In Equations 7.66a and 7.70, let us exchange primed for unprimed variables so that

$$\vec{\nabla}'^2\psi(\vec{x}',t') - \mu\varepsilon\partial^2\psi(\vec{x}',t')/\partial t'^2 = f(\vec{x}',t') \tag{7.114}$$

$$\left(\vec{\nabla}'^2 - \mu\varepsilon\partial^2/\partial t'^2\right)G(\vec{x}',t';\vec{x},t) = \delta(\vec{x}'-\vec{x})\delta(t'-t). \tag{7.115}$$

Now, we can multiply Equation 7.114 by $G(\vec{x}', t'; \vec{x}, t)$ and multiply Equation 7.115 by $\psi(\vec{x}', t')$ and subtract the last result from the first: that is,

$$\begin{aligned}
&G(\vec{x}', t'; \vec{x}, t)\vec{\nabla}'^2\psi(\vec{x}', t') - \mu\varepsilon G(\vec{x}', t'; \vec{x}, t)\partial^2\psi(\vec{x}', t')/\partial t'^2 \\
&\quad = G(\vec{x}', t'; \vec{x}, t)\, f(\vec{x}', t') \text{ minus} \\
&\quad \left(\psi(\vec{x}', t')\vec{\nabla}'^2 G(\vec{x}', t'; \vec{x}, t) - \mu\varepsilon\psi(\vec{x}', t')\partial^2 G(\vec{x}', t'; \vec{x}, t)/\partial t'^2\right) \\
&\quad = \psi(\vec{x}', t')\delta(\vec{x}' - \vec{x})\delta(t' - t)
\end{aligned}$$

and integrate over the volume of the domain and over time (t') from 0 to t^+, where t^+ is just greater than t. We will use the definition of the delta functions and note that

$$\begin{aligned}
&\partial[\psi(\vec{x}', t')(\partial G(\vec{x}', t'; \vec{x}, t)/\partial t') - G(\vec{x}', t'; \vec{x}, t)(\partial\psi(\vec{x}', t')/\partial t')]/\partial t' \\
&\quad = \psi(\vec{x}', t')\left(\partial^2 G(\vec{x}', t'; \vec{x}, t)/\partial t'^2\right) - G(\vec{x}', t'; \vec{x}, t)\left(\partial^2\psi(\vec{x}', t')/\partial t'^2\right). \quad (7.116)
\end{aligned}$$

The integral involving the terms on the right can be integrated, but, because at the upper limit, $t = t'$, we must use Equation 7.113 to conclude

$$\left.\begin{aligned}
G(\vec{x}', t'; \vec{x}, t) &= 0 \\
\partial G(\vec{x}', t'; \vec{x}, t)/\partial t' &= 0
\end{aligned}\right\} \text{ at } t' = t^+ > t + |\vec{x} - \vec{x}'|/c \quad (7.117)$$

Thus, by using Equation 7.113 inside all integrals,

$$\begin{aligned}
\psi(\vec{x}, t) &= \int_0^{t^+} \iiint_V G(\vec{x}, t; \vec{x}', t') f(\vec{x}', t')\, d^3x'\, dt' \\
&+ \sqrt{\mu\varepsilon} \iiint_V \left\{\psi(\vec{x}', 0)\frac{\partial}{\partial t'} G(\vec{x}, t; \vec{x}', 0) - G(\vec{x}, t; \vec{x}', 0)\frac{\partial}{\partial t'}\psi(\vec{x}', 0)\right\} d^3x' \\
&+ \int_0^{t^+} \oint_S \left\{G(\vec{x}, t; \vec{x}', t')\frac{\partial}{\partial n'}\psi(\vec{x}', t') - \psi(\vec{x}', t')\frac{\partial}{\partial n'} G(\vec{x}, t; \vec{x}', t')\right\} ds'\, dt'. \quad (7.118)
\end{aligned}$$

Equation 7.118 is the answer to all E&M problems for which boundary conditions are specified on the boundary S to a volume V on which either the quantity $\psi(\vec{x}', t')$ or the quantity $\partial\psi(\vec{x}', t')/\partial n'$ is specified *and* either the initial condition $\psi(\vec{x}', 0)$ or $\partial\psi(\vec{x}', 0)/\partial n'$ are specified throughout the volume V.

Note that this solution even contains the answer for homogeneous problems, where $f(\vec{x}', t') = 0$. In this case, the first integral in Equation 7.118 is zero.

Furthermore, because the function and its derivative cannot both be specified (an overspecification of the problem), we can see the following conditions that must be imposed on the Green's function in Equation 7.118 to get a solution:

- If $\psi(\vec{x}', 0)$ is specified in V, we will choose $G(\vec{x}, t; \vec{x}', 0) = 0$ in V.
- If $\partial\psi(\vec{x}', 0)/\partial t'$ is specified in V, we will choose $\partial G(\vec{x}, t; \vec{x}', 0)/\partial t' = 0$ in V.
- If $\psi(\vec{x}', t')$ is specified on S, we will choose $G(\vec{x}, t; \vec{x}', t') = 0$ on S.
- If $\partial\psi(\vec{x}', t')/\partial t'$ is specified on S, we will choose $\partial G(\vec{x}, t; \vec{x}', t')/\partial t' = 0$ on S.

This is the solution that many numerical codes use to build in boundary conditions that are specified on the surface S surrounding a finite volume V and initial conditions that are specified in the closed volume V. Let us be in awe of a mathematician like George Green who could have solved such a complete boundary value problem *with initial conditions* in 1824.

7.14 MEMRISTORS

In section 1.8, the quaternion form of Maxwell's equations was described in terms of the magnetic vector potential, $\vec{A}$, which was a three-component, time-dependent, vector in physical space. But, unlike the mathematical convenient construction of $\vec{A}$ described in section 7.5, Maxwell considered the magnetic vector potential to be invariant to the chosen gauge and had physical meaning. Although Heavyside abandoned the physical interpretation, Herman Weyl[v,vi] considered the transport of vectors[vii] in a three-component physical and one-component time (with distance ct) space in which the magnetic flux, Φ_M, was separated into spacelike and timelike[6] quantities, as shown in Equation 7.119:

$$\Phi_M = \oint_C \vec{A} \cdot d\vec{l} + \int_t^{t'} \Delta\phi_e c dt. \tag{7.119}$$

Weyl made the first attempt to formulate a unified field theory in which both gravitation and electromagnetism were incorporated into a single geometrical structure of a space–time manifold. Among other things, Weyl postulated that the length of a four dimensional vector displaced in a parallel fashion around a path C in three space and over the time interval $(t' - t)$ would undergo a change in physical length. Einstein pointed out that a consequence of Weyl's theory would be that the frequencies of spectral lines emitted by atoms would depend on the electromagnetic history of the atoms. An experimental test[viii] of Weyl's postulate for the frequency change, $\Delta f/f$, of iron Mössbauer nuclei, as explained in section 8.3, showed that such effects (if they exist at all) would be proportional to the dimensionless quantity, Φ_M/e, to less than $\pm 2 \times 10^{-48}$. The importance of this discussion here is that magnetic flux can be constructed[7] from a time integral of scalar potential difference, as shown in Equation 7.119.

In 1971,Leon Chua[ix] considered the ratio $d\Phi_M/dq$ to be characteristic of a fourth kind of circuit element he called the *memristor* (short for *memory resistor*) and noted that either term in Equation 7.119 would describe a passive two-terminal functional relationship between the integrals of magnetic vector potential or voltage and current because Stokes theorem gives

[6] Maxwell referred to scalar electric potential difference in Equation 1.58 as $\Delta\phi_e$ rather than scalar electric potential, V, relative to ground as used in Equation 7.85.

[7] One justification of color coding Maxwell's equations is that it emphasizes that time derivatives and time integrals of electric quantities (charge density, scalar potential, electric field intensity, electric flux density) give rise to magnetic quantities (effective magnetic monopole density, current, magnetic flux density, magnetic vector potential) and vice versa.

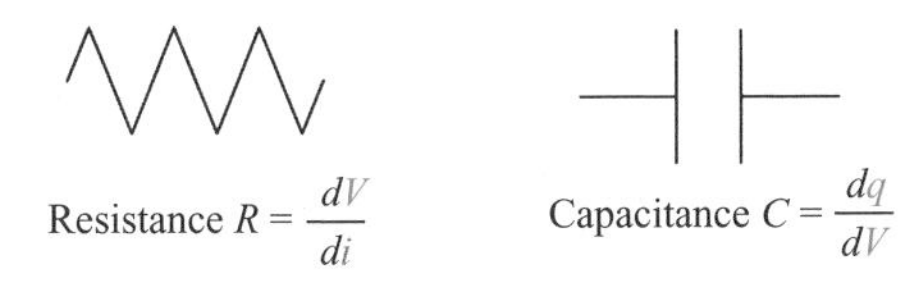

Resistance $R = \frac{dV}{di}$ Capacitance $C = \frac{dq}{dV}$

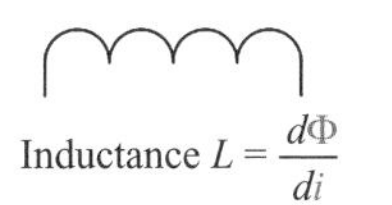

Inductance $L = \frac{d\Phi}{di}$

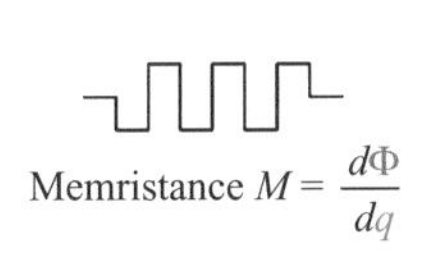

Memristance $M = \frac{d\Phi}{dq}$

Figure 7.7 Symmetry of *voltage*, *current*, *charge*, and *magnetic flux* linked to one another by fundamental material properties: *resistance*, *capacitance*, *inductance*, and *memristance*.

$$\oint_C \vec{A} \cdot d\vec{l} = \iint_S \vec{\nabla} \times \vec{A} \cdot d\vec{s} = \iint_S \vec{B} \cdot d\vec{s}.$$

Chua further argued that the relationship between magnetic flux and charge change completed a symmetry with the *resistor*, *inductor*, and *capacitor* that defined relations between four circuit variables *voltage*, *current*, *charge*, and *magnetic flux*; the device that linked the flux and the charge being the *memristor*, as shown in Figure 7.7.

Chua argued that the powerful part of this symmetry is that not only are instantaneous time varying quantities like *voltage* and *current* linked by material properties of *resistance*, but the *time integral of voltage* and the *time integral of current* (i.e., *charge*) are linked by material properties of *memristance*. Thus, in nonlinear devices, a hysteresis of the voltage versus current curves could be permanently retained until the charge was reversed in a complete cycle.

In 2008,[x] a team at Hewlett-Packard (HP) led by R. Stanley Williams announced the discovery of such a nonlinear device based on the resistive properties of titania.[8] In mildly reducing atmospheres, TiO_2 tends to lose oxygen and become substoichiometric TiO_{2-x}. For example, in its stoichiometric form at 25°C, TiO_2 has a resistivity of 10^{10} Ωm comparable to that of a good insulator, but, when raised to 700°C, its resistivity decreases to 2.5×10^2 Ωm comparable to that of semiconductors. The HP team assumed that the boundary between TiO_2 and TiO_{2-x} migrated under the influence of an electric field, as shown in Figure 7.8 to achieve a resistance versus charge behavior that persists even when the electric field was removed. However, the boundary moves only nanometers, and the chemistry of the material is uncertain. For example, it is unknown if there is an additional absorption or depletion of oxygen atoms in either of the two titania materials or if the concentration of the oxygen depleted volume changes to $TiO_{2-x'}$.

Because the titania material provides both capacitive and resistive properties, the voltage, $V(t)$, that moves the boundary by an amount, y, is a combination of the two charging mechanisms. Depending on the character of the voltage or current

[8] Titania is often used in the construction of ceramic capacitors because it has a high dielectric strength of 4 kV/mm, a high relative dielectric constant of $\varepsilon'_r = 85$, and a low $\tan\delta$ of 5×10^{-4} at 1 MHz.

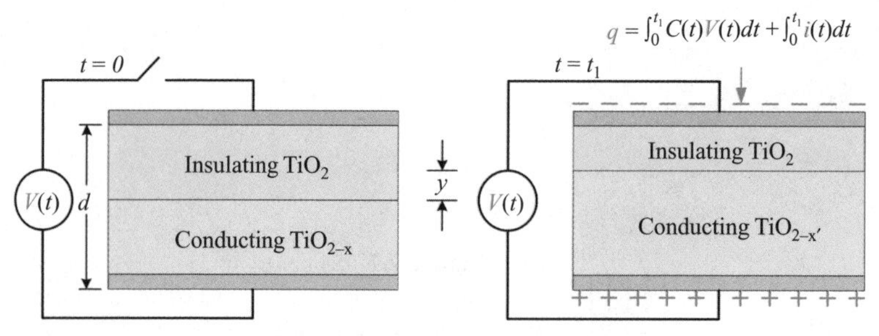

Figure 7.8 Williams's interpretation of the migration of the boundary between insulating TiO_2 and conducting TiO_{2-x} under the influence of a voltage differential.

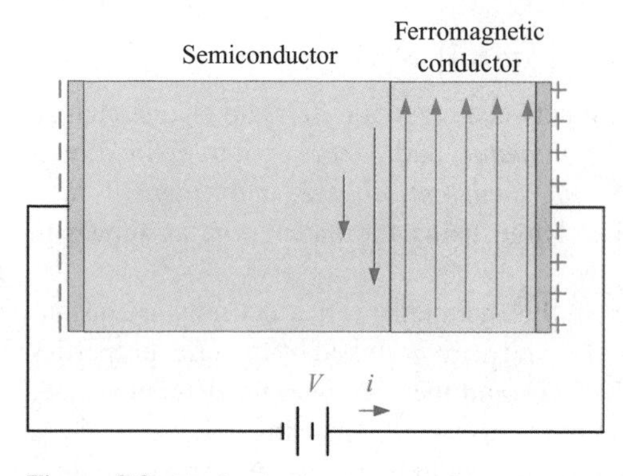

Figure 7.9 Device for the transport of spin-up electrons from a semiconductor into a spin-up conduction band of a conducting ferromagnet.

supply and the speed of the migrating boundary, one of these effects might dominate the other in a given frequency range. However, the memory mechanism (the resistance of the system) depends upon the boundary to remain in its last configuration once the voltage has been removed. The material would thus need to have a very low thermal diffusion constant to retain this information for long periods, as proposed in memory storage applications such as crossbar architectures.

Spintronic *Memristor*

Another proposed[xi] type of *memristor* is based on the spin of conduction electrons in ferromagnetic conductors, as shown in Figure 7.9.

As we have seen in section 6.6 on atomic magnetism, spin-dependent conduction bands in magnetic materials are often preferentially filled. In the case of ferromagnets, we conclude that one band (say the spin-up band) can have most or all of the conduction electrons at the expense of a nearly empty band (say, the spin-down band). This condition is shown on the right-hand side of in Figure 7.9 as a ferromagnetic conductor region with red arrows pointing up to indicate that the conduction electrons in this region dominantly have spins that point up; that is, the current density of conduction electrons have proportion $n_\uparrow$ that is much greater than the proportion $n_\downarrow$. Electrons in the nonmagnetic region on the lefthand side are in a semiconductor region with equal populations of spin-up and spin-down electrons ($n_\uparrow \approx n_\downarrow$) involved in current transport (propagating to the right in a random sequence of scattering events, as shown in section 5.7). At the contact boundary between the semiconductor and the ferromagnetic conductor, the spin-up electrons are permitted to continue their propagation to the right, while the spin-down electrons will be reflected back into the semiconductor. This will lead to a buildup of the population of spin-down electrons just to the left of the contact boundary, as shown by the decreasing length down arrows to indicate that the scattered spin-down electrons undergo spin relaxation back to the equal distribution condition. The local current density caused by the reflection of the spin-down electrons depends on the gradient of the population as $\vec{J}_{\uparrow(\downarrow)} = \sigma\vec{E} + eD\vec{\nabla}n_{\downarrow(\uparrow)}$, where D is a diffusion coefficient. The additional term proportional to the change in spin populations alters the effective conductivity of the device and results in spin-dependent conductivity. Device designs that can flip spin-up and spin-down region of the ferromagnet can cause a transition between a high and low resistance state at rates as fast as picoseconds. For reading a hard drive magnetized region, for example, we would want fast switching of the magnetic field, which would show up as a fast change in resistance. It is thought that the ferromagnetic regions might be contained in a nanometer of material that could lead to very high-density memory devices. The spintronic memristor, however, can suffer from the loss of memory of its magnetic state after power is turned off, but it is thought that a spin state can be held longer than a charge state.

*Mem*capacitors and *Mem*inductors

Circuit elements that store information without the need for a power source would represent a paradigm change in electronics, allowing for low-power computation and high-density storage. Ventra, Pershin, and Chua have speculated[xii] that many systems may belong to this class, including the thermistor (whose resistance depends on the temperature of the system). These researchers argue that the concept of a memory device is not necessarily limited to resistances but can be generalized to capacitive and inductive systems.

The exact character of the materials and several other candidate examples remains an area of speculation, but the application of such devices is worthy of a large research effort whether or not it represents a fourth fundamental circuit element.

7.15 ELECTRIC VECTOR POTENTIAL

In the last part of the twentieth century, some electrical engineers (e.g., Harrington and Balanis) modified the set of Maxwell equations to make them symmetric, as shown in Table 7.8 by including the magnetic monopole charge density and the magnetic monopole current density. The symbols used for magnetic charge density

Table 7.8 Symmetric Form of Maxwell's Equations

Differential form	Integral form	Name
$\vec{\nabla} \times \vec{E} = -\vec{J} - \partial \vec{B}/\partial t$	$\oint_C \vec{E} \cdot d\vec{l} = -I - \iint_S (\partial \vec{B}/\partial t) \cdot d\vec{s}$	Faraday's law
$\vec{\nabla} \times \vec{H} = \vec{J} + \partial \vec{D}/\partial t$	$\oint_C \vec{H} \cdot d\vec{l} = I + \iint_S (\partial \vec{D}/\partial t) \cdot d\vec{s}$	Ampere's law
$\vec{\nabla} \cdot \vec{D} = \rho_V$	$\oiint_S \vec{D} \cdot d\vec{s} = Q$	Gauss's law for electric charge
$\vec{\nabla} \cdot \vec{B} = \rho_V$	$\oiint_S \vec{B} \cdot d\vec{s} = Q$	Gauss's law for magnetic charge

and magnetic current density are dependent on the particular author, but the equations in Table 7.3 are consistent with the color-coding introduced earlier.

Maxwell's equations represent the same vector field quantities given in Table 7.2 with the additional quantities:

$\vec{J}$ = Magnetic current *density* (volts/meter^2)
ρ_V = Magnetic charge *density* (weber/meter^3)

with the units of the new field quantities in SI units shown in parentheses. Likewise, we can develop an equation of continuity for both electric and magnetic charge density using conservation of charge to write the symmetric forms shown in Equations 7.120 and 7.121:

$$\vec{\nabla} \cdot \vec{J} = -\partial \rho_V / \partial t \tag{7.120}$$
$$\vec{\nabla} \cdot \vec{J} = -\partial \rho_V / \partial t. \tag{7.121}$$

We can again choose to use vector and scalar potentials as mathematical tools to solve the symmetric form of Maxwell's equation; that is, we have shown in Chapter 3 that the divergence of the curl of *any* vector field is identically zero;

$$\vec{\nabla} \cdot (\vec{\nabla} \times \vec{A}) = 0. \tag{7.122}$$

Thus, in a *source-free* region of space, both $\vec{D}$ and $\vec{B}$ are solenoidal, ($\vec{\nabla} \cdot \vec{B} = 0$) and ($\vec{\nabla} \cdot \vec{D} = 0$), so we can thus assume that $\vec{B}$ may be written in terms of another vector field, $\vec{A}$, called the *magnetic vector potential*:

$$\vec{B} = \vec{\nabla} \times \vec{A} \tag{7.123}$$

and that $\vec{D}$ may be written in terms of another vector field, $\vec{A}$, called the *electric vector potential:*

$$\vec{D} = \vec{\nabla} \times \vec{A}. \tag{7.124}$$

To specify a *unique* definition of the vector fields, $\vec{A}$ and $\vec{A}$, we will need to make an additional gauge restriction on *both*.

Substituting Equation 7.123 into ($\vec{\nabla} \times \vec{E} = -\vec{J} - \partial \vec{B}/\partial t$) for free space (with $\vec{J} = 0$), we can write

$$\vec{\nabla} \times \vec{E} = -\partial\left(\vec{\nabla} \times \vec{A}\right)/\partial t \quad \text{or} \quad \vec{\nabla} \times \left(\vec{E} + \partial\vec{A}/\partial t\right) = 0. \tag{7.125}$$

Substituting Equation 7.124 into ($\vec{\nabla} \times \vec{H} = \vec{J} + \partial\vec{D}/\partial t$) for free space (with $\vec{J} = 0$), we can write

$$\vec{\nabla} \times \vec{H} = -\partial\left(\vec{\nabla} \times \vec{A}\right)/\partial t \quad \text{or} \quad \vec{\nabla} \times \left(\vec{H} + \partial\vec{A}/\partial t\right) = 0. \tag{7.126}$$

We have also shown in Chapter 3 that $\vec{\nabla} \times (-\vec{\nabla}V) = 0$ for *any* scalar field. Thus, because the curl of the vector field shown in parentheses in Equations 7.125 and 7.126 is zero, then that field can be written as the negative gradient of another scalar field that we will successively call the ***electric scalar potential***, V and the ***magnetic scalar potential***, V, with

$$\vec{E} + \partial\vec{A}/\partial t = -\vec{\nabla}V \quad \text{or} \quad \vec{E} = -\vec{\nabla}V - \partial\vec{A}/\partial t \tag{7.127}$$

$$\vec{H} + \partial\vec{A}/\partial t = -\vec{\nabla}V \quad \text{or} \quad \vec{H} = -\vec{\nabla}V - \partial\vec{A}/\partial t. \tag{7.128}$$

We can see from Equation 7.127 that the electric field intensity, $\vec{E}$, can be written in terms of the electric scalar potential, V, and the time derivative of the magnetic vector potential, $\vec{A}$. As long as these scalar and vector potentials are unique, the electric field intensity produced by them will also be unique.

NOTE In the special case of static (time-independent) fields and potentials, $\partial\vec{A}/\partial t = 0$, and we can see that Equation 7.127 reduces to $\vec{E} = -\vec{\nabla}V$ as we found in Chapter 6 for static electric fields.

We can see from Equation 7.128 that the magnetic field intensity, $\vec{H}$, can be written in terms of the magnetic scalar potential, V, and the time derivative of the electric vector potential, $\vec{A}$. As long as these scalar and vector potentials are unique, the magnetic field intensity produced by them will also be unique.

NOTE In the special case of static (time-independent) fields and potentials, $\partial\vec{A}/\partial t = 0$, and we can see that Equation 7.128 reduces to $\vec{H} = -\vec{\nabla}V$, as Maxwell originally proposed (see Equation 1.68).

For homogeneous media in time-varying fields, we can substitute $\vec{B} = \mu\vec{H}$ and $\vec{D} = \varepsilon\vec{E}$ into the symmetric forms to yield $\vec{\nabla} \times \vec{B} = \mu\vec{J} + \mu\varepsilon\partial\vec{E}/\partial t$ or $\vec{\nabla} \times (\vec{\nabla} \times \vec{A}) = \mu\vec{J} + \mu\varepsilon\partial\vec{E}/\partial t$ or

$$\vec{\nabla} \times \vec{\nabla} \times \vec{A} = \mu\vec{J} + \mu\varepsilon\partial\left(-\vec{\nabla}V - \partial\vec{A}/\partial t\right)/\partial t \tag{7.129}$$

and, using the identity $\vec{\nabla} \times \vec{\nabla} \times \vec{A} = \vec{\nabla}(\vec{\nabla} \cdot \vec{A}) - \vec{\nabla}^2\vec{A}$ in Equation 7.129, we see

$$\vec{\nabla}\left(\vec{\nabla} \cdot \vec{A}\right) - \vec{\nabla}^2\vec{A} = \mu\vec{J} - \vec{\nabla}(\mu\varepsilon\partial V/\partial t) - \mu\varepsilon\partial^2\vec{A}/\partial t^2$$

or

$$\vec{\nabla}^2\vec{A} - \mu\varepsilon\partial^2\vec{A}/\partial t^2 = -\mu\vec{J} + \vec{\nabla}\left(\vec{\nabla} \cdot \vec{A} + \mu\varepsilon\partial V/\partial t\right). \tag{7.130}$$

Likewise, using the symmetric form $\vec{\nabla} \times \vec{E} = -\vec{J} - \partial\vec{B}/\partial t$ or $\vec{\nabla} \times (\vec{\nabla} \times \vec{A}) = -\varepsilon\vec{J} - \varepsilon\partial\vec{B}/\partial t$ or,

$$\vec{\nabla} \times \vec{\nabla} \times \vec{A} = \varepsilon\vec{J} + \mu\varepsilon\partial\left(-\vec{\nabla}V - \partial\vec{A}/\partial t\right)/\partial t \tag{7.131}$$

and, using the identity $\vec{\nabla} \times \vec{\nabla} \times \vec{A} = \vec{\nabla}(\vec{\nabla} \cdot \vec{A}) - \vec{\nabla}^2 \vec{A}$ in Equation 7.131, we see

$$\vec{\nabla}(\vec{\nabla} \cdot \vec{A}) - \vec{\nabla}^2 \vec{A} = \varepsilon \vec{J} - \vec{\nabla}(\mu\varepsilon \partial V/\partial t) - \mu\varepsilon \partial^2 \vec{A}/\partial t^2$$

or

$$\vec{\nabla}^2 \vec{A} - \mu\varepsilon \partial^2 \vec{A}/\partial t^2 = -\varepsilon \vec{J} + \vec{\nabla}(\vec{\nabla} \cdot \vec{A} + \mu\varepsilon \partial V/\partial t). \tag{7.132}$$

Now, the definition of a ***unique*** vector field $\vec{A}$ or $\vec{A}$ requires an additional restriction or gauge. One way to provide this restriction is to specify their divergence; that is, although the curl of $\vec{A}$ or $\vec{A}$ is designated or $\vec{B}$ or $\vec{D}$, we are still at liberty to choose the divergence of $\vec{A}$ or $\vec{A}$. As we noted in Equation 7.28, we can use the Lorenz gauge to write[1]

$$\vec{\nabla} \cdot \vec{A} + \mu\varepsilon \partial V/\partial t = 0 \tag{7.133}$$

$$\vec{\nabla} \cdot \vec{A} + \mu\varepsilon \partial V/\partial t = 0, \tag{7.134}$$

that choice reduces Equations 7.130 and 7.132 to second-order, linear, inhomogeneous PDEs:

$$\vec{\nabla}^2 \vec{A} - \mu\varepsilon \partial^2 \vec{A}/\partial t^2 = -\mu \vec{J} \tag{7.135}$$

$$\vec{\nabla}^2 \vec{A} - \mu\varepsilon \partial^2 \vec{A}/\partial t^2 = -\varepsilon \vec{J} \tag{7.136}$$

for which Equation 7.135 is the inhomogeneous **wave equation** for the *magnetic vector potential* and Equation 7.136 is the inhomogeneous **wave equation** for the *electric vector potential.*

Equations 7.135 and 7.136 need only the current density, $\vec{J}$ or $\vec{J}$, to solve for $\vec{A}$ or $\vec{A}$. Some electrical engineering texts choose the Lorentz gauge for $\vec{A}$ and $\vec{A}$ to separate the terms into single PDEs that can be solved by integration techniques that were developed in section 7.9. Those texts then find a corresponding wave equation for the electric scalar potential by using Gauss's law $\vec{\nabla} \cdot \vec{D} = \rho_V$ and Equation 7.127:

$$\vec{\nabla} \cdot \vec{E} = \rho_V/\varepsilon \Rightarrow \vec{\nabla} \cdot (\vec{\nabla} V + \partial \vec{A}/\partial t) = -\rho_V/\varepsilon, \tag{7.137}$$

which leads to

$$\vec{\nabla}^2 V + \partial(\vec{\nabla} \cdot \vec{A})/\partial t = -\rho_V/\varepsilon, \tag{7.138}$$

and, using the Lorenz gauge ($\vec{\nabla} \cdot \vec{A} + \mu\varepsilon \partial V/\partial t = 0$), we see that the *electric scalar potential*, V, also satisfies the inhomogeneous wave equation:

$$\vec{\nabla}^2 V - \mu\varepsilon \partial^2 V/\partial t^2 = -\rho_V/\varepsilon \tag{7.139}$$

Equation 7.139 needs only ρ_V to solve for the electric scalar potential, V.

Likewise, those texts then find a corresponding wave equation for the magnetic scalar potential by using Gauss's law $\vec{\nabla} \cdot \vec{B} = \rho_V$ and Equation 7.128:

$$\vec{\nabla} \cdot \vec{H} = \mu\rho_V \Rightarrow \vec{\nabla} \cdot (\vec{\nabla} V + \partial \vec{A}/\partial t) = -\mu\rho_V, \tag{7.140}$$

which leads to

$$\vec{\nabla}^2 V + \partial\left(\vec{\nabla} \cdot \vec{A}\right)/\partial t = -\mu \rho_V \tag{7.141}$$

and, using the Lorentz gauge ($\vec{\nabla} \cdot \vec{A} + \mu\varepsilon\partial V/\partial t = 0$), we see that the *magnetic scalar potential*, V, also satisfies the inhomogeneous wave equation:

$$\vec{\nabla}^2 V - \mu\varepsilon\partial^2 V/\partial t^2 = -\mu\rho_V, \tag{7.142}$$

Equation 7.142 needs only ρ_V to solve for the magnetic scalar potential, V.

Conclusion

With a prior knowledge of ρ_V , ρ_V , $\vec{J}$ and $\vec{J}$, we can separate the x, y, and z components of the wave equations and solve for V and V and each component of $\vec{A}$ and $\vec{A}$ independent of the others. All three of these equations are of in the form of the same inhomogeneous wave equation and they are independent of one another. Thus, given the electric charge density, the magnetic charge density, the vector electric current density, and the vector magnetic current density, we can solve the inhomogeneous wave equation (subject to boundary conditions specified by a particular application) to find the potentials V, V, $\vec{A}$ and $\vec{A}$ from which we can then find all of the components of the electric field intensity and magnetic field intensity.

Equations 7.139, 7.142, 7.135, and 7.136, for V, V, $\vec{A}$ and $\vec{A}$ form a set of four equations equivalent in all respects to the symmetric Maxwell's equations (subject to the restriction of the Lorenz gauge). However, unlike Maxwell's equations, these four inhomogeneous PDEs are independent of one another, so they will be much easier to solve.

NOTE Using the electric vector potential and the magnetic vector potential results in electric and magnetic field that originates from

$$\vec{B} = \vec{\nabla} \times \vec{A} \tag{7.123}$$

$$\vec{D} = \vec{\nabla} \times \vec{A}. \tag{7.124}$$

$$\vec{E} = -\vec{\nabla} V - \partial\vec{A}/\partial t \tag{7.127}$$

$$\vec{H} = -\vec{\nabla} V - \partial\vec{A}/\partial t. \tag{7.128}$$

The resulting total electric and magnetic field is the vector sum due to both potentials:

$$\vec{E}_{total} = -\vec{\nabla} V - \partial\vec{A}/\partial t + \vec{\nabla} \times \vec{A}/\varepsilon \tag{7.143}$$

$$\vec{H}_{total} = -\vec{\nabla} V - \partial\vec{A}/\partial t + \vec{\nabla} \times \vec{A}/\mu. \tag{7.144}$$

Engineers sometimes use electric vector potential and magnetic vector potential to develop solutions because they are easier to find via the inhomogeneous wave equations with boundary conditions, as shown in section 7.9. The solutions can be chosen to have boundary conditions so that one part of the solution yields a transverse electromagnetic, transverse electric, or transverse magnetic solution in a particular coordinate system. The technique suffers from the fact that there are two vector potentials to find and from the fact that magnetic charge density may be approximated by Equation 6.74 and magnetic current density may be approximately by Equation 6.64. However, we are always mindful that this approximation is poor when considering fields in the microscopic near-field regime such as that shown in Figure 6.7.

CONCLUSION While it is more straightforward to obtain a solution, the two vector potential technique will not suffice for the analysis of crystal field effects or fields internal to atoms or molecules.

The physics community usually assumes that there is no such thing as magnetic charge density or a magnetic current density so that $\rho_V = 0$ and $\vec{J} = 0$. In this formalism, Maxwell's equations in Table 7.8 are equivalent to their asymmetric form in Table 7.2. Because we will often evaluate near fields, the asymmetric form of Maxwell's equations is used in *The Foundations of Signal Integrity*[xiii] to find solutions to applied problems in Signal Integrity.

ENDNOTES

i. L. V. Lorenz, "Eichtransformationen, und die Invarianz der Felder unter solchen Transformationen nennt man Eichinvarianz," *Philosophical Magazine Series 4,* 34 (1867): 287–301.

ii. A treatment of the computational techniques is found in Y. Chen, Q. Cao, and R. Mittra, *Multiresolution Time Domain Scheme for Electromagnetic Engineering* (Hoboken, NJ: John Wiley & Sons, 2005).

iii. Paul G. Huray, *The Foundations of Signal Integrity* (Hoboken, NJ: John Wiley & Sons, 2009), Chapter 3.

iv. J. D. Jackson, *Classical Electrodynamics*, 3rd ed. (Hoboken, NJ: John Wiley & Sons, 2001), 242.

v. H. Weyl, *Gravitation und Elektrizität* (Sitzungsberichte der Königlich Preußischen Akademie der Wissenschaften zu Berlin, 1918), 465–478.

vi. H. A. Lorentz, A. Einstein, H. Minkowski, and H. Weyl, *The Principle of Relativity* (New York: Dover, 1962).

vii. H. Weyl, *Space, Time, Matter* (New York: Dover, 1950).

viii. E. G. Harris, P. G. Huray, F. E. Obenshain, J. O. Thompson, and R. A. Villecco, "Experimental Test of Weyl's Gauge-Invariant Geometry," *Physical Review D* 7 (1973): 2326–2330.

ix. L. O. Chua, "Memristor—The Missing Circuit Element," *IEEE Transactions on Circuit Theory CT* 18, no. 5 (1971): 507–19.

x. D. B. Strukov, G. S. Snider, D. R. Stewart, and R. S. Williams, "The Missing Memristor Found," *Nature* 453 (2008): 80–3.

xi. Y. V. Pershin and M. Di Ventra, "Current-Voltage Characteristics of Semiconductor/Ferromagnet Junctions in the Spin-Blockade Regime," *Physical Review B* 77 (2008): 73301–73305.

xii. M. Di Ventra, Y. V. Pershin, and L. O. Chua, "Circuit Elements with Memory: Memristors, Memcapacitors and Meminductors," *Mesoscale and Nanoscale Physics* (2009): 1–6. Available at http://physics.uscd.edu/~diventra/memdevicespub.pdf, accessed August 20, 2009.

xiii. Huray, *Foundations of Signal Integrity.*

Appendix A

Measurement Errors

THE BINOMIAL DISTRIBUTION

Measurements can be made only a **finite** number of times and the precision of the measurement is predetermined by the accuracy of the measuring instrument: a **discrete** measure. In any single measurement, there will be a probability associated with a particular outcome (value obtained in that measurement). A sense of the mathematics associated with these probabilities can be obtained by conducting a measurement where the probabilities are known for a given event. For example, in tossing a die, only integer values can be measured: 1, 2, 3, 4, 5, or 6. In some measurements, the probability of obtaining a particular value will not change from one measurement to the next. Such events are said to be *independent*. Such is the case for tosses of a die: $p = 1/6$ is the probability that a particular integer will occur in any given toss, and $q = 1 - p = 5/6$ is the probability that a particular number will not occur. The probability that a particular integer will occur exactly x times in n tosses (and $n - x$ failures will occur) is given by the binomial distribution or the probability density function (PDF):

$$f(x) = \binom{n}{x} p^x q^{n-x} = \frac{n!}{x!(n-x)!} p^x q^{n-x} \tag{A.1}$$

This discrete probability function is called the binomial distribution because the coefficients and powers are the same as those of the xth term in a binomial expansion:

$$(q+p)^n = q^n + \binom{n}{1} q^{n-1} p + \binom{n}{2} q^{n-2} p^2 + \ldots + p^n = \sum_{x=0}^{n} \frac{n!}{x!(n-x)!} p^x q^{n-x}. \tag{A.2}$$

The mean of the binomial distribution is $\bar{x} = np$, the variance is $\sigma^2 = npq$, and the standard deviation is $\sigma = \sqrt{npq}$. The binomial probability distribution function for the example of 20 tosses of a die is shown in Figure A.1.

Figure A.1 shows that, on 20 throws of the six-sided die, the probability that a given value will occur x_1 times is given by $f(x_1)$. The probability that a given value

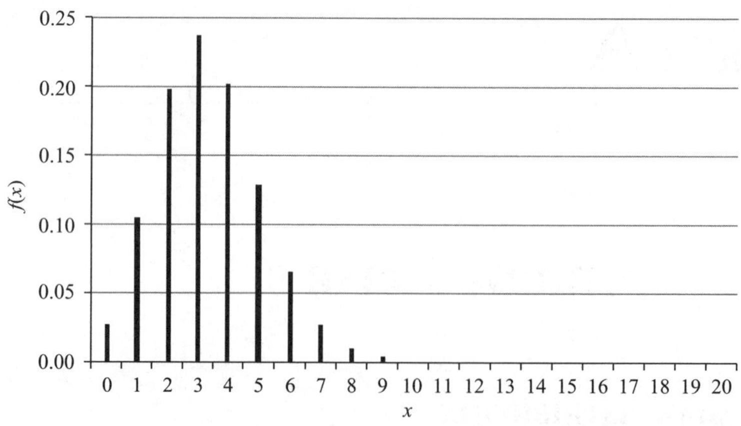

Figure A.1 Binomial probability density function for the case of $p = 1/6$ and $n = 20$.

will occur x_1 **or** x_2 times is given by $f(x_1) + f(x_2)$. The probability that a given value will occur 0 or 1 or 2 or 3, or ... , or 20 times is $\sum_{i=0}^{n} f(x_i)$. By replacing $q = 1 - p$, it can be seen from Equation A.2 that this sum is 1; that is, the sum of all of the discrete heights of bars in Figure A.1 is 1. The binomial probability density function is thus said to be normalized.

THE GAUSSIAN DISTRIBUTION

For an **infinite** set of measurements of a **continuous** quantity in which the measurements are *independent* of one another, a continuous PDF called the normal distribution, the bell-shaped curve, or the Gaussian distribution, is given by

$$f(x) = \frac{1}{\sigma}\frac{1}{\sqrt{2\pi}}e^{-\frac{(x-\bar{x})^2}{2\sigma^2}} \quad (-\infty < x < \infty) \tag{A.3}$$

or

$$f(t) = \frac{1}{\sqrt{2\pi}}e^{-\frac{t^2}{2}} \quad \text{where} \quad t = \frac{(x-\bar{x})}{\sigma}. \tag{A.4}$$

The Gaussian PDF is shown plotted in Figure A.2.

The coefficients in the Gaussian PDF are chosen so that the area under the curve is 1; that is, the probability that *some* value of t is obtained in a given measurement is 1. The quantity σ is called the standard deviation, and t is a variable that expresses $x - \bar{x}$ in terms of the number of standard deviations.

In contrast to the discrete binomial PDF, the continuous Gaussian PDF gives the probability, P, of finding a measurement between x_1 and x_2 as an integral:

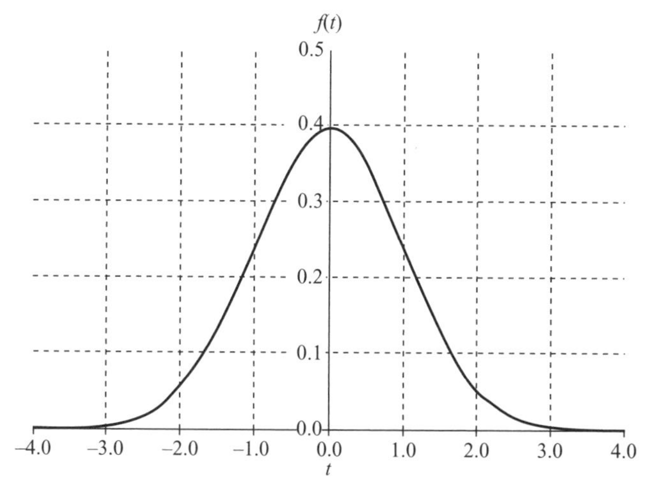

Figure A.2 Gaussian probability density function for a continuous variable, *t*.

$$P_{x_1,x_2} = \int_{x_1}^{x_2} f(x)\,dx = \int_{x_1}^{x_2} \frac{1}{\sigma}\frac{1}{\sqrt{2\pi}} e^{-\frac{(x-\bar{x})^2}{2\sigma}}\,dx \tag{A.5}$$

or

$$P_{t_1,t_2} = \int_{t_1}^{t_2} f(t)\,dt \quad \text{where } t_1 = \frac{(x_1 - \bar{x})}{\sigma} \text{ and } t_2 = \frac{(x_2 - \bar{x})}{\sigma} \tag{A.6}$$

The integral of a normalized Gaussian PDF (i.e., the area under the curve) gives the probability of finding a given measurement between x_1 and x_2 (or equivalently t_1 and t_2). The probability of finding a given measurement between $-\infty$ and $+\infty$ is 1. This can be seen by programming a calculator to integrate the Normalized Gaussian PDF or by finding the areas in Figure A.3, which gives the area under the curve between $t = 0$ and t.

Here, an integral from $x = \bar{x}$ or $t = (x - \bar{x})/\sigma = 0$ to $x = \infty$ or $t = (x - \bar{x})/\sigma = \infty$ is 0.5000.

Because the normalized Gaussian PDF is a symmetric function about $t = 0$, the integral from $x = -\infty$ or $t = (x - \bar{x})/\sigma = -\infty$ to $x = \bar{x}$ or $t = (x - \bar{x})/\sigma = 0$ is also 0.5000. By adding the two integrals (areas under the curve) $P_{-\infty,\infty} = 0.5000 + 0.5000 = 1.0000$ as expected (i.e., there is a probability of 1 that a given measurement will be found between $-\infty$ and $+\infty$).

EXERCISE

A.1 What is the probability that the next precise measurement will lie between

a. $\bar{x}$ and $(\bar{x} + \sigma)$?
b. $(\bar{x} - \sigma)$ and $(\bar{x} + \sigma)$?
c. $(\bar{x} + \sigma)$ and $(\bar{x} + 2\sigma)$?

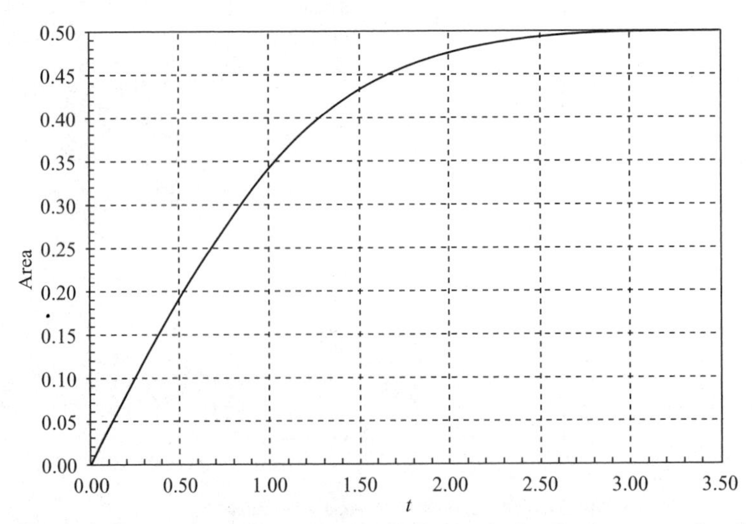

Figure A.3 Area under the Gaussian probability density function between $t = 0$ and some other value of t on the horizontal axis.

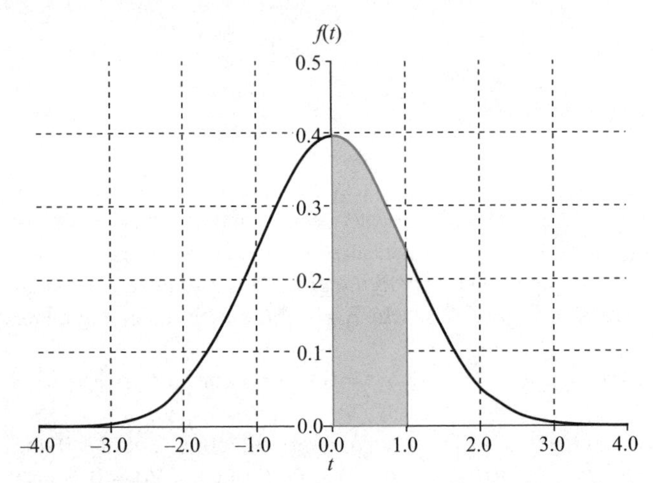

Figure A.4 The area under the Gaussian probability density function between $x_1 = \bar{x}$ and $x_2 = \bar{x} + \sigma$ or $t_1 = 0$ and $t_2 = 1$ is $P_{0,1} = 0.3413$.

ANSWERS $P_{x_1,x_2} = \int_{x_1}^{x_2} f(x)dx$ = area under the $f(x)$ curve between x_1 and x_2. Defining then $P_{t_1,t_2} = \int_{t_1}^{t_2} f(t)dt$ = area under the $f(t)$ curve between t_1. and t_2. Values of $f(t) = \left(1/\sqrt{2\pi}\right)e^{-\frac{t^2}{2}}$ are found by using an integrating calculator (or the values plotted in Figure A.3).

a. For $x_1 = \bar{x}$ and $x_2 = \bar{x} + \sigma$ or $t_1 = 0$ and $t_2 = 1$, the probability is given by the area under the PDF curve, as shown in Figure A.4.

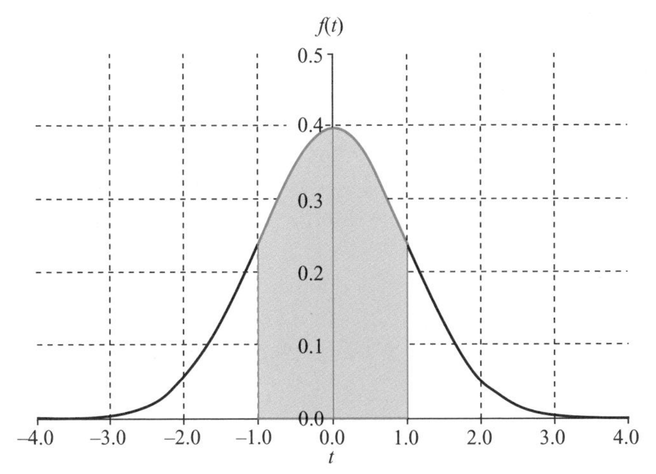

Figure A.5 The area under the Gaussian probability density function between $x_1 = \bar{x} - \sigma$ and $x_2 = \bar{x} + \sigma$ or $t_1 = -1$ and $t_2 = 1$ is $P_{-1,1} = 2P_{0,1} = 2(0.3413) = 0.6826$.

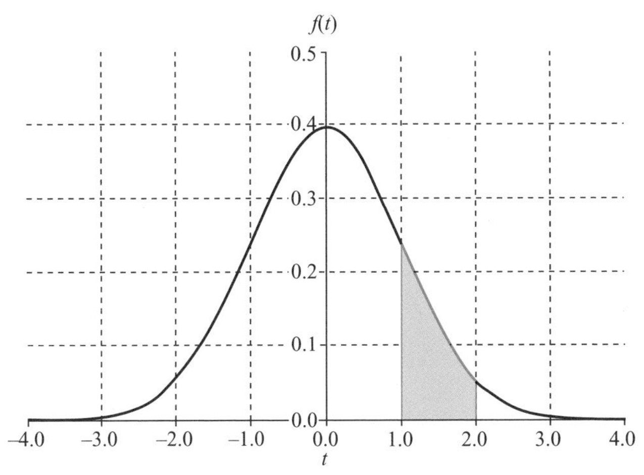

Figure A.6 The area under the Gaussian probability density function between $x_1 = \bar{x} + \sigma$ and $x_2 = \bar{x} + 2\sigma$ or $t_1 = 1$ and $t_2 = 2$ is $P_{1,2} = P_{0,2} - P_{0,1} = 0.4773 - 0.3413 = 0.1360$.

b. For $x_1 = \bar{x} - \sigma$ and $x_2 = \bar{x} + \sigma$ or $t_1 = -1$ and $t_2 = 1$, the probability is given by the area under the PDF curve, as shown in Figure A.5.

c. For $x_1 = \bar{x} + \sigma$ and $x_2 = \bar{x} + 2\sigma$ or $t_1 = 1$ and $t_2 = 2$, the probability is given by the area under the PDF curve, as shown in Figure A.6.

RELATION BETWEEN THE BINOMIAL AND GAUSSIAN PDF

The normalized Gaussian PDF applies strictly *only* to an infinite set of precise measurements of the continuous quantity x. In practice, an infinite set of measurements is never obtained, and measuring devices are never infinitely precise but are discrete. Nevertheless, it is convenient to have a standard shorthand scheme that allows one to report measurements of x to others without giving them all of the N measurements x_r (where $r = 1$ to N) and without explaining the limits of our ability to make each measurement. For this to be carried out, a property of the binomial PDF is used that it becomes closely approximated by the Gaussian PDF in the limit as n becomes large if neither p nor q is too close to zero. Textbooks on probability and statistics prove that this characteristic is valid.[1]

In a sense, this procedure is accomplished by comparing the discrete variable plot of the binomial PDF $f(x_1)$ of Figure A.1 with the plot of the continuous variable $f(x)$ of the Gaussian PDF of Figure A.2. This comparison is shown in Figure A.7,

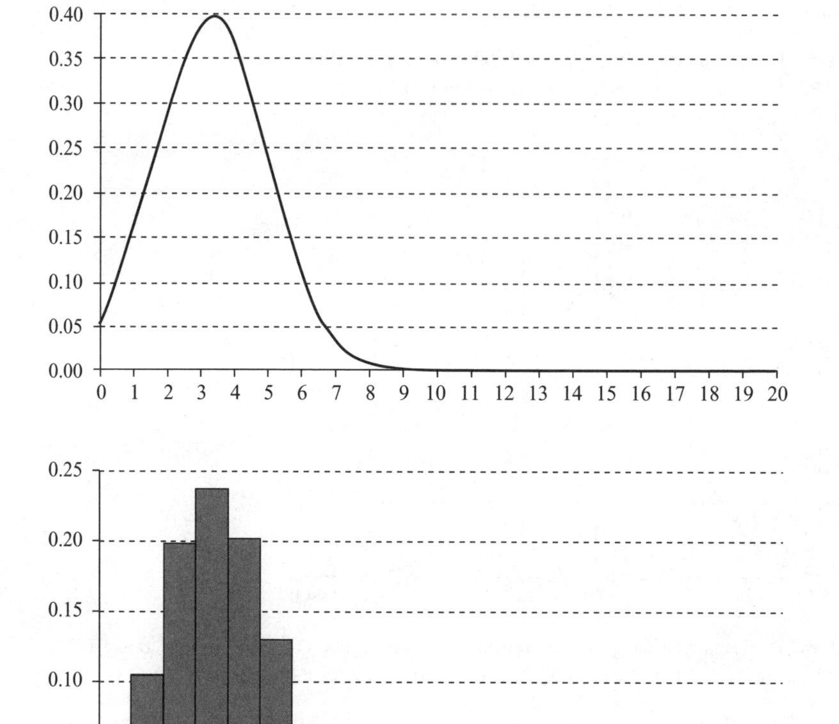

Figure A.7 Plot of the Gaussian probability density function with $\bar{x} = 3.333$ and $\sigma = 1.6667$ above a plot of the binomial probability density function of Figure A.1 (with bars of width $\Delta x = 1$).

[1] For example, Murray R. Spiegel and Larry J. Stephens, *Schaum's Outlines of Theory and Problems of Statistics*, 4th ed. (McGraw-Hill, 2008), p. 172.

where the discrete plot values of chart of Figure A.1 have been changed to bars of width $\Delta x = 1$. In this plot, the value of $\bar{x}$ for the Gaussian PDF has been chosen as $\bar{x} = np = (20)(1/6) = 3.3333$, and the standard deviation, σ, has been chosen as $\sigma = \sqrt{npq} = \sqrt{(20)(1/6)(5/6)} = 1.6667$. The values of $\bar{x}$ and σ are thus the same for both plots, but one (the Gaussian) PDF is for the continuous variable, and the other (the binomial) PDF is for the discrete variable.

AVERAGE If a statistical process is followed for a finite number of discrete measurements, x_i, that can be compared with the infinite continuous set of Gaussian values and can be reported as the average of measurements, $\bar{x}$, according to the formula,

$$\bar{x} = \frac{1}{N}\sum_{i=1}^{N} x_i = \frac{1}{N}\sum_{r=1}^{r=n\max} n_r x_r. \tag{A.7}$$

Here, the n *different* values of x_r (each of which was reported n_r times) have been summed, and the result has been divided by N (the total number of measurements).

Standard Deviation

For the standard deviation, σ, to be reported, the variance, μ, given by

$$\mu = \frac{1}{N}\sum_{i-1}^{N} (x_i - \bar{x})^2 = \frac{1}{N}\sum_{r=1}^{r=n\max} n_r (x_r - \bar{x})^2 \tag{A.8}$$

is defined. The quantity μ is the average square of the deviation of the rth measurement from the average, $(x_r - \bar{x})$. The square is desirable because the deviation is a positive number no matter whether x_r is smaller or larger than $\bar{x}$. The positive square root of the quantity μ thus gives an indication of the average absolute deviation of our measurements from the average, $\sigma = \sqrt{\mu}$.

NOTE Some textbooks define the value $s = \sqrt{\mu}$ to indicate that it was calculated by using a finite number of measurements and σ for a continuous distribution.

NOTE The values of $\bar{x}$ and σ_N often have units associated with them (e.g., cm or s). Unless the answer includes a number and a unit for both quantities, it is incorrect.

Measurement Precision

Measuring devices often have some inherent precision limits. For example, a meter stick may be incremented only to the nearest millimeter. While it may be possible to interpolate a given length measurement to the nearest tenth of a millimeter, it is questionable that giving a report to the nearest hundredth of a millimeter makes any sense. It might be better to choose to make all reports only to the nearest millimeter and say that, if the measure looks like it is closest to one value, that value will be reported (e.g., if the measure looks like it lies between 32 and 33 mm but closer to 32 mm than 33 mm, the measure will be reported as 32 mm). This is a personal convention but one that is often used. Measurements could then be reported only to the nearest *mm*, but consecutive measurements might yield different values for the reported length (e.g., the first measurement might be reported as 32 mm, and the next one might be reported as 33 mm).

Had the same precision convention been used for a continuous variable, any value that falls between 31.5 and 32.5 mm would be reported as 32 mm. This allows us to interpret measurements in terms of a normalized Gaussian PDF (that applies only to an infinite number of exactly precise measurements of a continuous variable). Thus, the probability that the next measurement in a series will be reported as 32 mm is approximately the area under a normalized Gaussian PDF curve between 31.5 and 32.5 mm. Of course, in order to compute this probability, it must be known which values of $\bar{x}$ and σ are to be used for the normalized Gaussian PDF. It will be agreed that those values will be determined by the measurement convention above for a set of discrete measurements.

Below are sample problems. If they can be answered correctly, the reader probably understands measurement error concepts well enough to report them in a publication.

EXERCISE

A.2 Assuming that the precision of a set of measurements can be expressed only to the nearest whole *cm* and, that using Equations A.7 and A.8, previous measurements have yielded the value $x = 32\ \text{cm} \pm 0.707\ \text{cm}$, what is the probability that the next measurement will be

a. 32 cm?

b. 33 cm?

c. 33 cm or greater?

ANSWERS $\bar{x} = 32$ cm and $\sigma = 0.707$ cm

Because the measurements are expressed only to the nearest whole *cm* and the curve $f(t) = \left(1/\sqrt{2\pi}\right)e^{-\frac{t^2}{2}}$ is a function of a continuous variable, t,

a. We can interpret any value of the continuous variable x between 31.5 and 32.5 cm as being reported as 32 cm with a device that measures to a precision of 1 cm. Thus, the probability that the next measurement will be between 31.5 cm, or

$$t_1 = \frac{(x_1 - \bar{x})}{\sigma} = \frac{(31.5\,\text{cm} - 32.0\,\text{cm})}{0.707\,\text{cm}} = -0.707,$$

and 32.5 cm, or

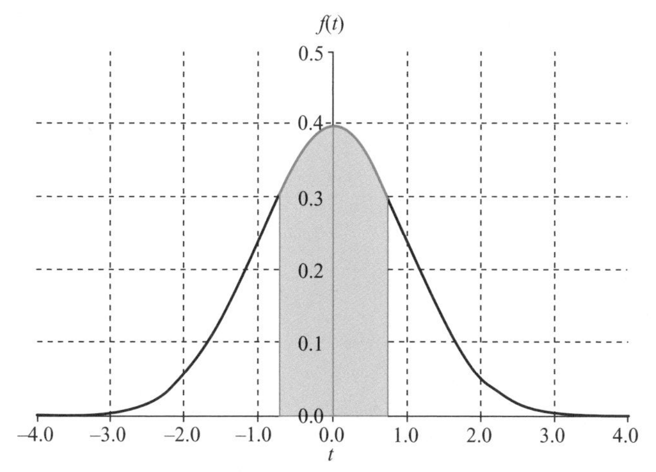

Figure A.8 Area under the Gaussian probability density function between $t_1 = -0.707$ and $t_2 = +0.707$.

$$t_2 = \frac{(x_2 - \bar{x})}{\sigma} = \frac{(32.5\,\text{cm} - 32.0\,\text{cm})}{0.707\,\text{cm}} = 0.707,$$

as shown in Figure A.8, is $P_{-0.707,+0.707} = 2P_{0,0.707} = 2(0.2602) = 0.5204$.

b. Any value of x between 32.5 and 33.5 cm can be interpreted as being measured as 33 cm. The probability that the next measurement will be between 32.5 cm, or

$$t_1 = \frac{(x_1 - \bar{x})}{\sigma} = \frac{(32.5\,\text{cm} - 32.0\,\text{cm})}{0.707\,\text{cm}} = 0.707,$$

and 33.5 cm, or

$$t_2 = \frac{(x_2 - \bar{x})}{\sigma} = \frac{(33.5\,\text{cm} - 32.0\,\text{cm})}{0.707\,\text{cm}} = 2.122,$$

is

$$P_{0.707,2.122} = P_{0,2.122} - P_{0,0.707} = 0.4830 - 0.2602 = 0.2228,$$

as shown in Figure A.9.

c. Any measurement between 32.5 cm, or $t_1 = \frac{(x_1 - \bar{x})}{\sigma} = \frac{(32.5\,\text{cm} - 32.0\,\text{cm})}{0.707\,\text{cm}}$ $= 0.707$, and ∞, or $t_2 = \infty$, will be reported as 33 cm or greater. Thus, the probability that the next measurement will be between $t_1 = 0.707$ and ∞ is

$$P_{0.707,\infty} = P_{0,\infty} - P_{0,0.707} = 0.5000 - 0.2602 = 0.2398,$$

as shown in Figure A.10.

A.3 Six decay rate measurements yield 200, 206, 204, 204, 206, and 204 counts per min. Plot a bar graph that shows the results, compare this to a theoretical

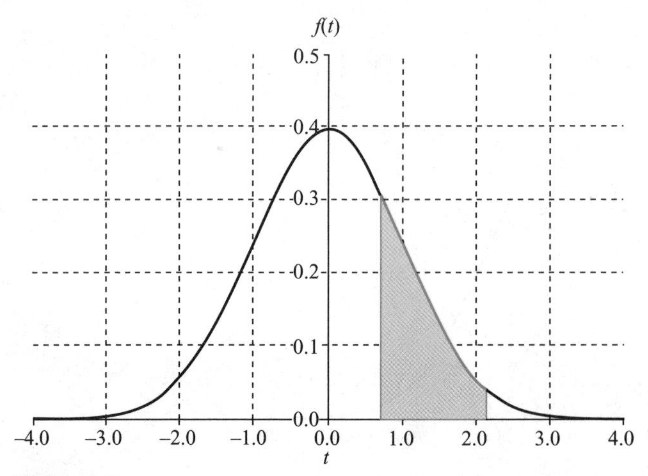

Figure A.9 Area under the Gaussian probability density function between $t_1 = +0.707$ and $t_2 = +2.122$.

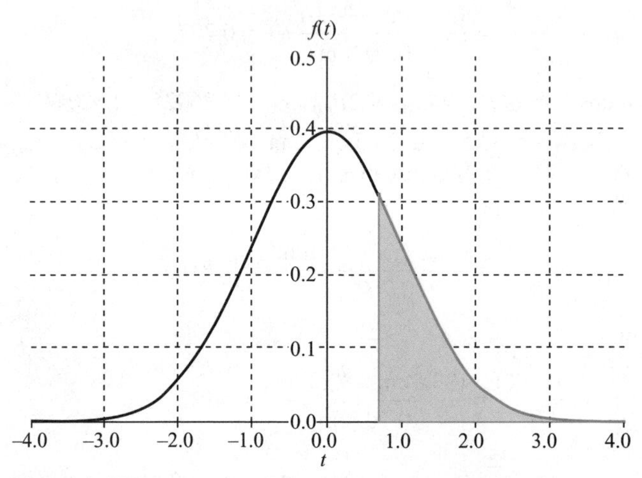

Figure A.10 Area under the Gaussian probability density function between $t_1 = +0.707$ and $t_2 = \infty$.

graph of a Gaussian distribution with the same value of $\overline{x}$ and σ, and indicate on the graph an area that shows the probability that the next measurement will be 204 counts per min.

ANSWER There are three different values of count rate, r, measured:

200 c/min	204 c/min	206 c/min
r_1	r_2	r_3
$n_1 = 1$	$n_2 = 3$	$n_3 = 2$

Thus,

$$N = \sum_{i=1}^{i=3} n_i = 6,$$

and

$$\bar{r} = \frac{1}{N}\sum_{i=1}^{i=3} n_i r_i = \frac{1}{6}[(1)(200\,\mathrm{c/min}) + (3)(204\,\mathrm{c/min}) + (2)(206\,\mathrm{c/min})] = 204\,\mathrm{c/min}$$

$$\mu = \frac{1}{N}\sum_{i=1}^{i=3} f_i(r_i - \bar{r})^2 = \frac{1}{6}\left[(1)(-4\,\mathrm{c/min})^2 + (2)(2\,\mathrm{c/min})^2\right] = 4\,\mathrm{c/min},$$

so

$$\sigma = \sqrt{\mu} = 2\,\mathrm{c/min}.$$

A bar graph for these measurements is shown in Figure A.11.
The probability that the next report will be 204 c/min = $P_{204\ \mathrm{c/min}}$ is

$$P_{204\mathrm{c/min}} = \int_{203.5\mathrm{c/min}}^{204.5\mathrm{c/min}} f(x)\,dx = \int_{-0.5\mathrm{c/min}/2\mathrm{c/min}}^{0.5\mathrm{c/min}/2\mathrm{c/min}} f(t)\,dt = \int_{-0.25}^{0.25} f(t)\,dt$$

or

$$P_{204\mathrm{c/min}} = 2\int_{0}^{0.25} f(t)\,dt = 2(0.0987) = 0.1974,$$

as shown in Figure A.12.

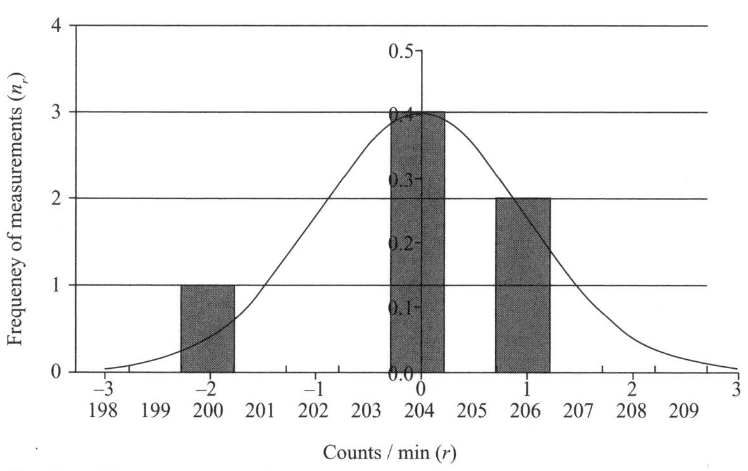

Figure A.11 Frequency of discrete decay rate measurements reported. Superimposed on the discrete measurements bar graph is a Gaussian probability density function, with the same mean and standard deviation as that produced by the discrete measurements.

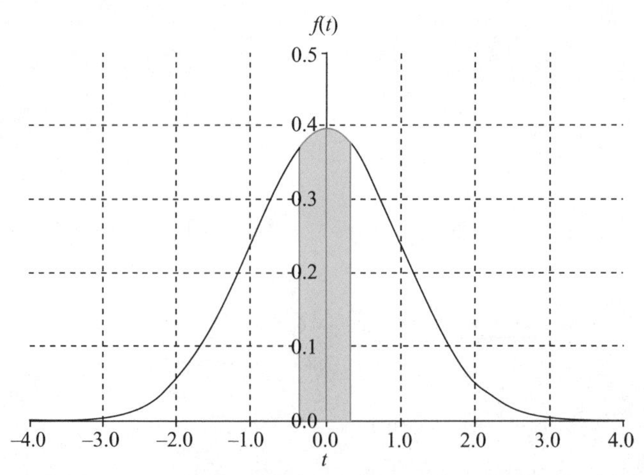

Figure A.12 Area under the Gaussian probability density function between $t_1 = -0.25$ and $t_2 = +0.25$.

The measurement bar graph can be compared to a normalized Gaussian PDF which has an average of 204 c/min ($t = 0$), and a standard deviation of 2 c/min ($\sigma = 1$) on a scaled horizontal axis, as shown to the right but, it is recognized that a Gaussian PDF applies only to a continuous set of precise measurements.

Sometimes, the vertical scale of the Gaussian distribution is also scaled to make the comparison on the superimposed graphs, as shown in Figure A.11. Here, the vertical scaling is chosen to make the height of the Gaussian the same as the height of the bar at $t = 0$. With a larger number of measurements, the areas could be chosen equal.

SUMS AND PRODUCTS OF MEASURED QUANTITIES

Sums

If we add one measured number $x_1 \pm \sigma_1$ to another measured number $x_2 \pm \sigma_2$, how would the sum be expressed? For an infinite number of measurements of the quantities, there would be two Gaussian distributions to be added, so their sum would be another Gaussian distribution centered at $x_1 + x_2$. However, the sum of two uncertain numbers has an even broader distribution because of the uncertainty of each. A mathematical analysis of the sum of two Gaussian distributions shows the answer to be $\sigma_{Sum} = \sqrt{\sigma_1^2 + \sigma_2^2}$.

Products

If one measured number ($x_1 \pm \sigma_1$) is multiplied by another measured number ($x_2 \pm \sigma_2$), how would the product be expressed? For a continuous set of measurements of the quantities, there would be two Gaussian PDFs. The product of two uncertain numbers would be expected to have an even broader distribution because of the uncertainty of each. A mathematical analysis of the sum of two Gaussian distributions shows the answer to be

$$(x_1 \pm \sigma_1)(x_2 \pm \sigma_2) = x_1 x_2 \left[1 + \sqrt{\sigma_1^2/x_1^2 + \sigma_2^2/x_2^2}\right].$$

Appendix B

Graphics and Conformal Mapping

GRAPHIC INTERPRETATIONS

Functions of a complex variable z are usually written, $f(z)$, or w. For example,

$$\begin{aligned} f(z) &= z^2 = (x+jy)^2 = (x^2-y^2)+j2xy \\ f(z) &= z^2 = (re^{j\theta})^2 = r^2e^{j2\theta} = r^2[\cos 2\theta + j\sin 2\theta] \\ f(z) &= z^2 = (r^2\cos 2\theta)+j(r^2\sin 2\theta) \\ f(z) &= z^2 = (\text{real part})+j(\text{imaginary part}), \end{aligned} \tag{B.1}$$

which is conventionally renamed as

$$w = u(x, y) + jv(x, y) \tag{B.2}$$

where

$$u(x, y) = (x^2 - y^2) = r^2\cos 2\theta$$

and

$$v(x, y) = 2xy = r^2 \sin 2\theta$$

EXAMPLES

B.1 Equations with z can also be represented by points in the w-plane, for example,

$$f(z) = z^2 = 1 \tag{B.3}$$

means

$$(x^2 - y^2) = 1 \quad \text{and} \quad 2xy = 0$$

or

$$u = 1 \quad \text{and} \quad v = 0$$

Maxwell's Equations, by Paul G. Huray

This equation can be graphically represented by showing that the point *1* in the z-plane with an arrow and the function $w = z^2$ pointed toward a w-plane, where the corresponding point $u = 1$, $v = 0$ is located. This figure is in every regard equivalent to Equations B.3. It could be said that the point $z = 1$ (point P') transforms to the point $w = 1$ (point P) under the transformation $w = z^2$. Equation B.3 and Figure B.1 are equivalent; given one, we can draw or write the other.

B.2 Equations of z can also be solved for z; for example,

$$f(z) = z^2 = 2.25 \tag{B.4}$$

can be solved as

$$z = 2.25^{1/2}$$

or

$$z = \left(2.25e^{j0}\right)^{1/2} \quad \text{and} \quad z = \left(2.25e^{j2\pi}\right)^{1/2}$$

or

$$z = 1.5e^{j0} \quad \text{and} \quad z = 1.5e^{j\pi}$$

or

$$z = 1.5 \quad \text{and} \quad z = -1.5$$

This inverse relationship can be equivalently represented graphically by showing the point 2.25 (point P) in the w-plane with an arrow and the function $z = w^{1/2}$ pointed toward a z-plane, where the corresponding points $x = 1.5$, $y = 0$ (point P') and $x = -1.5$, $y = 0$ (point P'') are located. Figure B.2 is in every regard equivalent to Equations B.4. It can be said that the point $w = 2.25$ transforms to the point $z = 1.5$ under the transformation $z = w^{1/2}$. Because Equation B.4 and Figure B.2 are equivalent; given one, the other can be drawn or written.

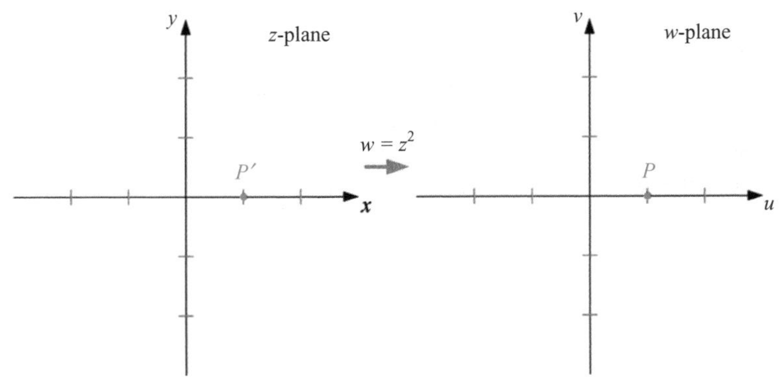

Figure B.1 Maping of the point P' from the z-plane onto point P in the w-plane under the transformation $W = z^2$.

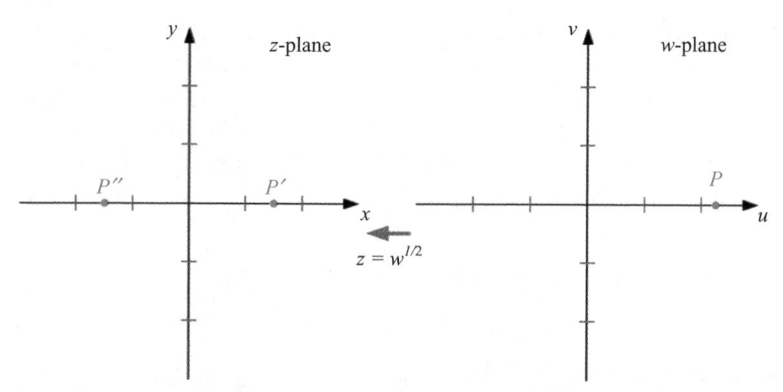

Figure B.2 Maping of the point P from the w-plane onto point P' and P'' in the z-plane under the transformation $z = w^{1/2}$.

Figure B.2 also points out a problem that there is not a one-to-one correspondence between points in the w-plane and points in the z-plane. By labeling the point (2.25, 0) in the w-plane as the point P, we could say that P maps from the w-plane under the transformation $z = w^{1/2}$ into the *two* points P' and P'' in the z-plane. The inverse way to say this is that both of the points P' and P'' in the z-plane map to the single point P in the w-plane under the transformation $w = z^2$.

B.3 Find the values of z that satisfy the equation,

$$z^2 = -2.25 \tag{B.5}$$

Solution:

$$z = (-2.25)^{1/2} = \left(2.25e^{j\pi}\right)^{1/2} \quad \text{and} \quad \left(2.25e^{j3\pi}\right)^{1/2}$$

or

$$z = 1.5e^{j\frac{\pi}{2}} \quad \text{and} \quad 1.5e^{j\frac{3\pi}{2}}$$

The two points Q' and Q'' in the z-plane can be represented by the transformation $w = z^2$ by using an arrow pointing from the z-plane to the w-plane, where the corresponding point Q is located. Of course, this equation inversely implies that the point Q in the w-plane maps onto the two points Q' and Q'' in the z-plane under the transformation $z = w^{1/2}$. All of these statements need not be made if Figure B.3 is shown; the figure says it all!

B.4 Suppose

$$z^2 = 2.25j. \tag{B.6}$$

then,

$$z = (2.25j)^{1/2} = \left(2.25e^{j\frac{\pi}{2}}\right)^{1/2} \quad \text{and} \quad \left(2.25e^{j\frac{5\pi}{2}}\right)^{1/2}$$

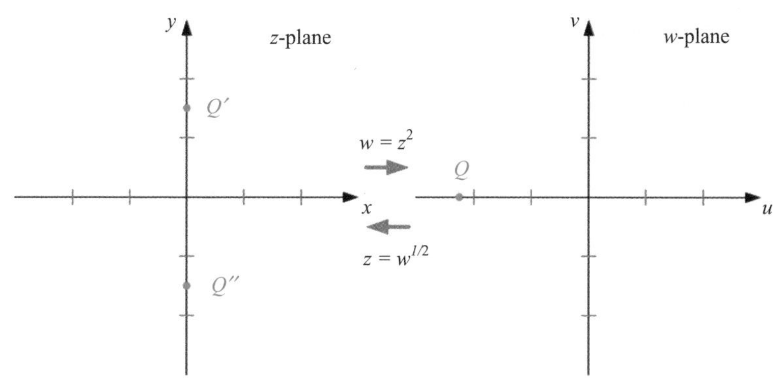

Figure B.3 Maping of the point Q from the w-plane onto point Q' and Q'' in the z-plane under the transformation $z = w^{1/2}$.

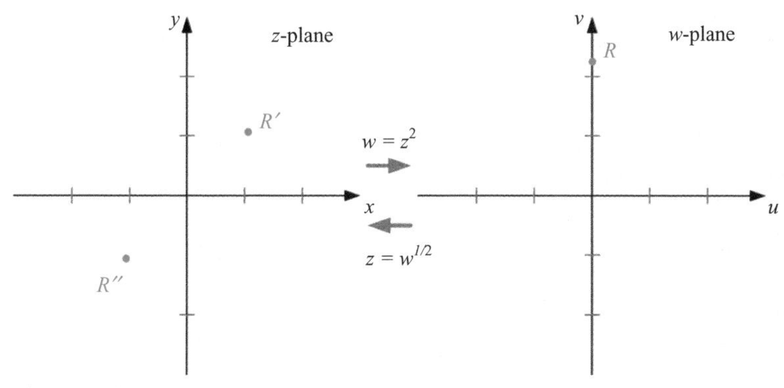

Figure B.4 Mapping of the point R in the w-plane onto points R' and R'' in the z-plane under the transformation $z = w^{1/2}$.

or

$$z = 1.5e^{j\frac{\pi}{4}} \quad \text{and} \quad 1.5e^{j\frac{5\pi}{4}},$$

as shown in Figure B.4.

B.5 Let us put all of the points P, R, and Q on the same graph, as shown in Figure B.5. Can you locate the image z-plane points that correspond to the point S in the w-plane?

ANSWER The point S in the w-plane is at $2.25e^{j\frac{\pi}{4}}$, S' is at $1.5e^{j\frac{\pi}{8}}$, and S'' is at $1.5e^{j\frac{9\pi}{8}}$ in the z-plane.

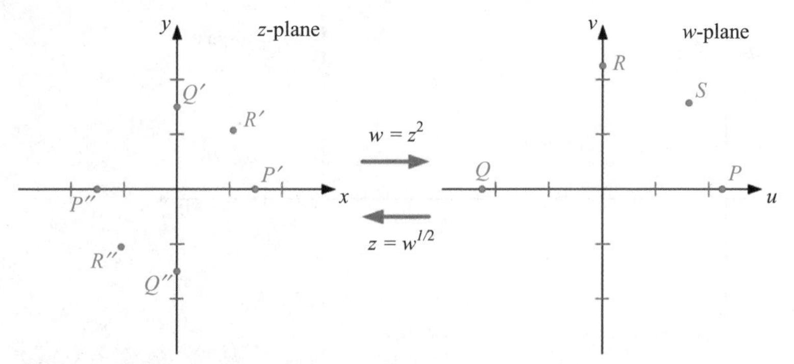

Figure B.5 Mapping of points P, R, and Q in the w-plane onto points P', R', and Q' and the points P'', R'', and Q'' in the z-plane under the transformation $z = w^{1/2}$.

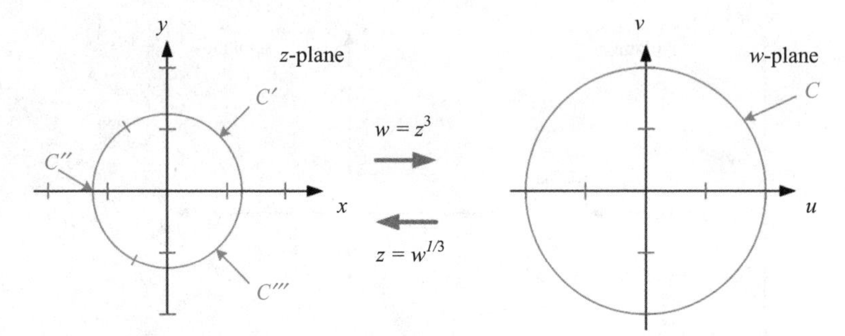

Figure B.6 Mapping of all points on the curve C in the w-plane onto points on the curve C', C'', and C''' in the z-plane under the transformation $z = w^{1/3}$.

B.6 Show how the points on the path C shown in Figure B.6 in the w-plane map onto corresponding points in the z-plane under the transformation $f(z) = z^3$.

ANSWER The image points are shown in Figure B.6 as three curve segments labeled C', C'', and C'''. It is understood that there are an infinite number of points on the path C, and there are also an infinite number of points on the paths C', C'', and C'''.

B.7 For the confusion of having a 3 : 1 correspondence between points in the z-plane and image points in the w-plane to be avoided, there is a common mathematical construct called a "branch cut" that restricts the values of θ in the z-plane to lie in the range $0 \leq \theta \leq 2\pi/3$ (sometimes called the principal

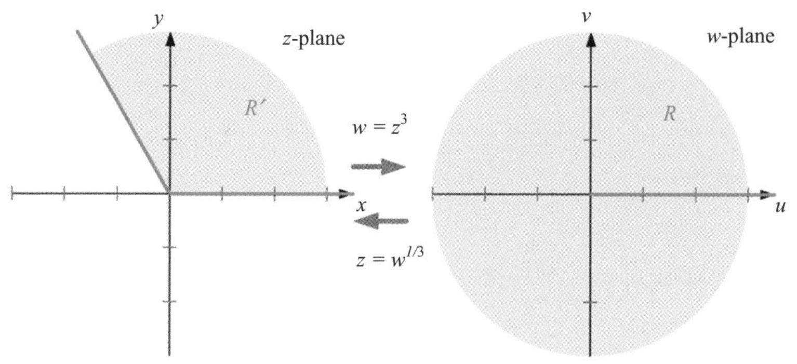

Figure B.7 Branch cut in the z-plane and the image branch cut in the w-plane that avoids the mapping of points onto multiple image points in the other plane.

branch). This is shown graphically by making a boundary line (shown in Figure B.7 as a heavy red line).

With such a branch cut, we will then agree never to cross the heavy red line with points on a contour (e.g., the segment C'). There is a corresponding boundary in the w-plane given by an "image branch cut" along the u-axis. With this "no-crossing" rule, it can now be shown that there is a one-to-one correspondence between points in the region R' of the z-plane and points R in the entire w-plane.

CONFORMAL MAPPING

It is clear that a graphic analysis of complex functions can help our understanding in solving problems quickly. Let us try to categorize some common functions along with their graphic interpretations.

Addition of a Constant

Figures B.8 and B.9 show that addition of a constant (even a complex constant like $a + jb$) corresponds to a *graphic translation* of the point P in the z-plane to the corresponding image point, P', in the w-plane. Likewise, the point Q or R or S in the z-plane corresponds to a graphic translation to the points Q', R', and S' in the w-plane. It is also clear that any point, z, on the line between P and Q, or Q and R, or R and S, or S and P in the z-plane shown in Figure B.10 translates by the same amount in moving to the w-plane. Furthermore, we can argue rigorously that any point, z, within the boundary formed by P, Q, R, S, and P in the z-plane translates into the region within the boundary formed by P', Q', R', S', and P' in the w-plane.

Note that the function, $f(z) = z + a + jb$, in Figure B.10 is not needed beside the arrow of transformation. Once the points P, Q, R, and S are shown in the z-plane and their corresponding images in the w-plane, it is clear that translation has occurred

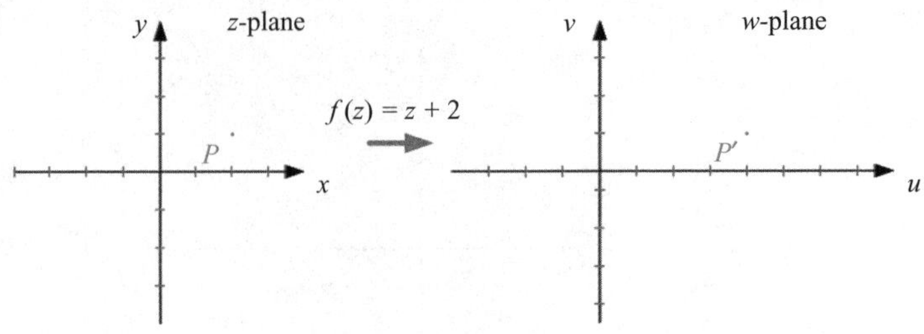

Figure B.8 Mapping of the point P in the z-plane onto the point P' in the w-plane under the addition of a real constant.

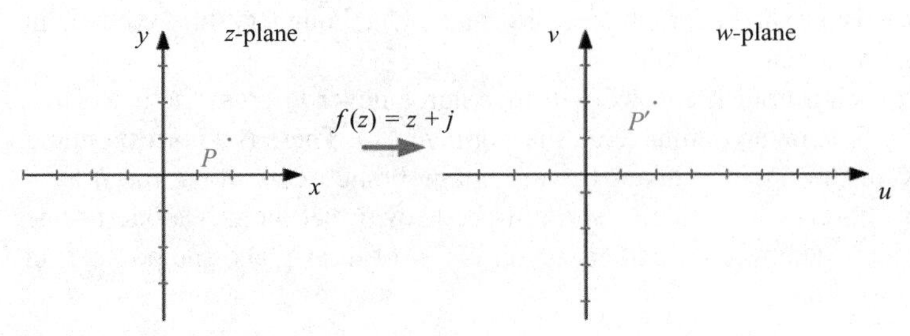

Figure B.9 Mapping of the point P in the z-plane onto the point P' in the w-plane under the addition of an imaginary constant.

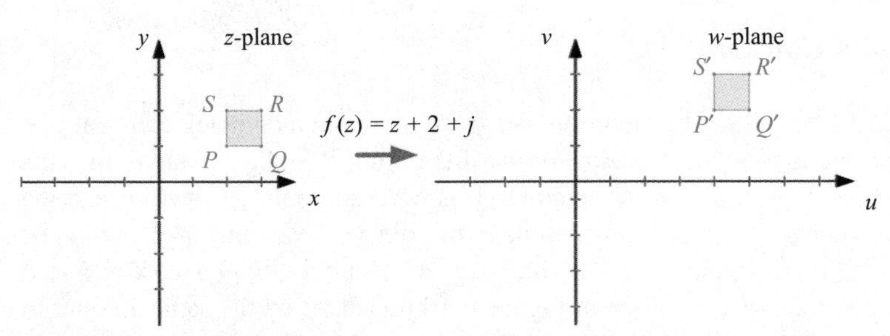

Figure B.10 Mapping of all points interior to the box defined by P, Q, R, and S in the z-plane onto points within the box P', Q', R', and S' in the w-plane.

and, hence, addition is implied. Once that is understood, the inverse transformation $z = w - a - jb$ is also implied. Formally, we do not need to see more than the points on the corners of a boundary in the two planes to understand that addition of a constant is the function that has produced the transformation; nevertheless the function or its inverse is conventionally stated somewhere on the page. We say that we have mapped points in the region bounded by P, Q, R, and S onto the translated region bounded by P', Q', R', and S' by the addition of a constant (i.e., addition by a constant = a graphic translation).

Multiplication by a Constant

Figure B.11 shows that multiplication by a constant, for example, $f(z) = r_0 e^{j\varphi} z$, corresponds to a rotation and a magnification of the region bounded by P, Q, R, S, and P in the z-plane into its corresponding region in the w-plane; that is, multiplication by a constant → a graphic rotation and magnification.

Complex Conjugation

Figure B.12 shows that the function, $f(z) = z^*$, produces a reflection of points bounded by P, Q, R, and S in the z-plane onto its corresponding region in the w-plane; that is, complex conjugation = a graphic reflection through the x-axis.

Combinations of Operations

We can combine a series of operations (e.g., complex conjugation, multiplication by a complex constant, addition of a complex constant), as graphically shown in Figure B.13, to produce an algebraic result like $f(z) = 2z^* e^{j\pi/2} - 3.5 - 2.75j$. Note that we would need the whole series of figures to unambiguously infer that the function that produced the conformal mapping was a combination of the specific functions used.

Through a series of mathematical transformations corresponding to reflection, rotation, magnification, and translation, as shown in Figure B.13, we can map the area bounded by P, Q, R, and S onto points within the area bounded by P''', Q''', R''', and S'''. We may also use graphic transformations found by others (e.g., *Schaum's Outlines on Complex Variables*[i]) in a series of transformations where it might be useful for solving two-dimensional (2-D) problems in which Laplace's equation is known to be satisfied. As long as all of the transformations in the series are *analytic*, then we can be sure that Laplace's equation is satisfied in any of the transformed regions.

More Complicated Transformations

Mathematicians have studied many more complicated transformation equations than those mentioned above. Some of these are cataloged in tables like *Schaum's Outlines*,

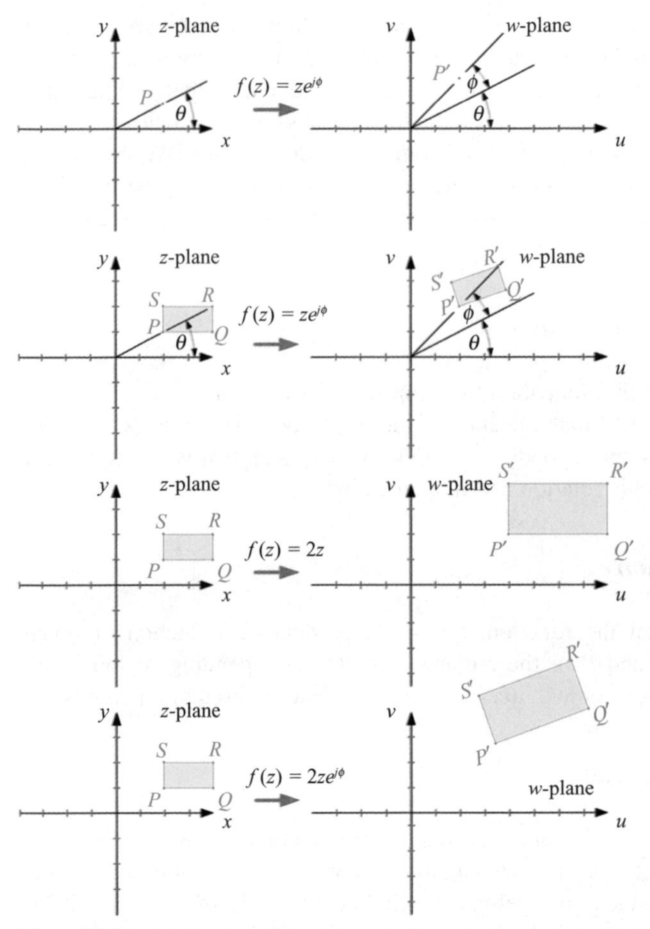

Figure B.11 (a) Mapping of the point P in the z-plane onto the point P' in the w-plane under multiplication by a phase factor $e^{j\phi}$; (b) mapping of points within the rectangle formed by P, Q, R, and S in the z-plane onto points within $P'Q'R'S'$ in the w-plane under multiplication by a phase factor $e^{j\phi}$; (c) mapping of points within the rectangle formed by P, Q, R, and S in the z-plane onto points within $P'Q'R'S'$ in the w-plane under multiplication by the real number 2; (d) mapping of points within the rectangle formed by $PQRS$ in the z-plane onto points within $P'Q'R'S'$ in the w-plane under multiplication by a magnitude and a phase factor, $2e^{j\phi}$.

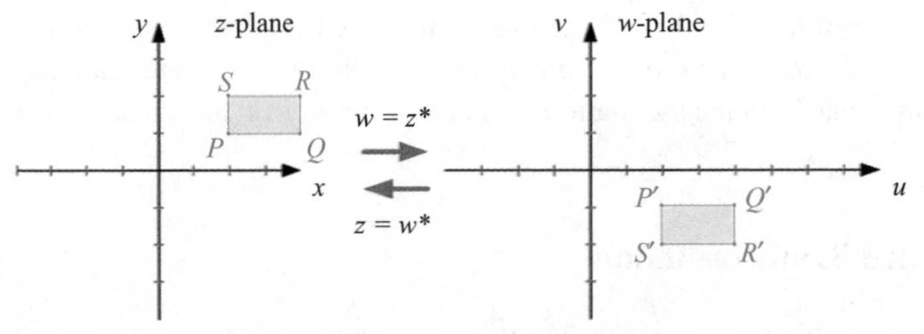

Figure B.12 Mapping of points within the rectangle formed by P, Q, R, and S in the z-plane onto points within $P'Q'R'S'$ in the w-plane under complex conjugation.

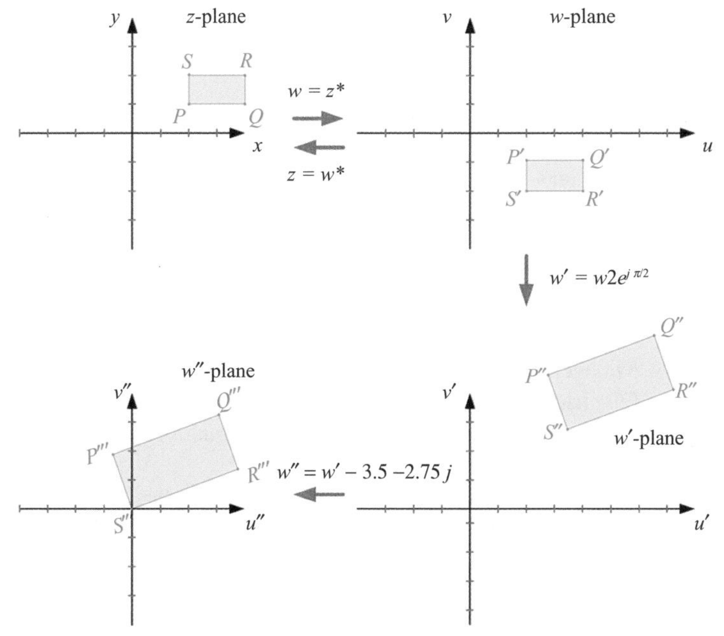

Figure B.13 Mapping of points within the rectangle formed by P, Q, R, and S in the z-plane onto points within $P'Q'R'S'$ in the w-plane under a combination of operations; complex conjugation, multiplication by a magnitude, and a phase factor $e^{j\phi}$, and then addition of a complex constant.

which shows a variety of functions that might appear in an electrical engineering problem (e.g., z^m, $e^{\pi z/a}$, $\sin \pi z/a$, $\cosh \pi z/a$, $a/2[z + 1/z]$, $[1 + z]^2/[1 - z]^2$, $\ln z$). It is possible for the user of such mappings to trust the mathematician that she has carried out the mapping correctly and to use the results in a 2-D problem solution, as shown below.

The mathematicians Cauchy and Riemann **defined** single-valued functions $f(z)$, where z is a complex number, to be *analytic in R* if their derivative $f'(z) \equiv \lim_{\Delta z \to 0} \dfrac{f(z+\Delta z) - f(z)}{\Delta z}$ exists in a region R and is independent of the manner in which $\Delta z \to 0$. They showed that a necessary and sufficient condition for analyticity is that the conditions $\partial u/\partial x = \partial v/\partial y$ and $\partial u/\partial y = -\partial v/\partial x$ be met. These are called the *Cauchy–Riemann* conditions. The word *conformal* in conformal mapping means that transformations are analytic in the region bounded by a specified area in the z-plane or in the w-plane and that $f'(z) \neq 0$. Points where $f'(z) = 0$ are called *critical points* of the transformation. Conformal mapping preserves the angle of intersection of two curves C_1 and C_2 at a point (x_0, y_0) in the z-plane when the two curves are mapped into curves C_1' and C_2' and intersect at the point (u_0, v_0) in the w-plane.

USE OF CONFORMAL MAPPING IN SOLVING 2-D ELECTRIC POTENTIAL PROBLEMS

One of the consequences of the Cauchy–Riemann conditions is that we may take their derivative to produce the properties $\partial^2 u/\partial x^2 + \partial^2 u/\partial y^2 = 0$ and $\partial^2 v/\partial x^2 + \partial^2 v/\partial y^2 = 0$. Thus, the real and imaginary parts of an analytic function satisfy Laplace's equation in 2-D, $\nabla^2 u = 0$ and $\nabla^2 v = 0$ at all points where $f(z)$ is analytic. It can be shown that any function [$\vec{E}(x, y)$, $\vec{D}(x, y)$, $\vec{B}(x, y)$, $\vec{H}(x, y)$, $\vec{A}(x, y)$, or $V(x, y)$] that satisfies Laplace's equation in a region R of the z-plane also satisfies Laplace's equation in an image plane produced by an analytic transformation (Spiegel).

A consequence of Maxwell's equations in charge free space is that all of the electromagnetic fields (electric field intensity, magnetic field intensity, magnetic vector potential, and electrostatic potential) satisfy Laplace's equation.

CONCLUSION Electromagnetic fields in a 2-D region of space defined by an x–y plane (the z-plane) satisfy Laplace's equation in any plane to which the region may be transformed by an analytic transformation.

APPLICATION Given a boundary value problem of any electromagnetic quantity in which the boundary conditions are specified on the boundaries of a region R in the x–y plane, conformal mapping of that region may be made to a corresponding region of an image plane (by one or a series of transformations). For example, a transformation sequence that maps the region R in a problem into the corresponding region of a parallel plate capacitor R' with simple boundary conditions (BC) can then be solved by inspection in R'. For the solution in the original region R to be found, the inverse transformation can be used.

Before the advent of computers, conformal mapping was a technique to analytically solve 2-D boundary value problems in electromagnetic theory (and in fluid flow, mechanical strain, or any other problems that had a variable that satisfied Laplace's equation). The technique is still useful to check the results of a computational model in 2-D to be sure that it yields the correct solution before proceeding to more complicated boundary conditions.

EXAMPLE

B.8 Conformal mapping may be used to find the electrostatic potential, $V(x, y)$, in a region of space bounded by the wedge shown in Figure B.14 with boundary conditions $V(x, 0) = 0$ V and $V(r, \pi/m) = 100$ V.

SOLUTION

In Figure B.14, an open region of a wedge of angle π/m in the z-plane bounded by the lines connecting A, B, C, D, and E has been mapped onto the upper

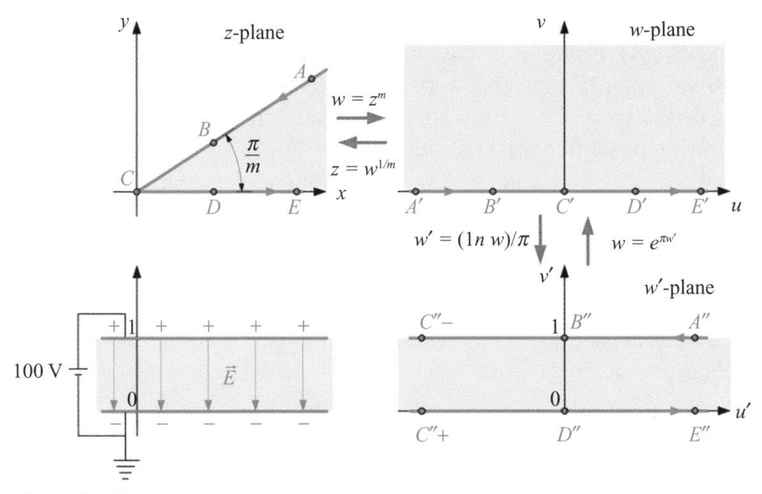

Figure B.14 Mapping of points within the triangle formed by the points $ABCDE$ in the z-plane onto points within the infinite slab $A''B''C''-C''+D''E''$ in the w'-plane under a combination of operations; $w = z^m$ and then $w' = \ln w/\pi$.

half of the w-plane by the transformation[1] $w = z^m$. Note that the image points A', B', C', D', and E' in the w-plane correspond to points on the u-axis. The arrows indicate that points A and E may be very far out on the boundary lines (even infinitely far) and that their corresponding images in the w-plane are also very far along the corresponding u-axis boundary line.

In a successive mapping corresponding to the transformation $w' = (\ln w)/\pi$, the region in the upper half of the w-plane has been mapped onto a strip of the w'-plane bounded by the u'-axis and the horizontal line $v' = 1$. The image points A'', B'', C'', D'', and E'' are also shown on the boundaries of the strip, with a notation that it depends on whether the point C' in the w-plane is an infinitesimal amount less than zero ($C'-$) or an infinitesimal amount more than zero ($C'+$). The image point of zero in the w-plane is recognized to be ambiguous when mapped to the w'-plane, so care should be taken in inferring any solutions to Laplace's equations at this *critical* point.

The application of this problem to an infinitely large parallel plate capacitor (a 2-D problem) is indicated by the configuration to the left of the w'-plane. Here, we have a boundary value problem in which the potentials $V(u', 1) =$ 100 V and $V(u', 0) = 0$ V and $V(u', v')$ satisfy Maxwell's equations. We will later show that, in a charge-free region of space, Maxwell's equations yield an electric potential, $V(u', v')$ that satisfies Laplace's equation $\nabla^2 V(u', v') = 0$ or $\partial^2 V/\partial u'^2 + \partial^2 V/\partial v'^2 = 0$ in 2-D. The second derivative term with respect to

[1] Murry R. Spiegel, *Schaum's Outlines on Complex Variables* (McGraw-Hill, 1999), 205.

the variable perpendicular to the w'-plane has been ignored because we can see that there is no variation in the potential in this direction. Furthermore, there is no variation in potential with respect to the variable u', so we can ignore this derivative as well, and Laplace's equation becomes $\partial^2 V/\partial v'^2 = d^2 V/dv'^2 = 0$. Solving this differential equation, we get $V(u', v') = a\,v' + b$. Applying the BC that $V(u', 0) = 0$ requires $b = 0$, and applying the BC that $V(u', 1) = 100$ V requires $a = 100$ V. Thus, $V(u', v') = 100$ V v' or $V(u', v') = 100$ V $\mathrm{Im}(w')$.

We can use the inverse transformation, $\ln w = \pi w'$, to write $V(u, v) = (100\ \text{volts}/\pi)\mathrm{Im}(\ln w)$ and thus find the potential in the region R' of the w-plane. Finally, we can use the transformation relationship $w = z^m$, to find the potential in the z-plane as

$$V(x, y) = (100\,\text{volts}/\pi)\,\mathrm{Im}\left(\ln z^m\right) = (100\,\text{volts}/\pi)\,\mathrm{Im}\,(m \ln z)$$

or

$$V(x, y) = (100\,\text{volts}/\pi)\,\mathrm{Im}\left[m\left(\ln re^{j\theta}\right)\right]$$

or

$$V(x, y) = (100\,\text{volts}/\pi)\,\mathrm{Im}\,[m \ln r + jm\theta] = (100\,\text{volts}/\pi)\,m\theta.$$

EXERCISE

Find the equipotential lines and electric field lines in all of the three regions, R, R', and R'' of space.

EXAMPLE

B.6 A long cylinder is split in two, and the halves are insulated from one another. One of the halves is grounded, and the other is raised to a potential V_0. Find the potential $V(x, y)$ at all points inside the cylinder.

SOLUTION The function $w = (1 + z)/(1 - z)$ maps the unit circle onto the half-plane, as shown in Figure B.15. Furthermore, in Example 1, it was shown that the function $w' = \ln w$ maps the half-plane onto a parallel plate capacitor.

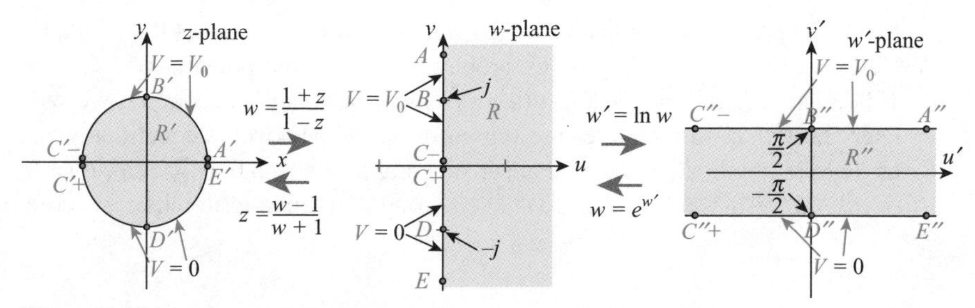

Figure B.15 Conformal mapping of a unit cylinder onto a parallel plate capacitor.

$\nabla^2 V = 0$ in any of the regions of the z-, w-, or the w'-plane because the mappings are conformal (the transformations are all analytic). In the w'-plane, we have a boundary value problem in which $\partial^2 V/\partial u'^2 + \partial^2 V/\partial v'^2 = 0$ subject to $V(u', \pi/2) = V_0$ and $V(u', -\pi/2) = 0$. The solution to this problem in the w' plane is

$$V(u', v') = (V_0/\pi)v' + V_0/2$$

or

$$V(u', v') = (V_0/2)\{1 + (2/\pi)\operatorname{Im}(\ln w)\}$$

or

$$V(u', v') = (V_0/2)\{1 + (2/\pi)\operatorname{Im}[\ln(u + jv)]\}$$

where

$$(u + jv) = re^{j\theta} = \sqrt{u^2 + v^2}\, e^{j\tan^{-1}\frac{v}{u}}.$$

Thus, $\operatorname{Im}\left[\ln\left(\sqrt{u^2+v^2}\right) + j\tan^{-1}(v/u)\right] = \tan^{-1}(v/u)$, so the solution in the w-plane is

$$V(u, v) = (V_0/2)\left\{1 + (2/\pi)\tan^{-1}(v/u)\right\}.$$

To obtain the solution in the z-plane, we recognize $w = (1 + z)/(1 - z)$, so

$$(u + jv) = \frac{1 + x + jy}{1 - x - jy}$$

or

$$(u + jv) = \frac{(1+x)(1-x) - y^2 + j2y}{1 - x^2 + y^2},$$

so

$$V(x, y) = (V_0/2)\left\{1 + (2/\pi)\tan^{-1}\left[2y/(1 - x^2 - y^2)\right]\right\}.$$

PROBLEMS

B.1 Show that equipotentials, $V = \alpha V_0$ $(0 \leq \alpha \leq 1)$, are circles with centers on the $\pm y$-axis and passing through the points $x = \pm 1$.

B.2 Find the potential $V(x, y)$ for all points *outside* the split cylinder.

HINT Consider the function $w = 1/z$.

ENDNOTE

i. Murry R. Spiegel, *Schaum's Outlines on Complex Variables* (McGraw-Hill, 1999), 205–11.

Appendix C

Vectors, Matrices, Orthogonal Functions

When using higher dimensions, it is easier to use a numerical designation for the Cartesian coordinate axes, as shown in Figure C.1.

In Figure C.1, the names of the base vectors $\hat{a}_x$, $\hat{a}_y$, $\hat{a}_z$ have been replaced with $\hat{a}_1$, $\hat{a}_2$, $\hat{a}_3$, and the names of the components of the vector $\vec{A}$ along the x_1, x_2, x_3 coordinate axes have been replaced with A_1, A_2, A_3. This convention is convenient for writing the vector $\vec{A}$ as

$$\vec{A} = A_1\hat{a}_1 + A_2\hat{a}_2 + A_3\hat{a}_3 = \sum_{i=1}^{i=3} A_i\hat{a}_i = A_i\hat{a}_i \tag{C.1}$$

where the *summation convention* has been used such that when an index subscript, i, is repeated, the sum over i is implied. This is shorthand in three dimensions, but it is crucial when we want to express a vector $\vec{A}$ in four dimensions or higher where it is impossible to visualize the four components of $\vec{A}$ but for which the algebra continues to apply.

n-DIMENSIONAL (*n*-D) VECTOR SPACE

In an *n-D* vector space, there are n linearly independent vectors in the space but not $n + 1$. Linear independence means that, for the n vectors in the set $\vec{x}_i$, there is no set of coefficients c_i such that $\sum_{i=1}^{n} c_i\vec{x}_i = 0$ except the set $c_i = 0$. If a set of vectors $\vec{a}_i$ forms a basis[1] (coordinate system) for a vector space, there exists a set of numbers x_i such that $\vec{x} = \sum_{i=1}^{n} x_i\vec{a}_i$ for any arbitrary vector $\boldsymbol{x}$ in the space.

[1] A basis set is any set of vectors $\vec{a}_i$ that "spans the space"; that is, given an arbitrary vector in the space, we may find a set of x_i so that $\vec{x} = \sum_{i=1}^{n} x_i\vec{a}_i$.

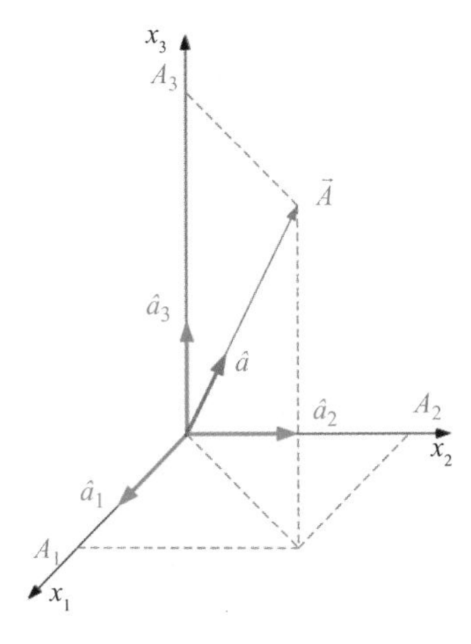

Figure C.1 Cartesian coordinate axes for three dimensions.

GRAM–SCHMIT ORTHOGONALIZATION

If we can define a scalar product (x, y) (called the inner product or dot product), we can create an orthonormal basis set, $\hat{a}_i$, from the set $\vec{a}_i$ by using a process called a Gram–Schmidt orthogonalization. The technique follows:

Given an arbitrary set of basis vectors $\vec{a}_i$,

1. Pick one (e.g., $\vec{a}_1$) and define $\hat{a}_1 = \dfrac{\vec{a}_1}{|\vec{a}_1|}$. (Note that $|\hat{a}_1|^2 = \dfrac{(\vec{a}_1, \vec{a}_1)}{|\vec{a}_1|^2} = 1$.)
2. Pick another (e.g., $\vec{a}_2$) and define $\hat{a}_2 = \dfrac{(\vec{a}_2 - c_1\hat{a}_1)}{|\vec{a}_2 - c_1\hat{a}_1|}$. Here, we will choose c_1 so that $(\hat{a}_1, \hat{a}_2) = 0$. We can see that this will be the case if $c_1 = (\hat{a}_1, \vec{a}_2)$, so we can write $\hat{a}_2 = \dfrac{(\vec{a}_2 - (\hat{a}_1, \vec{a}_2)\hat{a}_1)}{|\vec{a}_2 - (\hat{a}_1, \vec{a}_2)\hat{a}_1|}$.
3. Pick another (e.g., $\vec{a}_3$) and define $\hat{a}_3 = \dfrac{(\vec{a}_3 - c_2\hat{a}_1 - c_3\hat{a}_2)}{|\vec{a}_3 - c_2\hat{a}_1 - c_3\hat{a}_2|}$. Here, we will choose c_2 and c_3 so that $(\hat{a}_1, \hat{a}_3) = 0$ and that $(\hat{a}_2, \hat{a}_3) = 0$. We can see that this will be the case if $c_2 = (\hat{a}_1, \vec{a}_3)$ and $c_3 = (\hat{a}_2, \vec{a}_3)$, so we can write $\hat{a}_3 = \dfrac{(\vec{a}_3 - (\hat{a}_1, \vec{a}_3)\hat{a}_1 - (\hat{a}_2, \vec{a}_3)\hat{a}_2)}{|\vec{a}_3 - (\hat{a}_1, \vec{a}_3)\hat{a}_1 - (\hat{a}_2, \vec{a}_3)\hat{a}_2|}$.
4. The trend from the first three vectors in the orthonormal baisis set illustrates how each new basis vector is formed. However, it also illustrates that the choice of which vector is called $\vec{a}_i$, which is called $\vec{a}_2$, which is called $\vec{a}_3$ and so forth, is completely arbitrary.

Thus, even though the Gram-Schmidt process defines an orthonormal basis set it is not a unique set.

Note that we can teach a computer to do these operations for any n-D space[2] and we can even take the limit as $n \to \infty$ so it will always be possible to construct at least one orthonormal basis set.

Conclusion

Since it is always possible to construct an orthonormal basis set from an arbitrary basis set, we lose no generality in always working with orthonormal basis sets, where $(\hat{a}_i, \hat{a}_j) = \delta_{ij}$. We shall thus always assume an orthonormal basis set in this book.

SCALAR PRODUCT DEFINITIONS

1. If we have a set of vectors in *n-D* space with only *real* components, we see the following:
   - **a.** $(\vec{U},\vec{V}) = (\vec{V},\vec{U})$ commutative law
   - **b.** $(\vec{U},\vec{V} + \vec{W}) = (\vec{U},\vec{V}) + (\vec{U},\vec{W})$ associative law
   - **c.** $(a\vec{U},\vec{V}) = a(\vec{U},\vec{V})$ multiplication by a scalar
   - **d.** $(\vec{U},\vec{U}) \geq 0$ and $(\vec{U},\vec{U}) = 0$ iff $\vec{U} = 0$
2. If we have a set of vectors in n-D space with *complex* components, we see the following:
   - **a.** $(\vec{U},\vec{V}) = (\vec{V},\vec{U})^*$ the scalar product is not commutative
   - **b.** $(\vec{U},\vec{V} + \vec{W}) = (\vec{U},\vec{V}) + (\vec{U},\vec{W})$ associative law
   - **c.** $(a\vec{U},\vec{V}) = a^*(\vec{U},\vec{V})$ multiplication by a scalar
   - **d.** $(\vec{U},\vec{U}) \geq 0$ and $(\vec{U},\vec{U}) = 0$ iff $\vec{U}= 0$
3. If we have complex-valued functions $f(x)$ and $g(x)$ on the interval ($0 \leq x \leq 1$) we can consider the functions at a particular value of x to be a component of a vector. Because there are an infinite number of points, x, on the interval, we could consider the function to be an *infinite dimensional vector*, with a different component corresponding to each value of x. If the functions $f(x)$ and $g(x)$ have a real and imaginary components, then we can use the rules for an n-D space with complex components (as in [2] above) to form the rules for a scalar product but we will need to take the limit as $n \to \infty$. Breaking the interval into n equal parts, we can define a scalar product in the n-D space to be $(\vec{f},\vec{g}) = \sum_{i=1}^{n} f^*(x_i)g(x_i)$ (the vector symbol over f and g has been added for emphasis to indicate that we are thinking of the functions as vectors, but this is not a common notation, so we will drop the vectors hereafter). We now take the limit as $n \to \infty$ but we can see that, for

[2] Note that the basis set is not unique because the orthonormal basis set depends upon which of the basis vectors $\vec{a}_i$ we choose to call $\vec{a}_1$, which we choose to call $\vec{a}_2$, and so forth.

finite functions, the sum of the products would generally approach infinity. However, if we multiply the product by Δx, the space between two adjacent points of evaluation, we can then define $(\vec{f}, \vec{g}) = \sum_{i=1}^{n} f^*(x_i) g(x_i) \Delta x$ and, if we take the limit as $n \to \infty$, we must also take the limit as $\Delta x \to 0$, so we will generally produce a finite scalar product quantity defined as $(f, g) \equiv \lim_{n\to\infty} \sum_{i=1}^{n} f^*(x_i) g(x_i) \Delta x = \int_0^1 f^*(x) g(x) dx$. This is the traditional definition of an inner product of two complex-valued functions, and it obeys the following properties:

a. $(f, g) = (g, f)^*$ the scalar product is not commutative

b. $(f, g + h) = (f, g) + (f, h)$ associative law

c. $(af, g) = a^*(f, g)$ multiplication by a scalar

d. $(f, f) \geq 0$ and $(f, f) = 0$ iff $f = 0$ almost everywhere[3]

MATRIX CONVENTIONS

It is a common convention to specify a vector $\vec{u} = u_x\hat{a}_x + u_y\hat{a}_y + u_z\hat{a}_x$ in the numerical designation $\vec{u} = u_1\hat{a}_1 + u_2\hat{a}_2 + u_3\hat{a}_3$ because it is easy to extend the vector to n-dimensions. Using a definition of matrix multiplication, we can see that this is the same as

$$\vec{u} = (\hat{a}_1 \quad \hat{a}_2 \quad \hat{a}_3)\begin{pmatrix} u_1 \\ u_2 \\ u_3 \end{pmatrix}.$$

Most authors[4] use shorthand and simply write $\vec{u} = \begin{pmatrix} u_1 \\ u_2 \\ u_3 \end{pmatrix}$ with the understanding that the column matrix is to be multiplied by a row matrix $(\hat{a}_1 \quad \hat{a}_2 \quad \hat{a}_3)$.

[3] The condition that $f = 0$ *almost everywhere* is a condition mathematicians use when they study discontinuous functions. For example, if we have a function $f(x) = 0$ *except* at the points $x = ¼, ½$, and $¾$, where $f = 1$, then $\int_0^1 f^*(x) f(x) dx = 0$. The integral would also be equal to 0 for any *finite* number of points x_i on $(0 \leq x \leq 1)$ for which $f(x_i) = 1$.

[4] Physicists like to use the "bra"–"ket" notation for a vector in which the ket vector $|\vec{u}\rangle = \vec{u} = \begin{pmatrix} u_1 \\ u_2 \\ u_3 \end{pmatrix}$ (multiplication by the row base vectors is understood). In this convention, the bra vector $\langle\vec{u}| = (u_1^* \quad u_2^* \quad u_3^*)$. This is especially convenient for a scalar product of a bra and a ket vectors, where $\langle\vec{u}|\vec{v}\rangle = (u_1^* \quad u_2^* \quad u_3^*)\begin{pmatrix} v_1 \\ v_2 \\ v_3 \end{pmatrix} = u_1^* v_1 + u_2^* v_2 + u_3^* v_3$. Note that this *braket* convention holds for vectors in n-dimensions and also holds for the scalar product of two functions in the integral definition.

When we take the scalar product of two vectors with complex components, we can write

$$(\vec{u}, \vec{v}) = \langle \vec{u} | \vec{v} \rangle = (u_1^* \quad u_2^* \quad u_3^*) \begin{pmatrix} v_1 \\ v_2 \\ v_3 \end{pmatrix} = u_1^* v_1 + u_2^* v_2 + u_3^* v_3. \tag{C.2}$$

MATRIX MANIPULATION IN EXCEL

Microsoft Excel is set up to manipulate matrices[5] for the following operations:

Matrix Inversion

1. Put the matrix elements in separate cells; for example, put a 3 × 3 matrix in A1, B1, C1, A2, B2, C2, A3, B3, C3 (Excel calls this range A1:C3).
2. Select 3 × 3 cells anywhere; for example, select D1, E1, F1, D2, E2, F2, D3, E3, F3, and type "minverse(A1:C3)".
3. Then hit CTRL + SHIFT + ENTER. The inverse matrix of A1:C3 appears in D1:F3.
4. Note that you can now change any one of the elements in A1:C3, and its inverse will update automatically.

Matrix Multiplication

1. Select 3 × 3 empty cells for the final answers, for example, G1, H1, I1, G2, H2, I2, G3, H3, I3, and type "mmult(A1:C3,D1:F3)".
2. Then hit CTRL + SHIFT + ENTER. The product of matrix A1:C3 and matrix D1:F3 will appear in G1:I3.
3. Note you can change any element in the two matrices and the product will update automatically.

Square Matrix Times Column Matrix

1. Put the square matrix elements in separate cells; for example, put a 3 × 3 matrix into A1:C3.
2. Put the column matrix elements in separate cells; for example, put a 3 × 1 matrix into D1, D2 and D3.
3. Select 3 × 1 cells anywhere; for example, select E1, E2, E3 and type "mmult(A1:C3, D1:D3)".
4. Then hit CTRL + SHIFT + ENTER. The product of square matrix A1:C3 and column matrix D1:D3 will appear in E1:E3.

[5] Be sure that all cells in the Excel file used have a *number* format.

	A	B	C	D	E	F	G	H	I	J	K	L	M	N	O
1	Excel matrix inversion and multiplication examples														
2	Three by three matrix			Inverse matrix			Check								
3	1.00	2.00	3.00	−0.667	−1.333	1.000	1.000	0.000	0.000						
4	4.00	5.00	6.00	−0.667	3.667	−2.000	0.000	1.000	0.000						
5	7.00	8.00	10.00	1.000	−2.000	1.000	0.000	0.000	1.000						
6	Four by four matrix				Inverse matrix				Check						
7	1.00	1.00	2.00	4.00	1.000	−0.357	−0.357	1.000	1.000	0.000	0.000	0.000			
8	5.00	6.00	7.00	8.00	−2.000	−0.071	1.929	−3.000	0.000	1.000	0.000	0.000			
9	9.00	8.00	7.00	6.00	1.000	0.786	−2.214	3.000	0.000	0.000	1.000	0.000			
10	5.00	4.00	3.00	1.00	0.000	−0.286	0.714	−1.000	0.000	0.000	0.000	1.000			
11	Five by five matrix					Inverse matrix					Check				
12	1.00	2.00	3.00	4.00	1.00	0.000	−1.000	0.000	1.000	0.000	1.000	0.000	0.000	0.000	0.000
13	6.00	0.00	0.00	1.00	1.00	−0.250	9.000	3.000	−8.000	−3.750	0.000	1.000	0.000	0.000	0.000
14	2.00	3.00	4.00	5.00	6.00	0.250	−15.000	−2.000	13.000	2.750	0.000	0.000	1.000	0.000	0.000
15	7.00	0.00	0.00	1.00	1.00	0.250	7.000	0.000	−6.000	−0.250	0.000	0.000	0.000	1.000	0.000
16	1.00	2.00	3.00	4.00	5.00	−0.250	0.000	0.000	0.000	0.250	0.000	0.000	0.000	0.000	1.000
17	Three by three matrix			Column	Product										
18	1.00	1.00	2.00	1.00	9										
19	2.00	3.00	4.00	2.00	20										
20	5.00	6.00	7.00	3.00	38										

Figure C.2 Excel examples.

Excel Examples are shown in Figure C.2 for a

- three by three matrix inversion (with a check),
- four by four matrix inversion (with a check),
- five by five matrix inversion (with a check), and
- three by three matrix multiplied by a three by one column matrix.

n-D VECTOR SPACE

By expressing vector components as a column matrix, we can formulate a scalar product for an n-D vector space on the interval $(0 \le x \le 1)$ by taking the limit as $n \to \infty$ (and as $\Delta x \to 0$) to define a scalar product of two complex valued functions using the logic of the last section. Here,

$$(\vec{u}, \vec{v}) = \langle \vec{u} | \vec{v} \rangle \equiv \lim_{n \to \infty} \sum_{i=1}^{n} u^*(x_i) v(x_i) \Delta x = \int_0^1 u^*(x) v(x)\, dx \tag{C.3}$$

Treating row and column matrices as the complex and real components of a vector function on a continuous interval, we see that the analogy to an infinite dimensional vector space is synonymous to the inner product as an integral, and all of the properties stated in 3a, 3b, 3c, and 3d on page 277 are valid.

Note that we have arbitrarily chosen the interval for the functions to be $(0 \le x \le 1)$ but we could just as well have chosen the interval to be $(-1 \le x \le 1)$, $(-L \le x \le L)$, or $(a \le x \le b)$ and arrived at the same definitions. In addition, we chose to make the inner product finite by multiplying $f^*(x)g(x)$ by Δx and taking the limit as $\Delta x \to 0$, where we could have accomplished the same result by multiplying through by $w(x)\,\Delta x$. When a function $w(x)$ is chosen, it is called a *weighting function*, and it is useful in some applications, as we shall see.

ORTHOGONAL FUNCTIONS

One of the most useful applications of the infinite dimensional vector analysis is when we combine it with the Gram–Schmidt orthogonalization technique to create a series of orthogonal functions.

For example, we can generate a set of polynomial functions that are mutually orthogonal to one another on the interval $(-1 \le x \le 1)$ by beginning with the nonorthogonal powers of x: $1, x, x^2, x^3, x^4, \ldots$.

Using the Gram–Schmidt orthogonalization process, we can choose the first vector function to be $\phi_1(x) = c_{11}(1)$ and find a second function $\phi_2(x) = c_{21}(1) + c_{22}(x)$, and a third function $\phi_3(x) = c_{31}(1) + c_{32}(x) + c_{33}(x^2)$, and so forth. If the first two functions are to be orthogonal on the interval $(-1 \le x \le 1)$, then $\int_{-1}^{1} c_{11}(1)[c_{21}(1) + c_{22}(x)]dx = 0$ from which we conclude that $c_{11}[2c_{21}] = 0$ from which $c_{21} = 0$ since $c_{11} \ne 0$. In order that $\phi_1(x)$ and $\phi_2(x)$ be normalized on the interval $(-1 \le x \le 1)$, we require $\int_{-1}^{1} (c_{11})^2\, dx = 1$ and $\int_{-1}^{1} (c_{22}x)^2\, dx = 1$ from which $c_{11} = \pm\sqrt{1/2}$ and $c_{22} = \pm\sqrt{3/2}$. Thus, we find the first two orthogonal functions to be $\phi_1(x) = \pm\sqrt{1/2}$ and $\phi_2(x) = \pm\sqrt{3/2}x$.

PROBLEMS

C.1 **a.** Show that the coefficients $c_{31} = \pm\sqrt{5/8}$, $c_{32} = 0$, and $c_{33} = \mp 3\sqrt{5/8}$ so that $\phi_3(x) = \pm\sqrt{5/2}\,[3x^2 - 1]/2$.

b. Show that $\phi_4(x) = \pm\sqrt{7/2}\,[5x^3 - 3x]/2$.

c. Show that $\phi_5(x) = \pm\sqrt{9/2}\,[35x^4 - 30x^2 + 3]/8$.

C.2 Find the factor that makes these orthogonal functions equivalent to the ***Legendre Polynomials***:

$$P_0(x) = 1, \quad P_1(x) = x, \quad P_2(x) = [3x^2 - 1]/2,$$
$$P_3(x) = [5x^3 - 3x]/2, \quad P_4(x) = [35x^4 - 30x^2 + 3]/8,$$
$$P_5(x) = [63x^5 - 70x^3 + 15x]/8$$

Note the Legendre polynomials satisfy $P_n(1) = 1$, $P_n(-1) = (-1)^n$ and $\int_{-1}^{1} [P_n(x)]^2\, dx = 2/(2n+1)$

C.3 Given the set of functions, 1, x, x^2, x^3, x^4, ... , find an orthonormal set on the interval $(-1 \le x \le 1)$ if the weighting function is $w(x) = x$. These are called the *normalized Legendre Polynomials*.

C.4 Given the set of functions, c_0, $c_1 + c_2\, x$, $c_3 + c_4\, x + c_5\, x^2$, ... , where c_i = constants, find the constants so that the functions form a mutually orthogonal set on the interval $(0 \le x \le 1)$.

C.5 Given the set of functions, c_0, $c_1 + c_2\, x$, $c_3 + c_4\, x + c_5\, x^2$, ... , where c_i = constants, find the constants so that the functions form a mutually orthogonal set on the interval $(a \le x \le b)$.

USE FOR AN ORTHONORMAL FUNCTION SET

By analogy to the set of orthogonal basis vectors, $\phi_i(x)$, we can see that an arbitrary function, $f(x)$, can be expanded as a set of mutually orthonormal functions; $f(x) = \sum_{n=1}^{\infty} c_n \phi_n(x)$ on the interval $(a \le x \le b)$. Furthermore, we can see that the coefficients c_n are unique and that $c_n = \int_a^b f(x)\phi_n(x)dx$. The orthonormal series so defined are a generalization of the Fourier series, and the coefficients c_n are called the generalized Fourier coefficients.

As we will see later when we study spherical coordinates, it is often convenient to expand an arbitrary function of $\mu = \cos\theta$ on the interval $(-1 \le \mu \le 1)$ in terms of the Legendre Polynomials.

Other sets of orthogonal functions (e.g., Bessel, Laguerre, Hermite, Chebyshev) can be used to describe arbitrary functions in electromagnetic problems on a given interval and/or with respect to a given weighting function. We will wait until we encounter those problems, but the rationale for their use is the same as the Gram–Schmidt process described above.

STURM–LIOUVILLE EQUATION

One of the most elegant and beautiful proofs in the annals of mathematical physics lies in the solution to the Sturm-Liouville equation:

$$\frac{d}{dx}\left[p(x)\frac{dy}{dx}\right] + [q(x) + \lambda r(x)]y = 0 \tag{C.4a}$$

with some given boundary conditions in the general form

$$\alpha_1 y(a) + \alpha_2 y'(a) = 0, \quad \beta_1 y(b) + \beta_2 y'(b) = 0, \tag{C.4b}$$

where $p(x)$, $q(x)$, $r(x)$, α_1, α_2, β_1, β_2 are real and λ or y may be complex.

Quantities such as the electric field intensity, the magnetic field intensity, the electric vector potential, or the magnetic vector potential obey Maxwell's equations subject to some given boundary conditions, and these four coupled partial differential equations can usually be rewritten in terms of a set of uncoupled partial differential equations. Our choice of coordinate system (e.g., Cartesian, cylindrical, spherical, elliptic cylindrical, parabolic cylindrical, bipolar, prolate spheroidal, oblate spheroidal, parabolic, toroidal, bispherical, confocal ellipsoidal, conical, confocal parabolic) will depend on the specified boundary conditions. Typically, we are given Dirichlet boundary conditions $y(a)$, Neumann boundary conditions $y'(a)$, or mixed boundary conditions $\alpha_1 y(a) + \alpha_2 y'(a) = 0$ on the boundary a of some geometric shape. Applying the boundary conditions is made easier with a choice[6] of a particular coordinate system. In this book, we will limit ourselves to Cartesian, cylindrical, and spherical coordinate systems.

An excellent technique for solving the partial differential equations that come from combining Maxwell's equations is using separation of variables, as explained in detail in section 7.8. The ordinary differential equations that pertain to any one of the variables are in almost all cases in the form of the Sturm-Liouville equation. While the individual differential equations may be coupled through their use of a common parameter or constant, the solution to the Sturm-Liouville equation gives the function that will typically be multiplied by other functions to uniquely specify the solution to electromagnetic problems. Fortunately, Sturm-Liouville theory assures us of a complete orthogonal set of functions (called *eigenfunctions*) that span the space, so, once we have solved that differential equation, we can just write down the solution within some arbitrary constants whenever we encounter that problem in the future. The final solution to the problem is thus choosing the constants to meet the given boundary conditions.

EIGENVALUES

To find the properties of the eigenvalues, λ, for the Sturm-Liouville equation, we consider the complex conjugate of Equations C.4a and C.4b:

$$\frac{d}{dx}\left[p(x)\frac{d\tilde{y}}{dx}\right]+\left[q(x)+\tilde{\lambda}r(x)\right]\tilde{y}=0 \tag{C.5a}$$

$$\alpha_1\tilde{y}(a)+\alpha_2\tilde{y}'(a)=0, \quad \beta_1\tilde{y}(b)+\beta_2\tilde{y}'(b)=0 \tag{C.5b}$$

where the tilde over the eigenfunction y or the eigenvalue λ denotes the complex conjugate. We can multiply Equation C.4a by $\tilde{y}$ and Equation C.5a by y and subtract to see

$$\frac{d}{dx}[p(x)(y\tilde{y}'-\tilde{y}y')]=\left(\lambda-\tilde{\lambda}\right)r(x)y\tilde{y} \tag{C.6}$$

and integrating from a to b,

$$\left(\lambda-\tilde{\lambda}\right)\int_a^b |y|^2\, r(x)dx = p(x)(y\tilde{y}'-\tilde{y}y')|_a^b = 0 \tag{C.7}$$

[6] Nature does not know anything about coordinate systems, as that is a mathematical construction of humans. Thus, the answer to the problem cannot depend on our choice of coordinate system. Any choice will do, but some may lend themselves better to the application of a boundary condition.

upon using conditions C.4b and C.5b. However, because $r(x) \geq 0$ and is not identically zero everywhere on $(a \leq x \leq b)$, the integral on the left of Equation C.7 is not zero. Thus, the only possibility is that $\lambda = \tilde{\lambda}$ from which we conclude that *eigenvalues, λ, of Sturm-Liouville equations are real.*

EIGENFUNCTIONS

We can also show that eigenfunctions, y_1 and y_2, belonging to two different eigenvalues, λ_1 and λ_2, are orthogonal with respect to the weighting function $r(x)$ on the interval $(a \leq x \leq b)$. Here, we rewrite Equations C.4a and C.4b for the two functions:

$$\frac{d}{dx}\left[p(x)\frac{dy_1}{dx}\right]+[q(x)+\lambda_1 r(x)]y_1=0 \tag{C.8a}$$

$$\alpha_1 y_1(a)+\alpha_2 y_1'(a)=0, \quad \beta_1 y_1(b)+\beta_2 y_1'(b)=0 \tag{C.8b}$$

$$\frac{d}{dx}\left[p(x)\frac{dy_2}{dx}\right]+[q(x)+\lambda_2 r(x)]y_2=0 \tag{C.8c}$$

$$\alpha_1 y_2(a)+\alpha_2 y_2'(a)=0, \quad \beta_1 y_2(b)+\beta_2 y_2'(b)=0. \tag{C.8d}$$

If we now multiply Equation C.8a through by y_2 and Equation C.8c through by y_1 and subtract, we find

$$\frac{d}{dx}[p(x)(y_1y_2'-y_2y_1')]=(\lambda_1-\lambda_2)r(x)y_1y_2. \tag{C.9}$$

Integrating from Equation C.8a to C.8b and using Equation C.8b and C.8d,

$$(\lambda_1-\lambda_2)\int_a^b y_1y_2r(x)dx=p(x)(y_1y_2'-y_2y_1')|_a^b=0 \tag{C.10}$$

and because $\lambda_1 \neq \lambda_2$, it is required that

$$\int_a^b y_1y_2r(x)dx=0. \tag{C.11}$$

We have thus concluded that the eigenfunctions y_1 and y_2, belonging to two *different* eigenvalues, λ_1 and λ_2, are orthogonal with respect to the weighting function $r(x)$ on the interval $(a \leq x \leq b)$.

Completeness of Eigenfunctions

A final characteristic of eigenfunctions is that they form a *complete* set; that is, any arbitrary, single-valued function, $f(x)$, that is piecewise continuous on an interval $(a \leq x \leq b)$ can be exactly represented by the series

$$f(x)=\sum_{n=1}^{\infty}c_n\phi_n(x) \tag{C.12}$$

if the functions, $\phi_n(x)$, are eigenfuctions of a Sturm-Liouville equation. Specifically, we can say that the set $\phi_n(x)$ is complete because the limit of the mean square error vanishes as $m \to \infty$ as follows:

$$\lim_{m\to\infty} \int_a^b \left[f(x) - \sum_{n=0}^{m} c_n \phi_n(x) \right]^2 w(x)dx = 0 \tag{C.13}$$

In the language of a linear vector space, we can say the orthonormal functions, $\phi_n(x)$, on the infinite dimensional vector space interval $(a \le x \le b)$ form a basis that spans the space as long as we define the inner product according to Equation C.11. Such a space is called a *Hilbert space*. The concept is fully discussed in Arfken[i] and other basic texts on Mathematical Physics.

Summary Application

In section 7.8, we showed that Cartesian components of the electric field intensity, magnetic field intensity, electric flux density, magnetic flux density, scalar potential, or magnetic vector potential satisfy an inhomogeneous Helmholtz equation:

$$\vec{\nabla}^2 \psi - \mu\varepsilon \frac{\partial^2 \psi}{\partial t^2} = f(\vec{x}, t). \tag{C.14}$$

We chose to let $\psi(\vec{x},t) = \psi_s(\vec{x})T(t)$ so that the functions satisfy

$$\frac{\vec{\nabla}^2 \psi_S(\vec{x})}{\psi_S(\vec{x})} = -k^2 \quad \text{and} \quad \mu\varepsilon \frac{\partial^2 T(t)/\partial t^2}{T(t)} = -k^2. \tag{C.15}$$

We then let $\psi_s(\vec{x}) = X(x)Y(y)Z(z)$ so that $\partial^2 X(x)/\partial x^2 = -k_x^2 X(x)$, $\partial^2 Y(y)/\partial y^2 = -k_y^2 Y(y)$, and $\partial^2 Z(z)/\partial z^2 = -k_z^2 Z(z)$, with the condition that $k_x^2 + k_y^2 + k_z^2 = k^2$.

We now see that each of the spatial terms and the time-dependent term satisfy a Sturm-Liouville equation, so we are assured that the eigenvalues of the equations are real and that the eigenfunctions form a complete set of orthogonal functions. This means that we can expand the functions in a series $\psi(\vec{x}, t) = (1/2\pi)\int_{-\infty}^{\infty} \psi(\vec{x}, \omega) e^{-j\omega t} d\omega$, $f(\vec{x}, t) = (1/2\pi)\int_{-\infty}^{\infty} f(\vec{x}, \omega) e^{-j\omega t} d\omega$ in the time variable and in a similar series for each of the spatial variables. In section 7.8, we took it for granted that we could expand the individual space and time functions as Fourier transforms but we now see from the Sturm-Liouville theory that these expansions are assured to be orthogonal and complete, so we are rigorously justified in our adoption of the Fourier transform.

ENDNOTE

i. George Arfken, *Mathematical Methods for Physicists*, 3rd ed. (Burlington, MA: Elsevier Academic Press, 1985).

Bibliography

Abramowitz, Milton, and Irene A Stegun. *Handbook of Mathematical Functions*. National Bureau of Standards Applied Mathematics Series, 55, Superintendent of Documents, U.S. Washington, DC: Government Printing Office, 1964.

Arfken, George. *Mathematical Methods for Phyicists*. 3rd ed. New York: Academic Press, 1985.

Balanis, Constantine, A. *Advanced Engineering Electromagnetics*. Hoboken, NJ: John Wiley & Sons, 1989.

Chen, Yinchao, Qunsheng Cao, and Raj Mittra. *Multiresolution Time Domain Scheme for Electromagnetic Engineering*. Hoboken, NJ: John Wiley & Sons, 2005.

Cheng, David K. *Fundamentals of Engineering Electromagnetics*. Upper Saddle River, NJ: Prentice Hall, 1993.

Clark, Ronald C. *Einstein: The Life and Times*. New York: Harry N. Abrams, 1984.

Collins, Royal Eugene. *Mathematical Methods for Physicists and Engineers*. New York: Reinhold Book Corporation, 1968.

Davidson, David B. *Computational Electromagnetics for RF and Microwave Engineering*. Cambridge: Cambridge University Press, 2005.

DeBye, P. *Polar Molecules*. New York: Dover Publications, 1929.

Dwight, Herbert Bristol. *Tables of Integrals and Other Mathematical Data*. New York: Macmillan, 1961.

Edminister, Joseph A. *Electromagnetics*. 2nd ed. Schaum's Outline series. McGraw-Hill, 1993.

Feynman, Richard P., Robert B. Leighton, and Matthew Sands. *The Feynman Lectures on Physics*, vol. 2: *Electromagnetism and Matter*. Reading, MA: Addison-Wesley, 1964.

Feynman, Richard P., Robert B. Leighton, and Matthew Sands. *The Feynman Lectures on Physics*, vol. 3: *Quantum Mechanics*. Reading, MA: Addison-Wesley, 1965.

Ghosh, S. N. *Electromagnetic Theory and Wave Propagation*. 2nd ed. New Delhi: CRC Press, Narosa Publishing House, 1998.

Gradshteyn, S. I., and I. M. Ryzhik. *Table of Integrals, Series, and Products*. 4th ed. Translated by Yu. V. Geronimus and M. Yu. Tseytlin from Russian. New York: Academic Press, 1965.

Hall, Stephen H., and Howard L. Heck. *Advanced Signal Integrity for High-Speed Digital Designs*. Hoboken, NJ: John Wiley & Sons, 2009.

Hall, Stephen H., Garrett W. Hall, and James A. McCall. *High-Speed Digital System Design: A Handbook of Interconnect Theory and Design Practices*. Hoboken, NJ: John Wiley Interscience, 2000.

Halliday, David, Robert Resmick, and Jearl Walker. *Fundamentals of Physics*. 6th ed. New York: John Wiley & Sons, 2001.

Harrington, Roger F. *Time-Harmonic Electromagnetic Fields*. Hoboken, NJ: Wiley Interscience, John Wiley & Sons, 2001.

Hayt, William H., Jr. and John A. Buck. *Engineering Electromagnetics*. 7th ed. New York: McGraw-Hill, 2006.

Huray, Paul G. *The Foundations of Signal Integrity*. Hoboken, NJ: John Wiley & Sons, 2009.

Iles, George. *Flame Electricity and the Camera, Man's Progress from the First Kindling of Fire to the Wireless Telegraph and the Photography of Color*. New York: Doubleday & McClure, 1900.

Isaacson, Walter. *Einstein: His Life and Universe*. New York: Simon and Schuster, 2007.

Jackson, John David. *Classical Electrodynamics*. 3rd ed. Hoboken, NJ: John Wiley & Sons, 1999.

Johnson, Howard, and Martin Graham. *High-Speed Signal Propagation: Advanced Black Magic*, Upper Saddle River, NJ: Prentice Hall, 2003.

Kittel, Charles. *Introduction to Solid State Physics*. 7th ed. Hoboken, NJ: John Wiley & Sons, 1996.

Kittel, Charles. *Quantum Theory of Solids*. Hoboken, NJ: John Wiley & Sons, Inc., 1963.

Laudau, L. D., and E. M. Lifshitz. *The Classical Theory of Fields*. 4th revised English ed. Translated by Morton Hamermesh from Russian. London: Pergamon Press, 1975.

Landau, L. D., E. M. Lifschitz, and L. P Pitaevskii. *Electrodynamics of Continuous Media*. 2nd ed. Translated by J.B. Sykes, J.S. Bell, and M.J. Kearsley from the second edition of Elektrodinamika sploshnykh sred, Izdated'stvo 'Nauka', Moscow, 1982. Amsterdam, The Netherlands: Elsevier, Butterworth-Heinemann, 2006.

Mathews, Jon, and R. L. Walker. *Mathematical Methods of Physics*. New York: W. A. Benjamin, 1965.

Maxwell, James Clerk. "A Comparison of the Electric Units and on the Electromagnetic Theory of Light." *Philosophical Transactions, Royal Society of London* 158 (1868). Reprinted in Maxwell, Scientific Papers.

Maxwell, James Clerk. *A Treatise on Electricity & Magnetism*. Vol. 1. Unabridged, slightly altered, republication of the third edition, published by the Clarendon Press in 1891. New York: Dover, 1954.

Maxwell, James Clerk. *A Treatise on Electricity & Magnetism*. Vol. 2. Reprint of the original edition. Merchant Books, 2007.

Maxwell, James Clerk. *The Dynamical Theory of the Electromagnetic Field*. Reprint of the original. Edited by Thomas F. Torrance. Eugene, OR: Wipf and Stock, 1996.

Ohanian, Hans C. *Gravitation and Spacetime*. New York: W. W. Norton, 1976.

Orlando, Terry P., and Kevin A. Delin. *Foundations of Applied Superconductivity*. Reading, MA: Addison-Wesley, 1991.

Paul, Clayton R. *Electromagnetics for Engineers. with Applications to Digital Systems and Electromagnetic Interference*. Hoboken, NJ: John Wiley & Sons, 2004.

Pozar, David M. *Microwave Engineering*. 3rd ed. Hoboken, NJ: John Wiley & Sons, 2005.

Schelkunoff, S. A. *Electromagnetic Waves*. New York: D. Van Nostrand, 1943.

Schlesinger, Mordechay, and Milan Paunovic. *Modern Electroplating*. Electrochemical Society Series. 4th ed. New York: John Wiley Interscience, 2000.

Scott, William T. *The Physics of Electricity and Magnetism*. New York: John Wiley & Sons, 1962.

Spiegel, Murray R. *Schaum's Outline Series. Complex Variables: With an Introduction to Conformal Mapping and Its Applications*. McGraw-Hill, 1999.

Standage, Tom. *The Victorian Internet; The Remarkable Story of the Telegraph and the Nineteenth Century's On-Line Pioneers*. New York: Walker, 1998.

Tyndall, John Tyndall. *Fragments of Science*. 6th ed. Vols. I and II. New York: Appleton, 1897.

Wills, A. P. *Vector Analysis with an Introduction to Tensor Analysis*. New York: Dover, 1958.

Index